Production Logging—Theoretical and Interpretive Elements

A.D. Hill
Assoc. Professor
U. of Tcxas

SPE Monograph Series, Volume 14

Henry L. Doherty Memorial Fund of AIME
Society of Petroleum Engineers
Richardson, TX USA

Dedication

To Ellen, Austin, and Laura.

Disclaimer

Printed in the United States of America.

ISBN 978-1-55563-030-0

11 12 13 14 15 16 17 / 14 13 12 11 10 9 8 7

Society of Petroleum Engineers
222 Palisades Creek Drive
Richardson, TX 75080-2040 USA

http://store.spe.org
books@spe.org
1.972.952.9393

Preface

Production logging comprises a large number of measurements made throughout the life of a well. It is likely that some production logs are run on the majority of all wells that are completed. Yet there is a surprising dearth of literature on production logging relative to other well-testing procedures, such as openhole logging or pressure-transient testing. The production logging literature consists largely of a relatively small number of journal articles scattered over 50 years of literature and service-company manuals. By nature, each service-company manual is limited in scope and specific to a particular company's tools and procedures. The purpose of this book is to draw together in one reference current knowledge of production logging measurements and interpretation procedures. It is intended primarily as a source of information for the petroleum engineer faced with obtaining and interpreting production logs.

This book itself is limited in scope; it should be considered a starting point for the individual wishing to become an accomplished practitioner of production logging interpretation. Such an individual must become aware of more features in the logs than could possibly be presented in a book of this length. This monograph provides the fundamentals needed to develop expertise in production logging.

Throughout the monograph, guidelines are presented to aid the engineer in selecting the proper logging techniques for obtaining desired data. With each logging method, the theory of the measurement is presented in sufficient detail to allow the engineer to appreciate the accuracy and limitations of the data obtained and to understand interpretation procedures. General operational procedures are reviewed, primarily to point out the steps that must be followed to obtain high-quality logs. Interpretation methods are then detailed, with examples for illustration. I have tried to demonstrate the utility of a properly run and interpreted production log and to point out the limitations inherent in many of the measurements.

This book is not intended to be a training manual to teach production logging operations. For example, little mention is made of the surface electronics used to acquire and display production logs, yet detailed knowledge of this equipment is required for someone to run a production log. Because each vendor's tools are different, such information is best taught by the vendor.

Comprehension of the information in this monograph will enable an engineer (1) to know which log or combination of logs to run to accomplish given logging objectives, (2) to know which logs not to run, (3) to understand the logging procedures that must be followed to obtain the most information with each log, (4) to make an initial interpretation of all commonly used production logs, and (5) to appreciate the relative accuracies of the various logs in a wide range of well conditions.

The information sought with production logs is extremely valuable to the production or reservoir engineer. Yet in some instances current technology is inadequate in providing all the answers we seek. I hope that new production logging techniques will evolve in the near future that can conclusively diagnose even the most difficult well conditions.

A. Daniel Hill
Austin, Texas
May 1990

SPE Monograph Series

The Monograph Series of the Society of Petroleum Engineers was established in 1965 by action of the SPE Board of Directors. The Series is intended to provide authoritative, up-to-date treatment of the fundamental principles and state of the art in selected fields of technology. The Series is directed by the Society's Monograph Committee. A committee member designated as Monograph Editor provides technical evaluation with the aid of the Review Committee. Below is a listing of those who have been most closely involved with the preparation of this monograph.

Monograph Review Committee

Norman R. Carlson, Atlas Wireline Services, Monograph Editor
Richard M. McKinley, Exxon Production Research Co.
Gilbert L. Feather, Amoco Production Co.

Monograph Committee (1989)

John L. Gidley, John L. Gidley & Assocs. Inc., Chairman
Glenn P. Coker, Amoco Corp.
Satish K. Kalra, Arco Oil & Gas Co.
Medhat M. Kamal, Arco Oil & Gas Co.
Charles E. Konen, Amoco Production Co.
Raymond E. Roesner, Atlas Wireline Services
David R. Underdown, Arco Oil & Gas Co.
Robert R. Wood, Shell Western E&P Inc.

Acknowledgments

As author of a technical monograph, I owe a debt of gratitude to the many engineers and scientists who have developed the technology of production logging. I often feel I have served more as a collator than as an author, because many of the thoughts expressed in this book are those of others.

Many individuals have contributed to the production of this monograph. First, I recognize my former colleagues at Marathon Oil Co., particularly Bob Earlougher Jr., Paula Galloway, Irv Johnson, Hossein Kazemi, and Tim Oolman. Their collaboration and supervision piqued my interest in production logging.

My efforts to learn and understand production logging have been aided greatly by the many students who have taken my course on this subject at the U. of Texas. Also, I have had a number of excellent graduate students who have contributed significantly to the technology of production logging. I greatly appreciate the efforts of Jay Akers, Les Anthony, Kathryn Boehm, Chip Cox, Ross Headifen, Steve Morriss, Grant Scott, Song Shanhong, Ricardo Solares, Farah Zavareh, and Ding Zhu.

The ultimate value of this monograph will be due in large part to the efforts of the SPE Monograph Review Committee: Norm Carlson (Monograph Editor), Atlas Wireline Services; Gil Feather, Amoco Production Co.; and Mac McKinley, Exxon Production Research Co. I am indebted to these gentlemen for the many contributions they made to this book and to their companies for the large time commitment that was required.

I thank Joanna Castillo for the skillful preparation of many of the figures in this monograph and Joye Johnson for her assistance in the production of this book. I also recognize Julie Noe and Georgeann Bilich of the SPE staff for their considerable editorial efforts.

Finally, I thank my wife, Ellen, for her contributions as a word processor and for her understanding and encouragement throughout my efforts on this monograph.

Contents

Chapter 1
Introduction

1.1 Definition and Use of Production Logging

Production logging traditionally encompasses a number of well logging techniques run on completed injection or production wells, with the goal being to evaluate the well itself or the reservoir performance. In recent years, however, the role of production logging has expanded to include applications that start at the early stages of drilling and that last throughout the life of the well. The purpose of production logs is to evaluate fluid flow inside and outside pipe or, in some cases, to evaluate the well completion directly. The most common application of production logging is the measurement of the well's flow profile, the distribution of flow into or out of the wellbore. Wade *et al.*,[1] referring to production wells, state that production logging is used to answer the question "How much of what fluid is coming from where." Cased-hole formation-evaluation logs, such as pulsed-neutron logs, are sometimes regarded as production logs; for the purposes of this work, however, formation-evaluation logs will not be considered, except as they apply to measuring flow profiles. The primary logging methods that will be considered are temperature, radioactive-tracer, and spinner-flowmeter logs for single-phase flow; temperature, fluid-density, fluid-capacitance, and flowmeter logs in multiphase flow; and noise, cement-bond, and unfocused-gamma-ray-density logs as applied for well completion evaluation.

Production logging has traditionally been applied for problem well diagnosis or reservoir surveillance.[2] As more U.S. fields move into secondary and tertiary recovery, the need for production logging is increasing. In these advanced stages of production, reservoir sweep efficiency is often critical; production logging is one of the few means available to ascertain the vertical distribution of injected or produced fluids. Likewise, competent well completions are vital to efficient reservoir performance and, again, production logging is a primary method of well evaluation. As McKinley[2] illustrates, production logging techniques are being applied even during the drilling stage in certain instances.

Production logs, like most well tests, rely on indirect measurement to obtain desired results. For example, in a temperature log, the wellbore temperature as a function of depth is measured. The engineer then attempts to determine the injection or production intervals by applying an interpretation scheme. For this reason, log interpretation is of critical importance in production logging. Almost all production logging interpretation relies on an understanding of the fluid movements in and around the wellbore and of how these fluid movements affect the logging measurement. Thus, log interpretation will be emphasized in this book.

1.2 History of Production Logging

Production logging began with the use of temperature surveys to locate fluid entries in a wellbore.[3] Early workers in the field recognized that the cooling of gas as it expands caused low-temperature anomalies that located the sources of gas entries. Cool fluids were also injected to locate permeable zones from the temperature anomalies that remained after shut-in.[4]

In the 1940's, flow rate and pressure measurements were added to temperature surveys to obtain more information about well conditions.[5,6] The types of fluid in the well could be identified by measuring a pressure gradient. Flow measurements gave information about the quantities of fluid produced or injected.

Further development saw the introduction of surface-recording production logging instruments. These offered the obvious advantage of providing the operator more flexibility and control during logging. With the development of reliable grease-injection lubricators, surface-recording instruments became the standards of the industry.[2]

By the mid-1960's, other production logging instruments had been developed to gain further information about well conditions, particularly in multiphase flow. Density and capacitance (water holdup) meters were introduced to resolve complex multiphase flow behavior.[1] Also, cement-bond logging had achieved widespread use as a completion evaluation method.[7-9]

As new logging instruments became available, interpretation methods evolved for the more complex flow situations being encountered. New logging techniques and interpretation methods are still being tested today and no doubt will be for years to come as more is learned about flow characteristics in wellbores.

1.3 Current Technology

A large suite of production logging measurements is available today. In single-phase flows, temperature, radioactive-tracer, and spinner-flowmeter logs are commonly applied. When run and interpreted properly, these logs provide accurate measurements of flow profiles in most instances. In multiphase flows, fluid-density, fluid-capacitance, and deflector-flowmeter logs are used in addition to the logs used in single-phase flows. For completion evaluation, the engineer can choose from noise, cement-evaluation, and, in some instances, unfocused-gamma-ray-density logs.

Running all the production logging tools needed for a particular job on the same tool string is usually possible.[10] Tools that can make simultaneous measurements have recently been developed,[11] eliminating some of the error that can result from well instability. Because a suite of logs is often required, particularly in multiphase flows, the ability to run several tools in one string and to make measurements simultaneously is clearly advantageous.

Though production logging devices have been improved significantly over the past 40 years, the technology is still severely limited in some cases, particularly when applied in multiphase flows. Most flow profiling techniques, when applied properly, will provide accurate results in a single-phase-flow well. Notable exceptions are in low-flow-rate wells or irregular boreholes. In multiphase flow, conditions often are so complex and so different from the conditions for which the tools are designed that production logging measurements often are inaccurate, and, in the worst cases, yield a misleading or totally false picture of the well conditions. This should not be surprising, considering the difficult task we are attempting—to meter a two- or three-phase flow stream remotely with devices that often sample only a small portion of the flow stream. Yet, there are currently no means of accurately metering multiphase flow streams on the surface without first separating the phases, even though much more sophisticated instrumentation can be applied.

1.4 Future Needs

Improved production logging results in the future will require the development of new measurements because only incremental improvements appear possible with current tools and interpretation methods. A completely new approach may be required to improve significantly our ability to measure flow profiles, particularly in multiphase flows.

Current practices use measurements of the flow stream in the well to infer the locations and rates of fluid entering the well. For example, we might measure the average velocity and average density of the flow stream at several locations in the well. From these measurements, the volumetric flow rates of each phase are calculated by making assumptions about the multiphase flow. The rates of fluid entries into the wellbore are then calculated as the differences between wellbore flow rates at various depth locations. This approach is fraught with difficulty—the basic measurements are often inaccurate and the interpretation must rely on assumptions about the flow conditions, which introduces further error.

Significant improvement in our ability to measure flow profiles in multiphase flows will likely require more direct measures of the quantities of interest. Techniques to measure directly the average velocity of one or more of the phases present would greatly improve production logging results in multiphase flow. Alternatively, a direct measure of the flow rate issuing from a perforation, rather than inferring the flow from the formation from the changes in flow rate in the wellbore, may be a significant innovation in the future.

Production logging measurements and/or interpretation methods will need modification to be applicable to emerging applications of production logging. These logs can be valuable diagnostic tools during the drilling of exploration wells.[2] Production logs will be increasingly applied in enhanced recovery operations as these continue to expand. The different nature of the fluids in these new areas will require new approaches to production logging.

1.5 Scope and Objectives

This book is intended primarily as a source of information for the petroleum engineer faced with running and interpreting production logs. I have attempted to cover all the commonly applied production logs dealing with flow inside or outside of casing or completion evaluation; cased-hole formation-evaluation logs are not covered, except as applied for profile measurement.

Throughout the monograph, guidelines are presented to aid in the selection of proper logging techniques for obtaining the desired data. With each logging method, the theory of the measurement is presented in sufficient detail to allow the engineer to appreciate the accuracy and limitations of the log and to understand interpretation procedures. General operational procedures are reviewed, primarily to point out the steps that must be followed to obtain high-quality logs. Log interpretation methods are then detailed, with examples for illustration. Throughout the monograph, I have tried to demonstrate the utility of a properly run and interpreted production log, while pointing out the limitations inherent in many of the measurements.

1.6 Organization

This monograph is organized into four general sections. The first section is introductory and should serve as a roadmap for the reader by presenting the applications of production logging through the life of a well or reservoir. The reader will be directed to subsequent chapters for details on the logging methods and interpretations required for specific problems. The next three sections treat logging in single-phase flow, logging in multiphase flow, and logs for flow behind pipe or completion evaluation. Specific logging techniques are introduced in the earliest applicable section (e.g., temperature logging in the single-phase-flow section), then discussed again as they apply in other sections.

The first section of the monograph comprises Chaps. 1 and 2. Chap. 1 introduces the book, while Chap. 2 describes the use of production logging for well and reservoir diagnosis. For the reader with a well or reservoir problem, Chap. 2 provides guidelines on what logs, if any, should be run, how accurate the results might be, and what other well tests might complement production logs.

The second section, Chaps. 3 through 7, presents logging measurements as applied in single-phase flow. Effects of multiphase flow on some of these same measurements are discussed in subsequent chapters. Chap. 3 briefly reviews the fundamentals of flow in pipes. Chaps. 4, 5, and 6 cover temperature logging, radioactive-tracer logging, and spinner-flowmeter logging, respectively. The production logging techniques applied in single-phase flow are summarized in Chap. 7, with emphasis on the relationships between the various logging methods.

The difficult subject of production logging in multiphase flow is treated in the third section, Chaps. 8 and 9. Chap. 8 presents the background in multiphase flow in pipes needed to understand production logging measurements and interpretation procedures applied in this environment. Chap. 9 considers production logging in multiphase flow. Previously presented measurements (temperature, tracer, and spinner logs) are discussed as they pertain to multiphase systems, and measurements unique to multiphase flow are introduced. The chapter emphasizes the need for multiple logging measurements in multiphase flow and points out the limitations of current logging practices.

The final section of this monograph, Chaps. 10 through 12, discusses techniques aimed primarily at completion evaluation or flow outside the wellbore. Chap. 10 considers noise logging and Chap. 11 treats cement-bond and ultrasonic-pulse-echo logs for cement evaluation. Chap. 12 discusses other, less frequently applied logging techniques for flow behind casing. These include the unfocused-gamma-ray-density log, radial-differential-temperature logging, and the application of the pulsed-neutron log as a flow profiling method.

Throughout the monograph, I have tried to present sufficient theory of the logging measurements to enable the engineer to judge the merits of the logs in different situations. This theory is followed by examples to illustrate the common applications of production logging.

References

1. Wade, R.T. *et al.*: "Production Logging—The Key to Optimum Well Performance," *JPT* (Feb. 1965) 137–44.
2. McKinley, R.M.: "Production Logging," paper SPE 10035 presented at the 1982 SPE Intl. Petroleum Exhibition and Technical Symposium, Beijing, March 18–26.
3. Schlumberger, M., Doll, H.G., and Perebinossoff, A.: "Temperature Measurements in Oil Wells," *J. Inst. Pet. Technologists* (Jan. 1937) **23,** No. 159.
4. Millikan, C.V.: "Temperature Surveys in Oil Wells," *Trans.*, AIME (1941) **142,** 15–23.
5. Dale, C.R.: "Bottom Hole Flow Surveys for Determination of Fluid and Gas Movements in Wells," *Trans.*, AIME (Aug. 1949) **186,** 205–10.
6. Riordan, M.B.: "Surface Indicating Pressure, Temperature and Flow Equipment," *Trans.*, AIME (1951) **192,** 257–62.
7. Grosmangin, M., Kokesh, F.P., and Majani, P.: "A Sonic Method for Analyzing the Quality of Cementation of Borehole Casings," *JPT* (Feb. 1961) 165–71.
8. Riddle, G.A.: "Acoustic Wave Propagation in Bonded and Unbonded Oil Well Casing," paper SPE 454 presented at the 1962 SPE Annual Meeting, Los Angeles, Oct. 7–10.
9. Pickett, G.R.: "Acoustic Character Logs and Their Applications in Formation Evaluation," *JPT* (June 1963) 659–67.
10. Meunier, D., Tixier, M.P., and Bonnet, J.L.: "The Production Combination Tool—A New System for Production Monitoring," *JPT* (May 1971) 603–13.
11. Anderson, R.A. *et al.*: "A Production Logging Tool with Simultaneous Measurements," *JPT* (Feb. 1980) 191–98.

Chapter 2
Use of Production Logging for Well and Reservoir Diagnosis

2.1 Introduction

Production logging can be a powerful tool for evaluating the performance of a well or reservoir, but also can be a waste of time and resources when applied poorly or in inappropriate circumstances. An understanding of the capabilities and limitations of the various production logging measurements available and of the types of problems that can be addressed with production logs is essential to the efficient use of these services. This chapter summarizes the applications of production logs throughout the life of a well by discussing the well or reservoir problems usually investigated with production logs and by showing the logs applicable in each case.

Reservoir performance monitoring, well completion evaluation, and planning and evaluation of well workovers are the most common applications of production logging.[1-8] Production logs are sometimes run as part of a regular reservoir surveillance program; they are used more commonly to diagnose a problem implied by a well's performance. Surface conditions or a comparison of a well's behavior with that of nearby wells often suggests problems that can be pinpointed with production logs.

Production logs can be useful diagnostic tools from the time a well is being drilled to its abandonment. This chapter first describes applications of production logging during drilling and completion, followed by the more common applications during a well's production or injection stage. Uses of production logs in conjunction with well workovers are then described. Guidelines for planning a production logging job are presented, followed by tables that summarize the use of production logs for well or reservoir performance evaluation.

2.2 Production Logging Applications During Drilling and Completion

Though production logs traditionally have been thought of as for use in completed injection or production wells, they are increasingly being applied as a well is being drilled or completed. From the start of drilling operations, production logs can be used to locate lost-circulation zones or to find underground blowouts or the source of kicks. Virtually all completion operations—including cementing, gravel packing, perforating, and well stimulation—can be evaluated with production logs.

2.2.1 Lost-Circulation Zones. A lost-circulation zone is a region where drilling mud is entering the formation because of an excessive overbalance between the wellbore and the formation pressures, as illustrated in Fig. 2.1. In a lost-circulation zone, fluid is being injected from the wellbore into the formation, so some of the same techniques used to measure injection profiles in completed wells can be applied. The complicating factors in locating lost-circulation zones as opposed to measuring injection profiles are that the flow of interest is occurring in the annulus between the drillpipe and the borehole wall and that the drilling mud will often be viscous. Under these conditions, a log that can detect flow behind pipe is needed.

Temperature and radioactive-tracer logs can be used to locate lost-circulation zones. A shut-in temperature log would be interpreted in this case in the same manner as in an injection well—zones of fluid entry are identified by cool anomalies caused by the cooling of the formation by the injected fluid. While drilling mud is injected, the lost-circulation zone may be indicated by a break toward warmer temperatures moving down the well. This is because the drilling mud circulating up the annulus has been warmed by contact with the formation at lower depths; above the lost-circulation zone, less of this warm drilling mud will be flowing in the annulus, resulting in less heating of the wellbore above the lost-circulation zone. Fig. 2.2 illustrates this behavior. Run 1 was made with drilling mud circulating at a high rate; a slight shift to higher temperatures is observed at the location of the lost-circulation zone. For Run 2, the annulus was shut in and drilling mud was injected so that mud was being displaced into the lost-circulation zone. In this case, none of the warm drilling mud from the bottom of the well is circulating above the zone of loss. This results in a sharp break to higher temperatures at the location of the lost circulation.

A radioactive-tracer log can locate a zone of lost circulation as long as the zone is not too far off bottom. A slug of radioactive tracer ejected into the drilling mud at the bottom of the drillstring can be tracked back up the annulus with the tool's gamma ray detector. At the lost-circulation zone, the radioactivity will decrease as tracer is carried into the formation with the drilling fluid. Unless all the drilling fluid is being lost to the formation, some tracer will be circulated to the surface, thus requiring precautions in handling the radioactive fluid.

Interpretation of temperature and tracer logs to detect lost-circulation zones can be complicated by certain borehole conditions that lead to similar log responses—in particular, large washouts. The radioactive-tracer log is likely to yield more positive identification of the lost-circulation zone in this instance.

2.2.2 Well-Kick Source and Underground Blowouts. A well kick occurs when fluid enters the wellbore from the formation during drilling and reaches the surface. When fluid influx to the wellbore occurs, the rapid increase in wellbore pressure sometimes causes formations above the kick source to break down, resulting in an underground blowout. Production logs have proven useful in locating the source of fluids in well kicks and in locating underground blowouts, particularly with the improvements in noise logging that have been made in the past 15 years.[3,10,11]

Temperature and noise logs are the main diagnostic tools for locating the sources of kicks and underground blowouts. Figs. 2.3

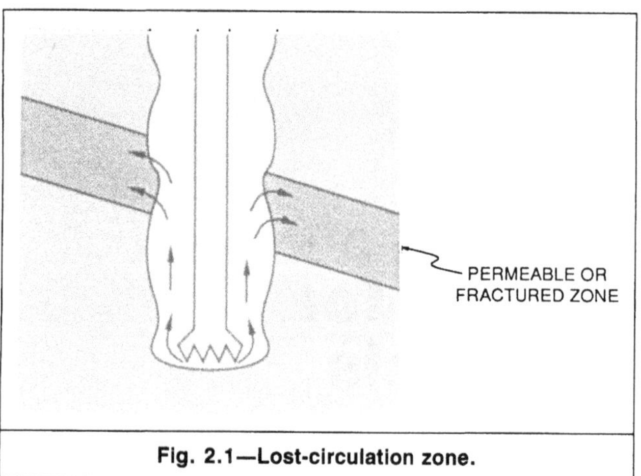

Fig. 2.1—Lost-circulation zone.

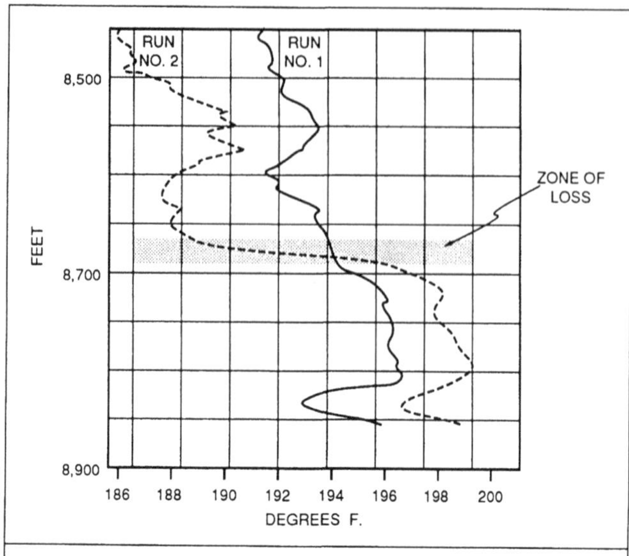

Fig. 2.2—Temperature log locates zone of lost circulation (from Ref. 9, courtesy PennWell Publishing Co.).

and 2.4 illustrate this application. This well was drilling ahead just below the casing set at 17,000 ft [5182 m] when a kick occurred and the well was shut in. Fig. 2.3 shows the temperature log that was run after 6 hours. The increase in temperature above the geothermal temperature from about 15,600 ft [4755 m] to total depth (TD) indicates that fluid from the kick source at TD is flowing upward and entering the formation at 15,600 ft [4755 m]. The temperature log does not resolve whether the fluid is flowing through the casing and exiting through a casing leak at 15,600 ft [4755 m] or whether the kick has broken down the cement so that the fluid is channeling through the cement to the location of the underground blowout. The noise log (Fig. 2.4) answers this question. The multiple peaks on the noise log indicate flow through a tortuous path, so the flow path is most likely through channels in the cement. The frequency characteristics of the noise peaks also indicate that the flow is single-phase. Because the wellbore contains drilling mud, the source fluid for the underground blowout must be a liquid and not a gas.

2.2.3 Cement Tops. One of the earliest applications of production logging was the location of the top of cement. Because curing cement is undergoing an exothermic reaction, the temperature of the wellbore generally will rise significantly in the cemented region, making a temperature log a means of detecting the top of the cement. Alternatively, the first part of the cement pumped can be tagged with radioactive material so that a gamma ray detection tool can find the cement top. A cement-bond log usually will also indicate the top of cement.

Locating cement tops from the heating caused by the curing cement dates from the earliest applications of temperature logging. There are many examples of the use of temperature logs to locate cement tops.[12-16] Fig. 2.5 illustrates the large temperature increase, indicating the top of cement, that can occur when a temperature log is run shortly after cementing. On the other hand, a temperature log will not always clearly pinpoint the top of cement, as seen in Temperature Log A in Fig. 2.6. This log may have been run too long after cementing for a clear temperature anomaly to persist.

A gamma-ray log can be used to locate the top of cement if the first portion of the cement is tagged with a radioactive material. Fig. 2.7 illustrates the high gamma ray level that identifies the final location of the cement that has been tagged with radioactivity. The hotspots below the top of cement apparently result from the accumulation of radioactive cement in washouts.

Finally, a cement-bond or ultrasonic-pulse-echo log should give a clear indication of the top of cement when the cement is bonded reasonably well to the casing. Chap. 11 discusses the responses of these logs to uncemented and cemented casing. An unfocused-

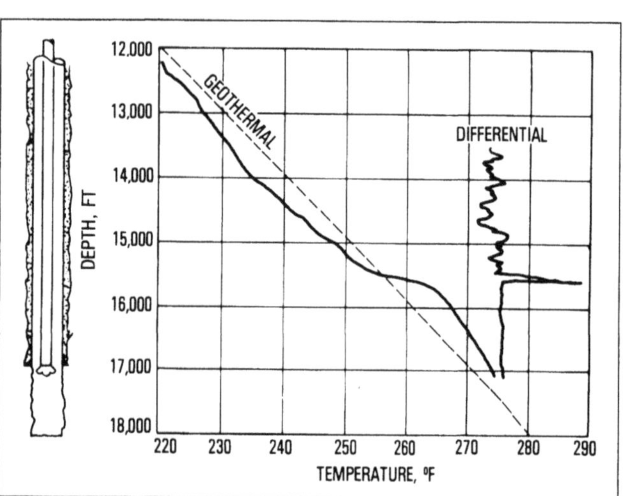

Fig. 2.3—Temperature log from well with underground blowout (from Ref. 3).

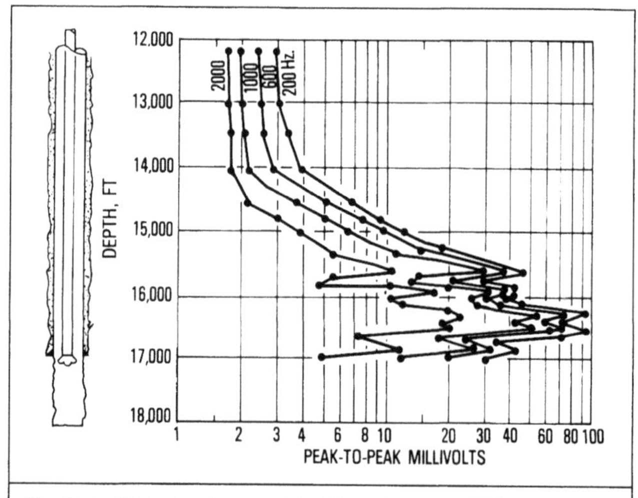

Fig. 2.4—Noise log from well with underground blowout (from Ref. 3).

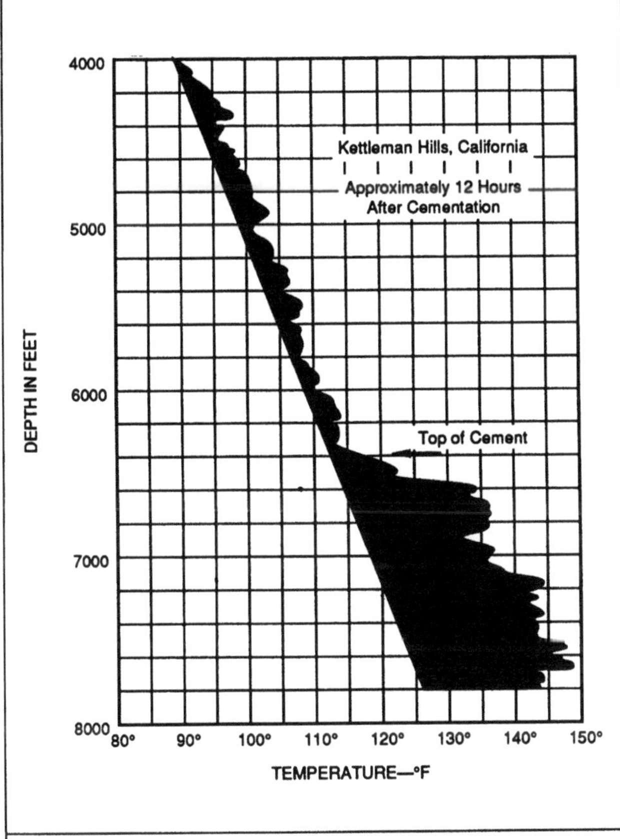

Fig. 2.5—Temperature log indicating top of cement (from Ref. 16, courtesy Gulf Publishing Co.).

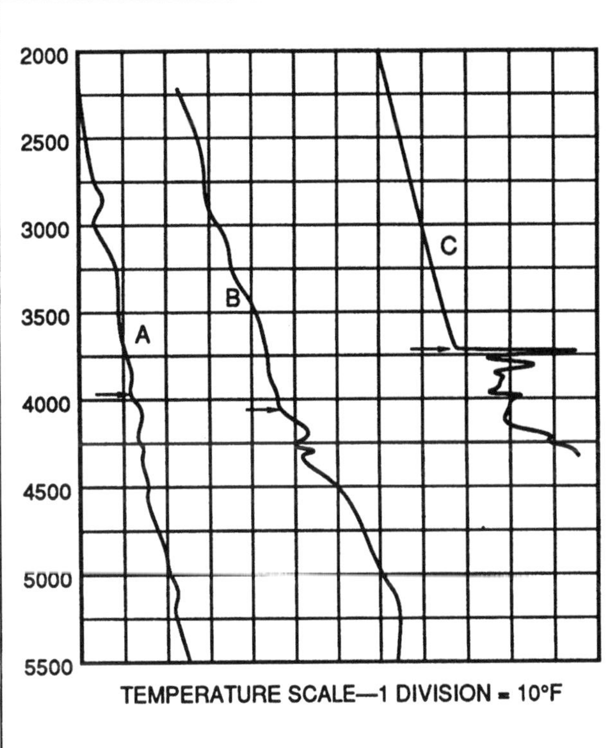

Fig. 2.6—Lack of response of temperature log to cement top—Well A (from Ref. 14).

gamma-ray-density log (Chap. 12) may also be used to locate cement tops.

2.2.4 Cement Quality. The need to determine the quality of cement has led to the development of logs aimed specifically at the evaluation of cement—the cement-bond and ultrasonic-pulse-echo logs. Applications of these logs to the diagnosis of cement quality are treated thoroughly in Chap. 11. The primary purpose of cementing is the hydraulic isolation of the various permeable zones penetrated by a wellbore. Other production logs—primarily temperature, radioactive-tracer, and noise logs—can be used to detect channeling in cement.

Channeling between the casing and the formation resulting from poor cement conditions can result in a number of problems—including excessive water or gas production or lost production to other permeable zones, as illustrated in Figs. 2.8 and 2.9. To identify channeling positively, production logs that respond to flow outside the casing and that measure flow inside casing must be run. The applications of temperature, radioactive-tracer, and noise logs to the detection of channeling in cement are described in the chapters on these logs.

2.2.5 Gravel-Pack Quality. Gravel-pack completions are commonly used to prevent the production of formation sand. Fig. 2.10 shows that, in a gravel-pack completion, sand is placed in the annulus between a slotted or perforated liner and the formation. This sand (the gravel) is of such a size that fluid flow is not significantly affected, but the formation sand is filtered by the gravel pack, thus preventing sand production. To function properly, the gravel pack must completely fill the annulus, yet uniform placement of gravel-pack sand is often difficult, particularly in deviated wells. Thus, a means of evaluating the effectiveness of a gravel pack is desirable.

The primary means of evaluating a gravel pack is with an unfocused-gamma-ray-density log. This log measures the density of the material behind the liner, and in this way can locate voids in a gravel pack. Unfocused-gamma-ray-density logging is discussed

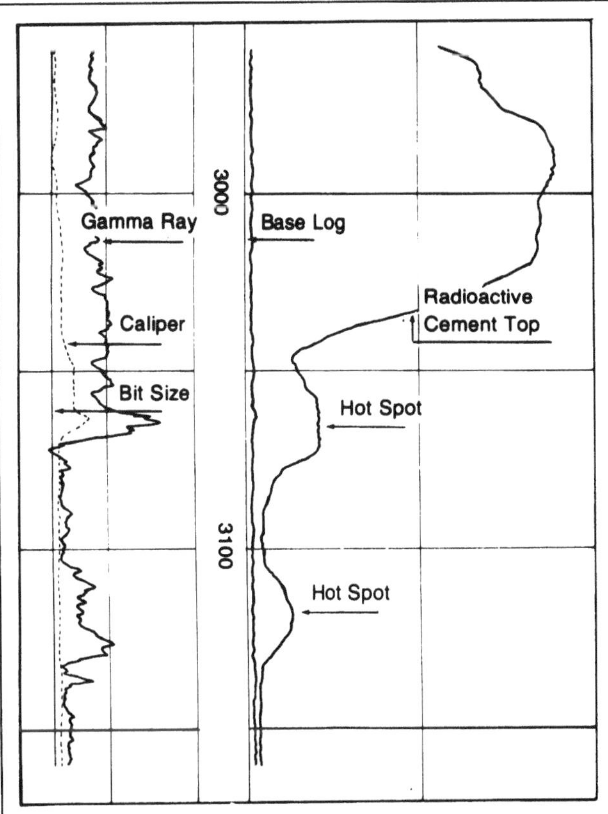

Fig. 2.7—Locating cement top with radioactive tracer log (from Ref. 17, courtesy Atlas Wireline Services, Western Atlas Intl. Inc.).

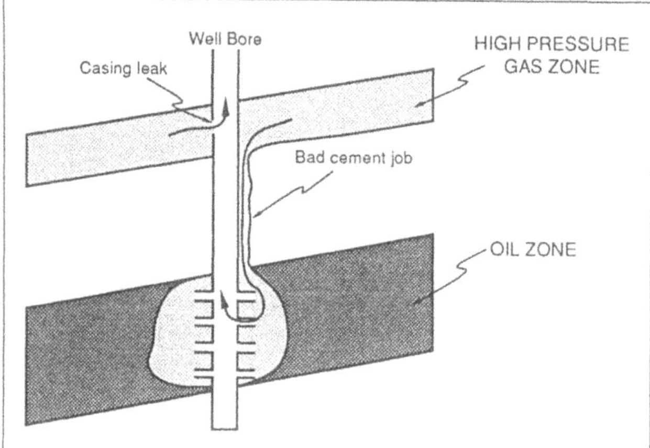

Fig. 2.8—Gas channeling. (From Ref. 18. Reprinted with permission of Edgell Communications. Copyright ©Sept. 1956, PETROLEUM ENGINEER INTERNATIONAL.)

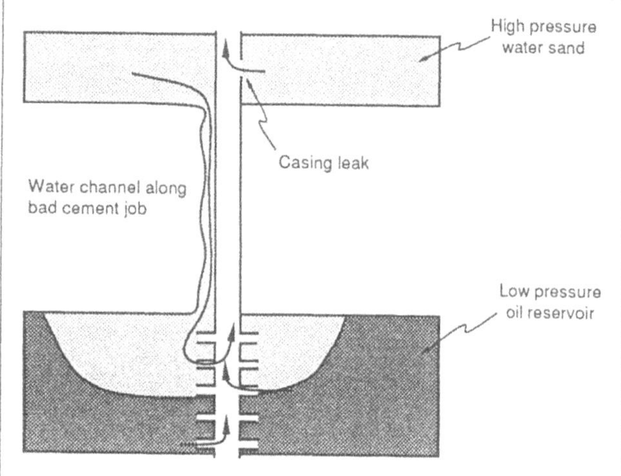

Fig. 2.9—Water channeling. (From Ref. 18. Reprinted with permission of Edgell Communications. Copyright ©Sept. 1956, PETROLEUM ENGINEER INTERNATIONAL.)

in Chap. 12. It may be possible to detect the locations of sand production, and thus regions of poor gravel-pack quality, with a noise log.[19] This technique is discussed in Chap. 10.

2.2.6 Perforation Location and Performance. In a cased completion, perforations provide the path for fluids from the formation to enter the wellbore. The perforations must be located at the depths intended and must be open to flow for optimum well performance. Production logs are often run immediately after perforating to evaluate the success of the perforating operation and to provide a baseline measurement of the production or injection profile of the well.

The locations of perforations can sometimes be seen directly with the casing-collar-locator log run with most production logs for depth correlation. More frequently, the locations and performance of perforations are inferred from the producing characteristics of the various perforated intervals in a well, as determined from a flow profile. The logging procedures used to measure flow profiles are a means of at least indirectly evaluating the performance of perforations. Use of production logs to measure flow profiles is treated in Sec. 2.3.

2.2.7 Well Treatment Evaluation. Many wells require stimulation upon initial completion, and in tight formations the well stimulation treatment may be the most crucial (and perhaps most expensive)

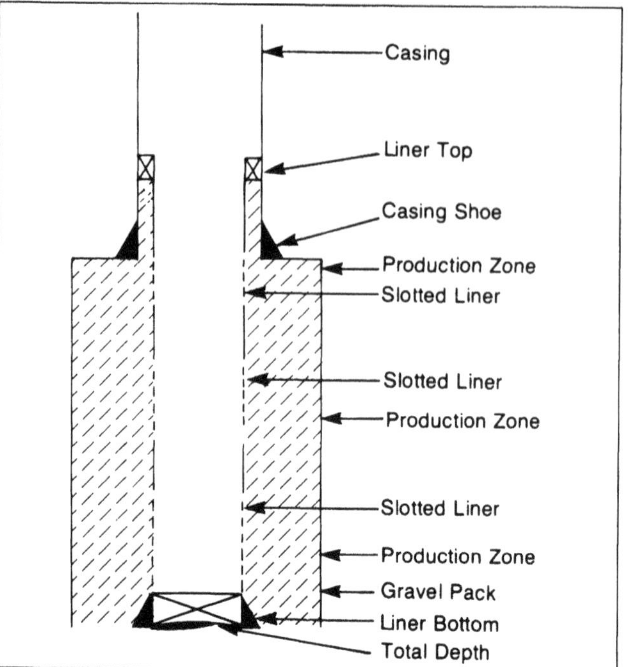

Fig. 2.10—Typical gravel pack completion (from Ref. 17, courtesy Atlas Wireline Services, Western Atlas Intl. Inc.).

phase of the well completion. Production logs can be used to evaluate the regions affected by the well treatment and the effect of the treatment on the productivity of different zones penetrated by the well. The use of production logs to diagnose well treatments is discussed in Sec. 2.4.

2.3 Applications of Production Logs During Subsequent Production or Injection

The primary applications of production logging are problem-well diagnosis and reservoir surveillance throughout the producing (or injecting) life of a well. This section describes the common uses of production logs during this phase of a well's life. The section is divided into single-phase and multiphase flow applications because the well problems or logging objectives and the logging procedures applied often differ for these two types of flows.

Deciding whether the flow consists of a single phase or multiple phases in the region of the well to be logged is not as trivial a problem as it may seem. Most production logging measurements are made below the tubing, in the region of the reservoir zones producing or receiving fluids. Most injection wells—water or gas—will be single-phase, though exceptions can exist, particularly when steam or CO_2 is injected. In production wells, the downhole flow conditions may be single-phase, but are more likely to be two- or three-phase. Examples of single-phase production wells would be an oil well producing no water with a bottomhole pressure (BHP) above the bubblepoint pressure of the crude oil or a dry gas well. Even in these cases, the flow conditions opposite the producing formations may be multiphase because water may be present in the wellbore even though there is no net production of water at the surface. In general, single-phase flow conditions are rare in producing wells, so the complications of multiphase flow should usually be considered when logging in production wells.

2.3.1 Single-Phase Flow Applications. Because single-phase flow conditions are most commonly encountered in injection wells, this section will describe the uses of production logs in injection wells, though it should be remembered that these same techniques can be applied in the occasional production well that is producing a single phase. Production logs are used in injection wells to monitor reservoir behavior and to evaluate problems observed with the individual injection well or the reservoir. Among the problems that may arise are abnormally low or high injectivity, abnormal pressure or fluid level in the annulus, and low productivity or high water production in offset producers. To evaluate such problems, production logs are run to measure the amount of flow to each reservoir interval, to check for interval isolation, to locate high-permeability zones, and to find leaks in the well equipment. Stratton *et al.*[6] concisely summarize these applications of production logs:

"The injection wells are generally receiving water as part of a secondary recovery, a pressure maintenance, or sometimes simply a disposal system. The injection medium may also be liquid hydrocarbons, gas (including air), or a combination of liquid and gas such as saturated steam. Whatever the injection fluid, the main purpose of the production log is generally to determine the injectivity profile, that is, to quantitatively assign volumes or percentages of fluid to each of the intervals taking significant amounts of fluid. While the profile is being obtained, it is often important to check also for casing failures, packer leaks, faulty cement jobs and interzonal migration."

Measurement of Flow Profiles. The fundamental information sought with a production log in an injection well is the flow profile, the amount of fluid being injected into each interval. The injection profile obtained with a production log is typically displayed as shown in Fig. 2.11, with the percent of total injection rate plotted vs. depth in the right track and a bar graph of the percent of total flow rate entering certain discrete intervals presented in the left track. Flow profiles are measured in injection wells with temperature, radioactive-tracer, and spinner-flowmeter logs. A temperature log will yield a qualitative indication of the formation injection intervals, while the spinner-flowmeter or radioactive-tracer logs more precisely define the amount of flow exiting the wellbore. The formation injection intervals and the fluid exit locations may not be contiguous; if channeling is occurring, they will not be. Thus, a combination of logs is often necessary to determine the profile of injection into the formation. Chaps. 4 through 7 treat the use of these logs for measuring flow profiles.

Determination of Interval Isolation. The flow profile shows where fluids leave the wellbore, but there is no guarantee that fluids are entering the formation at these same locations, because they may move through channels behind the casing and enter zones other than those intended. The capability of the well completion to isolate the injection zones from other formations is crucial to proper reservoir management and is thus an important property to be evaluated with production logs. Cement-bond or ultrasonic-pulse-echo logs indicate the possibility of channeling by measuring the mechanical properties of the cement behind the casing. To identify channeling positively, a production log that can respond to flow behind the casing is needed. Among the logs that will serve this purpose are temperature, radioactive-tracer, and noise logs.

Reasons for Anomalous Rate Changes. A change in rate and/or wellhead pressure often indicates a serious well or reservoir problem. Abnormally low injectivity or a marked drop in injection rate can result from formation damage around the wellbore, plugged perforations, restrictions in the casing or tubing, or scaling. An unusually high injection rate may be caused by leaks in the casing, tubing, or packer, by channeling to other zones, or by fracturing of the reservoir. Some of these problems can be clearly identified with production logs and application of techniques used for flow profiling or measuring interval isolation, while others may require additional well tests to confirm the cause of the abnormal behavior. For example, a pressure-transient test can measure a well's skin factor, a measure of restricted or enhanced flow around the wellbore; a flow profile obtained with production logs can indicate the vertical distribution of the damaged or stimulated regions around the wellbore.

The cause of a rate change in a well is often easier to diagnose if production logs have been run periodically throughout the life of the well. For example, the flow profile in a water injection well may change gradually throughout the life of the well as the saturation distribution changes in the different reservoir layers. Occasional logs would show this as being a natural progression in a waterflood. Without knowledge of this gradual change, a profile obtained several years after the start of injection might appear suffi-

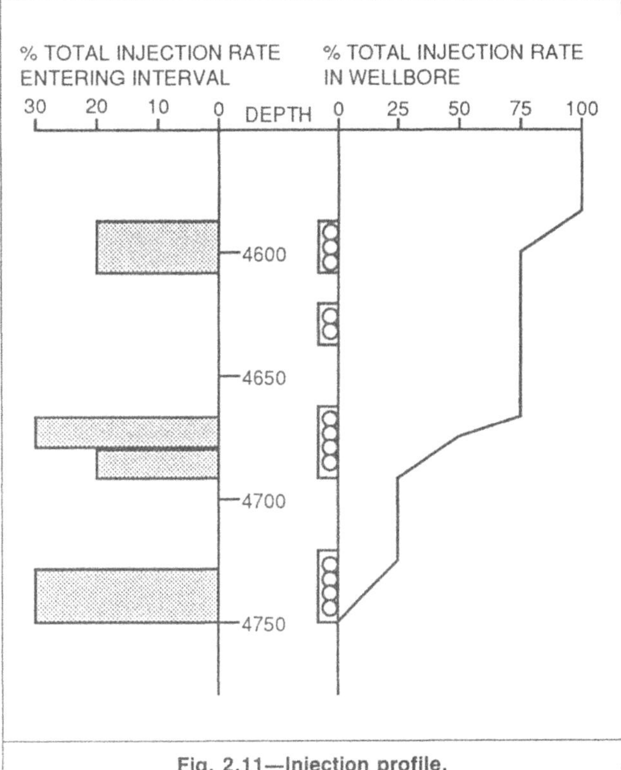

Fig. 2.11—Injection profile.

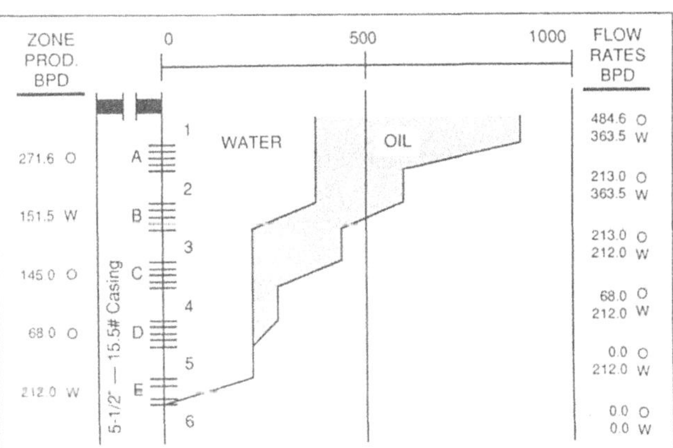

Fig. 2.12—Production well profile (from Ref. 20, courtesy Schlumberger).

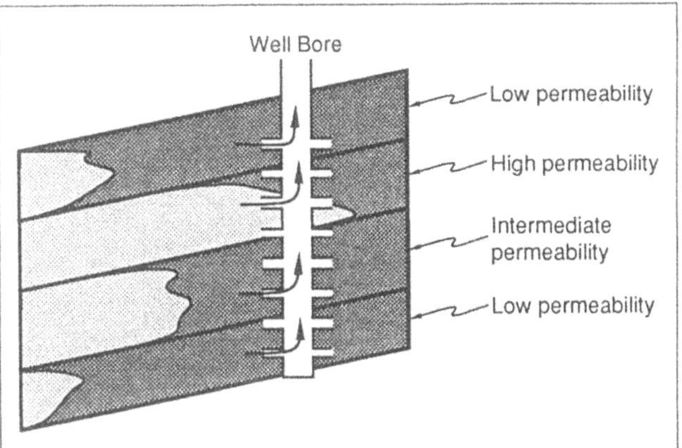

Fig. 2.13—Early water breakthrough in high-permeability layers. (From Ref. 18. Reprinted with permission of Edgell Communications. Copyright ©Sept. 1956, PETROLEUM ENGINEER INTERNATIONAL.)

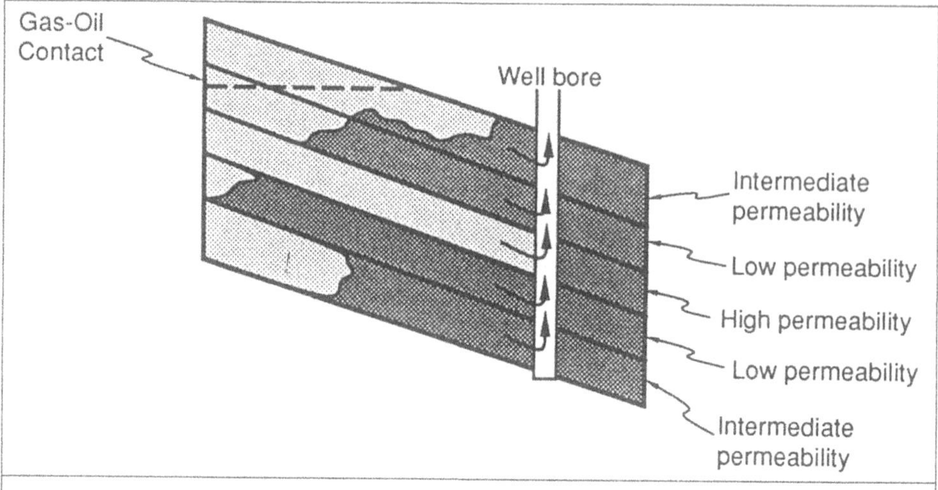

Fig. 2.14—Early gas breakthrough through high-permeability layers. (From Ref. 18. Reprinted with permission of Edgell Communications. Copyright ©Sept. 1956, PETROLEUM ENGINEER INTERNATIONAL.)

ciently different from the initial profile to lead an engineer to conclude that channeling had developed or some other drastic change had taken place.

2.3.2 Multiphase Flow Applications. At the downhole conditions where production logging measurements are made, the presence of more than one phase is likely in production wells, regardless of the surface production conditions. In oil production wells, water production is common, and if the BHP is below the bubblepoint pressure, free gas will be present in the wellbore. A gas well may have water or condensate present in the wellbore even when there is no liquid production at the surface. Thus, in any production well, the possibility of multiphase flow must be considered in planning a production logging job or in interpreting production logs.

As in a single-phase well, the fundamental information sought with a production log is the flow profile, but in the multiphase case, the locations and rates of the entries of each phase are needed. Fig. 2.12 shows a flow profile for a well producing oil and water. The water production rate and the total production rate are plotted vs. depth, with the oil rate shown as the difference between the two curves. To define the flow profile of more than one phase, a log or logs that can identify the amount of each phase present in the wellbore and logs that measure total flow rate must be run.[21] Fluid density or capacitance logs are generally used with some type of flow metering technique, such as a spinner-flowmeter, a flow-concentrating-flowmeter, or a radioactive-tracer log. The results of these combinations of logs are interpreted by making assumptions about the nature of the multiphase flow to yield the flow rate of each phase at different depth positions in the well, as discussed in Chap. 9. Including a temperature log as one of the production logs run in a multiphase well is always a good practice. For example, when flowmeters and fluid-identification logs are inaccurate because of the complex flow regime, a temperature log will often yield at least a qualitative measure of the flow profile.

Production logs in multiphase-flow wells are often aimed at the same problems as in single-phase wells, such as testing for interval isolation or determining the reason for anomalous rate changes. In addition, an anomalous rate of production of a particular phase is

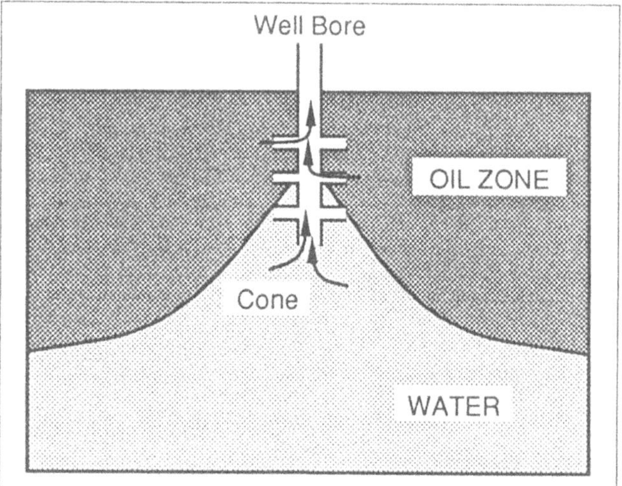

Fig. 2.15—Water coning. (From Ref 18. Reprinted with permission of Edgell Communications. Copyright ©Sept. 1956, PETROLEUM ENGINEER INTERNATIONAL.)

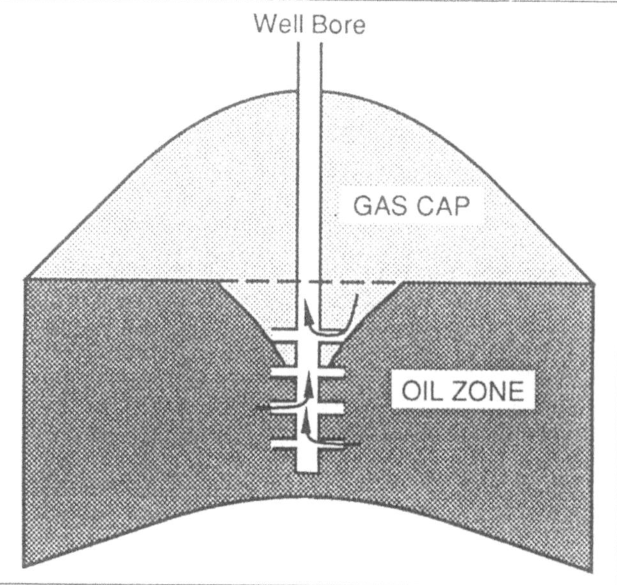

Fig. 2.16—Gas coning. (From Ref. 18. Reprinted with permission of Edgell Communications. Copyright ©Sept. 1956, PETROLEUM ENGINEER INTERNATIONAL.)

often the problem being investigated with a production log in a multiphase well. Excessive water or gas production is usually the difficulty addressed and can be caused by channeling, preferential flow through high-permeability zones in the reservoir, or coning.

Channeling between the casing and the formation caused by poor cement conditions (Figs. 2.8 and 2.9) is sometimes the cause of high water or gas production rates. As discussed in Sec. 2.2.4, logs that respond to flow behind casing—including temperature, radioactive-tracer, and noise logs—are the best means of identifying channeling.

Preferential flow of water or gas through high-permeability zones (Figs. 2.13 and 2.14) is an extremely pervasive problem that can sometimes be pinpointed with production logs. Excessive water production may result from injected water in a waterflood or from water encroaching from an aquifer. An accurate flow profile of the production well will identify the location of the high-permeability zone or zones contributing the high water production. Because the measurement and interpretation of logs in the multiphase flow in production wells are generally less accurate than those in a single-phase flow, water distribution in the reservoir is often monitored by measuring injection well profiles and assuming continuity of the reservoir layers between injectors and producers in waterflood operations.[22] Excessive gas production can similarly result from flow of injected gas or from gas moving through high-permeability layers from a gas cap. Again, a flow profile measured in a production well will identify the offending entry locations, or the high-permeability zones causing high gas production can be inferred from profiles measured in gas injection wells when gas is being injected into the reservoir.

Water or gas coning, illustrated in Figs. 2.15 and 2.16, is another possible source of excessive water or gas production. Water coning results when a well is completed near a water/oil contact and sufficient vertical permeability exists for water to migrate upward to the wellbore as the pressure is drawn down around the well. Similarly, gas can cone down from a gas cap if vertical permeability is high enough. Discussions of water or gas coning are presented by Frick and Taylor[23] and Timmerman.[24] Coning is a difficult phenomenon to identify conclusively with production logs. Consider a well experiencing water coning. A flow profile will indicate water production from the lower part of an interval. This water could result from channeling from below the perforated zone, flow through a high-permeability zone located in the lower part of the interval, or coning. A log responding to flow outside the casing, such as a noise log, might be able to eliminate channeling as the water source. Distinguishing between coning and flow through a high-permeability layer will be difficult with production logs alone. Other well tests, such as pressure-transient tests, or information from cores about the permeability distribution in the reservoir (including vertical permeability) will be needed to identify coning clearly. Another possible approach is to measure the flow profile at several different total production rates because coning is sensitive to the pressure drawdown of the well. An increasing water production rate as total production is increased and, in particular, the water production expanding upward as total rate increases would indicate coning.

2.4 Applications of Production Logs With Well Workovers

Production logs can often play a key role in planning and evaluating well workover operations. Among the workovers that can benefit from production logging are cement squeezing, reperforating, acidizing, fracturing, and water shutoff or profile modification treatments.

Information provided by production logs is often essential to the proper planning of a well workover. For example, before conducting a cement squeeze operation aimed at eliminating channeling, the channel must be located with a temperature, radioactive-tracer, or noise log. A flow profile measured with production logs will identify zones of low productivity or injectivity that may be corrected by reperforating, acidizing, or fracturing. The noise log shown in Fig. 2.17, for example, shows that only 19 out of several hundred perforations are producing significant amounts of fluid. The log suggests that reperforating or acidizing might significantly increase the productivity of this well.

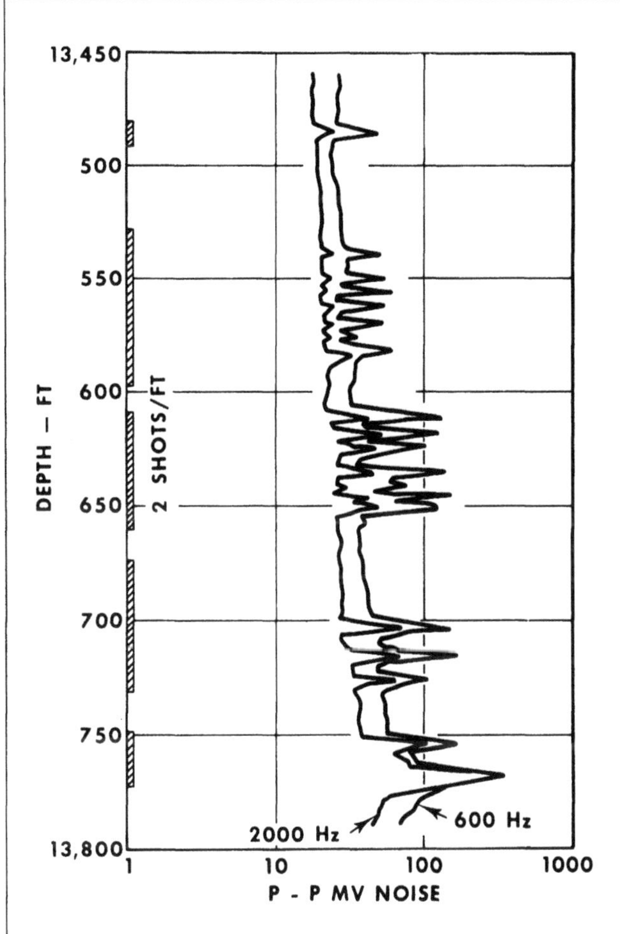

Fig. 2.17—Flow from individual perforations indicated by noise-amplitude peaks (from Ref. 19).

Production logs have a similar role in the evaluation of workover operations because they can indicate what regions of the well were affected by the workover and to what extent. Temperature logs are routinely used to measure the fracture height near the wellbore after fracturing operations.[25-27] Fracture height near the wellbore is also frequently measured by radioactively tagging the fracture proppant and then running a radioactive-tracer log after the treatment to locate the tagged proppant. A flow profile measured after a well stimulation or a profile modification treatment can often give the best indication of whether the desired effect of the workover was achieved.

2.5 Production Logging Job Planning and Reporting

2.5.1 Planning the Production Logging Job. The benefits of running a production log can be maximized by planning the job so that the appropriate tools are run properly and by conscientiously recording sufficient data for correct log interpretation. In this section, it will be assumed that the logging objective is to measure a flow profile because this is the most common goal of production logging. The first step in planning a production logging job is to determine whether the well being considered is suitable for production logging in general and then whether it is suitable for particular logging methods. The primary general requirement for applying production logs is that the logging tools can be placed in the well in the desired locations (no restrictions exist that would prevent logging). When logging a well that has not been logged recently, running a gauge bar on a slickline to test for wellbore accessibility is a prudent practice.

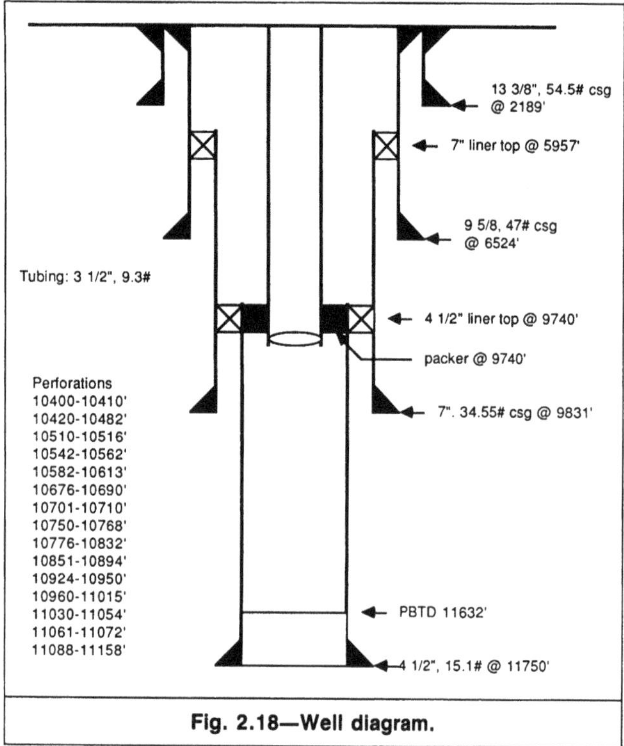

Fig. 2.18—Well diagram.

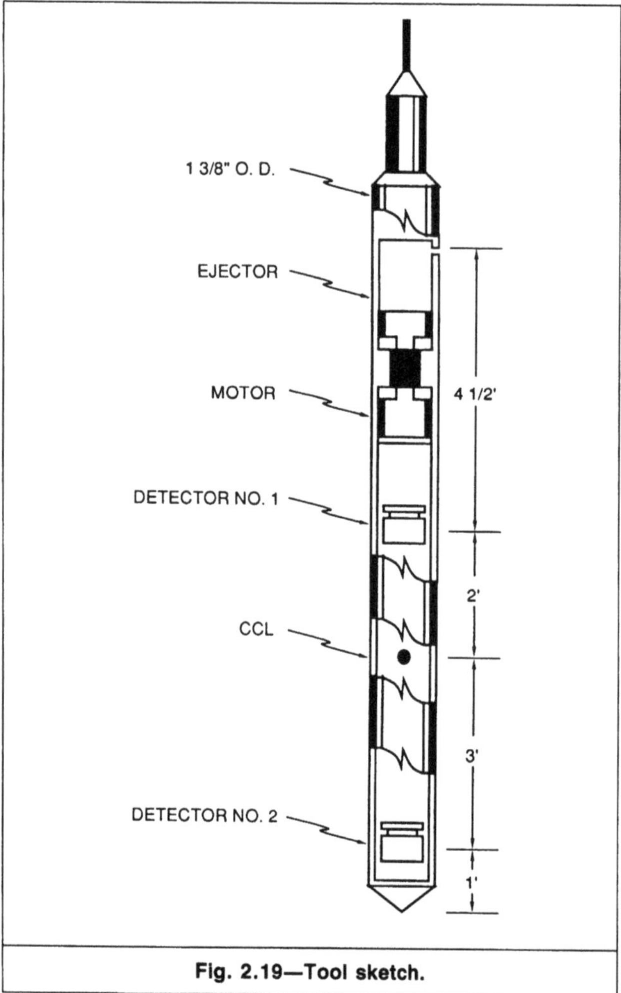

Fig. 2.19—Tool sketch.

If the well is suitable for production logging, the next decision to be made is which log or logs to run. This may depend on the particular logging objectives (flow profiling, channel identification, etc.), but in the typical case where the objective is to measure the flow profile and detect problems such as leaks or channels, logs that respond to flow both inside and outside casing are needed. In a single-phase well, a combination of a temperature and a radioactive-tracer log or a temperature and a spinner-flowmeter log should provide the desired information. The choice between the radioactive-tracer and the spinner-flowmeter logs might depend on many factors, including availability, cost, and expected accuracy. Chap. 7 presents comparisons of the different logging methods in single-phase flow.

The choice of logs for a multiphase flow well is more complicated because, in addition to measuring flow inside and outside the casing, logs to identify the composition of the flow stream in the wellbore are needed. Furthermore, the flow-rate-measuring logs are inherently less accurate in multiphase flow. A spinner flowmeter, for example, will often be unreliable, particularly if the well is deviated. In general, the engineer must rely on experience in a particular area in selecting logging suites in multiphase wells and in recognizing that often only qualitative information about the well is obtainable with production logs.

2.5.2 Record Keeping. Maintaining good records of well history and of previous production logs run on a well is important to sound production logging practices. Some logs, notably the temperature log, are sensitive to the production or injection history of a well, not just its current operating conditions. Comparisons of previous production logs on a well with a current log can often show things about the well's or reservoir's behavior that would not be evident on the basis of a single log. The benefits of production logging can be maximized by having previous logs readily available to the engineer interpreting a new production log on a well.

2.5.3 Production Logging Data Requirements. The value of a production log is greatly enhanced when it contains adequate supporting data along with the logs recorded. As a minimum, the log should contain *all* information needed for an independent interpretation of the log. This information includes (1) a description of the well completion, (2) production or injection conditions, (3) wellbore fluid properties, (4) description of logging tools—including dimensions and logging procedures, and (5) raw (uninterpreted) data from logging tools.

The required information about the well completion can be best presented with a well sketch, such as the one shown in Fig. 2.18. The sketch should include the sizes and bottom locations of all casing and tubing strings, the locations of packers, and the locations of all perforations in a cased-hole completion. Any special completion, such as a gravel pack, should be clearly described. In an openhole completion, the nominal hole size should be noted along with any known variations in hole diameter. In fact, if available, a caliper log should be included. The well diagram should also indicate the lowest depth that could be logged and the well deviation.

When a flow profile is measured, it is essential to be able to estimate the total rates of each phase at bottomhole conditions in a well. At a minimum, the surface production rates and the surface temperature and pressure should be indicated on the log. In addition, if any significant rate changes have occurred within the past year, the rate history should be given because some logs, particularly the temperature log, respond to regions of past injection or production as well as current conditions.

Physical properties of the wellbore fluids are needed to convert surface rates to downhole rates. Among the information that should be included are the FVF's for each fluid, solution GOR, the gravities of each phase, the surface and bottomhole temperatures and pressures, the bubblepoint of the oil in an oil production well, and the viscosities of all fluids at downhole conditions. If all this information is not available, as a minimum, the surface and bottomhole temperatures and pressures and the gravities of all fluids should be provided so that fluid properties can be estimated from corre-

TABLE 2.1—WELL DIAGNOSIS—DRILLING AND COMPLETIONS

Surface Indication of Poor Behavior/ Logging Objective	Possible Causes	Recommended Logs or Other Tests	Comments	References
Well treatment design and evaluation		Flow profiling logs—various combinations of temperature, radioactive-tracer, spinner, density, and capacitance (see Tables 2.2 and 2.3)	Many well treatments require flow profile before and after treatment for design and analysis.	Sec. 2.2.7 Chaps. 7, 9
Evaluation of cement quality		1. Cement-bond 2. Ultrasonic-pulse-echo 3. Temperature 4. Radioactive-tracer 5. Noise 6. Unfocused-gamma-ray-density	Cement-bond or ultrasonic-pulse-echo logs give general indication of quality of cement. To locate channels, temperature, radioactive-tracer, or noise logs can be more positive indicators.	Secs. 2.2.4, 4.4.2, 5.3.3, 10.3 Chaps. 11, 12
Fracture-height measurement		1. Shut-in temperature 2. Radioactive-tracer (multiple isotope spectroscopy)	Temperature log measures fracture height based on cooling by the fracture fluid. Radioactive-tracer log locates radioactive proppant and/or fluids used in treatment.	Secs. 2.2.7, 4.5, 5.1
Mud kick or lost circulation	Underground blowout	1. Temperature 2. Noise	Temperature log responds to flow into the formation similar to an injection well. Noise log can detect the blowout location if downhole velocities are sufficiently high.	Sec. 2.2.2 Chaps. 4, 10, particularly Sec. 10.3
Evaluation of gravel-pack quality		1. Unfocused-gamma-ray-density 2. Noise	Unfocused-gamma-ray-density log responds to density of material behind pipe—can locate voids in gravel pack. Noise log can sometimes locate source of sand production.	Secs. 2.2.5, 10.4, 12.3
Evaluation of perforation performance		1. Flow profiling logs—various combinations of temperature, radioactive-tracer, spinner, density, and capacitance (see Tables 2.2 and 2.3) 2. Noise 3. Horizontal-spinner	Performance of perforations is inferred from the productivity or injectivity of zones perforated and comparison with core or openhole log data on reservoir permeability. Flow from individual perforations can be measured with noise log if velocities are high enough. Horizontal-spinner log can give qualitative indication of flow from high-rate perforations.	Secs. 2.2.6, 2.3, 6.6.3, 10.4 Chaps. 7, 9
Locating cement top		1. Temperature 2. Radioactive-tracer 3. Cement-bond 4. Ultrasonic-pulse-echo 5. Unfocused-gamma-ray-density	Temperature log run shortly after cementing shows warm region opposite cement resulting from heat of hydration. Radioactive-tracer log locates tracer added to first few barrels of cement. Cement-bond or ultrasonic-pulse-echo distinguishes between cemented region and free pipe.	Secs. 2.2.3, 4.4.2, 5.1 Chap. 11
Reduced or completely lost drilling returns	Lost-circulation zone	1. Temperature 2. Radioactive-tracer	With a shut-in temperature log, lost-circulation zone indicated by cool anomaly. Temperature log with fluid circulating has a break to higher temperatures at lost-circulation zone. With a radioactive-tracer log, mud is tagged with tracer and movement followed up the annulus—lost-circulation zone indicated by location tracer is lost to the formation. Complications—certain borehole effects, particularly washouts—can give similar response on both shut-in temperature and tracer log.	Secs. 2.2.1, 4.4.2 Chap. 5
Gas flow to surface while drilling	Inflow of gas from the formation owing to underbalance	1. Temperature 2. Noise	Temperature log responds to inflow of fluid in same manner as in production well. Noise log can locate gas entry from characteristic noise of gas flowing in liquid.	Sec. 2.2.2 Chaps. 4, 10

TABLE 2.2—WELL DIAGNOSIS—SINGLE-PHASE (INJECTION) WELLS

Surface Indication of Poor Behavior/ Logging Objective	Possible Causes	Recommended Logs or Other Tests	Comments	References
Low injectivity	1. Low formation permeability 2. Plugged perforations 3. Near-wellbore formation damage 4. Wellbore restrictions	1. Temperature and radioactive-tracer or spinner 2. Pressure-buildup or -drawdown test 3. Caliper	Injection profile measured with spinner or radioactive-tracer logs will indicate low-injectivity intervals. Comparison with core data or openhole logs may distinguish between low permeability and plugged perforations or formation damage. Pressure-transient tests will also distinguish between low permeability and near-wellbore or perforation damage. Restrictions may be located by how easily tools run in well; otherwise, a caliper log can be run to locate restrictions in casing.	Sec. 2.3.1 Chaps. 4 through 7
Abnormally high injectivity	1. Channeling 2. Tubing leak 3. Casing leak 4. Packer leak 5. Fractured reservoir	1. Temperature and radioactive-tracer 2. Temperature and spinner 3. Noise 4. Cement-bond 5. Step-rate test	Temperature log can indicate channels or leaks. Radioactive-tracer log can distinguish between channeling, casing leaks, and packer leaks. Tubing leaks can be located with special radioactive-tracer log. Spinner can suggest channeling or leaks by shape of profile or measured fluid velocity. Noise log can be an excellent locator of channels or leaks. Cement-bond log indicates possible channel locations. A fractured reservoir is suggested by absence of channels or leaks as determined by logs. This can be confirmed with a step-rate test.	Secs. 2.3.1, 2.2.4, 4.4.2, 5.3.3, 10.3 Chap. 11
Abnormal annulus pressure or fluid level	1. Packer leak 2. Tubing leak	1. Radioactive-tracer 2. Noise	Radioactive-tracer log easily detects packer leaks and can be used to locate tubing leaks. Noise log is a good indicator of tubing leaks; injecting gas can enhance response when locating tubing leaks.	Sec. 2.3.1 Chaps. 5, 10
Low productivity in offset producers	1. Channeling, leaks 2. Problems in producers 3. Reservoir problems—directional permeability, barriers, poor pattern placement, etc.	Same as for abnormally high injectivity	If no problems are found in injection well, production wells should be tested with production logs and/or pressure-transient tests.	Secs. 2.3.1, 2.2.4, 4.4.2, 5.3.3, 10.3 Chap. 11
High water production in offset producers	1. High-permeability zone(s)	1. Spinner or radioactive-tracer 2. Temperature	Injection profile measured with spinner or radioactive-tracer should locate high-permeability zones. Temperature log may indicate a high-permeability zone in a young well.	Sec. 2.3.1 Chaps. 4 through 7
Routine reservoir surveillance		1. Temperature and radioactive-tracer 2. Temperature and spinner	Injection profile should be measured. Further tests for leaks or channels should be made if problems indicated.	Sec. 2.3.1 Chaps. 4 through 7

lations. A review of techniques to calculate bottomhole flow rates from surface conditions is given by Atlas Wireline Services.[17]

The description of the logging tools and procedures should begin with a sketch of the tool string, showing the dimensions of the tools used and their locations in the tool string. Fig. 2.19 is such a sketch for a radioactive-tracer logging tool. Other information about the tools used, such as time constants used in averaging tool responses, should be noted on the log. All surface calibrations performed should be included on the log. A brief summary of the logging procedures used, which should include the number of runs made, the cable speeds used, and calibration constants, is necessary for later log interpretation.

Finally, it is important that the raw responses from the logging tools, not just interpreted results, be presented on the log. A radioactive-tracer log, for example, should include the actual responses of the gamma ray detectors, not just a table of transit times taken from the detector responses. As will be seen in later chapters, there are often several interpretation procedures that can be applied to a given log; the purchaser of the logging service should have all information required for an independent interpretation presented on the log.

2.5.4 Quality Control. Quality control should be a major concern of both the logging service company running the log and the operator

TABLE 2.3—WELL DIAGNOSIS—MULTIPHASE (PRODUCTION) WELLS

Surface Indication of Poor Behavior/ Logging Objective	Possible Causes	Recommended Logs or Other Tests	Comments	References
Low productivity	1. Low formation permeability 2. Plugged perforations 3. Near-wellbore damage 4. Wellbore restrictions 5. Channeling or crossflow 6. Lift mechanism problem	1. Temperature and spinner 2. Temperature and flow-concentrating-flowmeter	Production profile measured with spinner- (usually qualitative) or flow-concentrating-flowmeter will indicate low-productivity intervals. Comparison with core data or openhole logs may distinguish between low permeability and plugged perforations or formation damage. Pressure-transient tests will also distinguish between low permeability and near-wellbore or perforation damage. Restrictions may be located by how easily tools run in well; otherwise, a caliper can be run to locate restrictions in casing. Channeling or crossflow may be indicated by temperature or flowmeter logs. Problems with lift mechanisms should be determined with other tests.	Sec. 2.3.2 Chaps. 4, 9
Excessive water production	1. High-permeability zone 2. Channeling 3. Coning	1. Temperature, flowmeter, and density 2. Temperature, flowmeter, and capacitance 3. Noise	A combination of a flowmeter (spinner- or flow-concentrating) and density or capacitance log can be used to locate water entries. Information will often be only qualitative. Density or capacitance alone is sometimes sufficient to locate water entries. Water channeling may be located with temperature or noise log. Coning is difficult to distinguish from bottom high-permeability zone; special tests or sequence of logs run at different times or with different well rates may confirm coning.	Sec. 2.3.2 Chaps. 4, 9
Excessive gas production	1. High-permeability zone 2. Production from gas cap 3. Channeling 4. Coning	1. Temperature, flowmeter, and density 2. Noise	Temperature log may indicate gas production by cool anomalies. Combination of flowmeter (spinner- or flow-concentrating) and density logs may locate gas entries, though information is often only qualitative. Density log alone may identify gas-productive zones. Channeling may be indicated by temperature log; noise log should be run to locate gas channels more positively. Distinguishing between high-permeability zones, gas-cap production, and coning is difficult. Other information that may help includes well rate history, reservoir pressure history, core data, or openhole logs.	Sec. 2.3.2 Chaps. 4, 9
Routine reservoir surveillance		Various combinations of temperature, spinner, flow-concentrating-flowmeter, density, and capacitance	Particular tool combination depends on well conditions. Temperature should be part of all logging suites. If flow is single-phase or pseudohomogeneous, spinner-flowmeter is suitable; otherwise, flow-concentrating-flowmeter should be used. Recommended tool combinations are (1) single-phase—temperature and spinner; (2) oil/water flow—temperature, spinner- or flow-concentrating-flowmeter, density or capacitance for high or low oil concentrations; (3) gas/oil or gas/water flow—temperature, spinner- or flow-concentrating-flowmeter, density; and (4) oil/water/gas flow—temperature, spinner- or flow-concentrating-flowmeter, density, and capacitance.	Sec. 2.3.2 Chaps. 4,9

of the well. Ultimately, however, quality control is the responsibility of the operator paying for the service. As such, the first step in ensuring a quality log is to be sure that the well conditions are appropriate for the logs to be run. This can be accomplished with joint prelog planning by the operator and the logging company.

Once a logging program has been chosen, the logging operation should be conducted to ensure that the best possible log is obtained. All logging tools should be checked on the surface for proper operation. Repeat runs should always be made, and, where feasible, downhole calibration checks should be conducted (e.g., a fluid-density tool should be run in static water at the bottom of the well when possible.) Finally, the operator and the logging company personnel should cooperate to see that all the data described in the previous section are included on the log.

2.6 Well Diagnosis Summary

Tables 2.1 through 2.3 summarize the applications of production logs to diagnose well and reservoir behavior.

References

1. Wade, R.T. *et al.*: "Production Logging—The Key to Optimum Well Performance," *JPT* (Feb. 1965) 137–44.
2. Petevello, B.G.: "Evaluation of Well Performance Through Production Logging," *Proc.*, CWLS Fifth Formation Evaluation Symposium, Calgary (1975).
3. McKinley, R.M.: "Production Logging," paper SPE 10035 presented at the 1982 SPE Intl. Petroleum Exhibition and Technical Symposium, Beijing, March 18–26.
4. Curtis, M.R.: "Flow Analysis in Producing Wells," paper SPE 1908 presented at the 1967 SPE Annual Meeting, Houston, Oct. 1–4.
5. Cocanower, R.D. and Morris, B.P.: "Selection of Profile Techniques Based on Well Conditions," *JPT* (May 1966) 576–88.
6. Stratton, R., Chase, R., and Schaller, H.E.: "Case Histories of Production Logging," *JPT* (Feb. 1970) 207–13.
7. Jones, L.D.: "Analytical and Interpretive Problems of Production Logging," *JPT* (Aug. 1967) 993–98.
8. Connolly, E.T.: "Resume and Current Status of the Use of Logs in Production," paper presented at the 1965 SPWLA Annual Logging Symposium, Dallas, May 4–7.
9. Goins, W.C. Jr. and Dawson, D.D. Jr.: "Temperature Surveys to Locate Zone of Lost Circulation," *Oil & Gas J.* (June 22, 1953) 170–71.
10. McKinley, R.M., Bower, F.M., and Rumble, R.C.: "The Structure and Interpretation of Noise from Flow Behind Cemented Casing," *JPT* (March 1973) 329–38.
11. Robinson, W.S.: "Field Results From the Noise-Logging Technique," *JPT* (Nov. 1976) 1370–76.
12. Schlumberger, M., Doll, H.G., and Perebinossoff, A.A.: "Temperature Measurements in Oil Wells," *J. Inst. Pet. Technologists* (Jan. 1937) **23,** No. 159, 1–25.
13. Deussen, A. and Guyod, H.: "Use of Temperature Measurements for Cementation Control and Correlations in Drill Holes," *AAPG Bulletin* (1937) **21,** No. 6, 789–805.
14. Millikan, C.V.: "Temperature Surveys in Oil Wells," *Trans.*, AIME (1941) **142,** 15–23.
15. Guyod, H.: "Temperature Well Logging, Part 7," *Oil Weekly* (Dec. 16, 1946) 38–40.
16. Halbouty, Michel T.: "Temperature Effects on Oil Well Drilling," *Oil Weekly* (Dec. 18, 1939) 10–16.
17. *Interpretive Methods for Production Well Logs*, second edition, Atlas Wireline Services, Western Atlas Intl. Inc., Houston (1982) 77.
18. Clark, N.J. and Schultz, W.P.: "The Analysis of Problem Wells," *The Petroleum Engineer* (Sept. 1956) **28,** B-30–38.
19. McKinley, R.M. and Bower, F.M.: "Specialized Applications of Noise Logging," *JPT* (Nov. 1979) 1387–95.
20. Curtis, M.R.: "Flow Analysis With the Gradiomanometer and the Flowmeter," Schlumberger (1966).
21. Carlson, N.R. and Johnston, M.: "Importance of Production Logging Suites in Multiphase Flow," paper presented at the 1983 SPWLA Annual Logging Symposium, Calgary, June 27–30.
22. Yoelin, S.D., Howald, C.D., and Holbert, D.R.: "Production Logging as Used To Solve Water Injection Problems in the Huntington Beach Offshore Field," *JPT* (Sept. 1970) 1083–88.
23. "Reservoir Performance Equations," *Petroleum Production Handbook, Volume II—Reservoir Engineering,* T.C. Frick and R.W. Taylor (eds.), SPE, Richardson, TX (1962) 43–46.
24. Timmerman, E.H.: *Practical Reservoir Engineering, Volume II,* PennWell Books, Tulsa (1982) 49–60.
25. Bundy, T.E.: "Prefracture Injection Surveys: A Necessity for Successful Fracture Treatments," *JPT* (May 1982) 995–1001.
26. Wages, P.E.: "Interpretation of Postfracture Temperature Surveys," paper SPE 11189 presented at the 1982 SPE Annual Technical Conference and Exhibition, New Orleans, Sept. 26–29.
27. Agnew, B.G.: "Evaluation of Fracture Treatments with Temperature Surveys," *JPT* (July 1966) 892–98.

SI Metric Conversion Factors

cycles/sec	× 1.0*	E+00	= Hz
ft	× 3.048*	E−01	= m
°F	(°F−32)/1.8		= °C
in.	× 2.54*	E+00	= cm

*Conversion factor is exact.

Chapter 3
Single-Phase Flow in Pipes

3.1 Introduction

Before considering the behavior of specific logging tools in single-phase flow, some understanding of the nature of the flow itself is necessary. Two factors—whether the flow is laminar or turbulent and the velocity profile, which is the axial velocity as a function of radial position—affect production logging tools used in single-phase flow, such as spinner flowmeters and radioactive tracers. We will consider single-phase flow in circular and annular geometries because these are two common configurations encountered in production logging.

3.2 Laminar and Turbulent Flow

Fluid flow is characterized as being either laminar or turbulent, depending on the value of a dimensionless group, the Reynolds number, N_{Re}, defined for a circular geometry as

$$N_{Re}=\frac{d_{ci}\bar{v}\rho}{\mu}, \quad (3.1)$$

where d_{ci}=casing ID, $\bar{v}$=average velocity, ρ=fluid density, and μ=fluid viscosity. Consistent units must be used in the evaluation of the Reynolds number so that N_{Re} is dimensionless. Table 1 shows a few sets of units that can be used to calculate N_{Re}.

Laminar and turbulent flows are distinguished by $N_{Re}<2{,}000$ and $N_{Re}>4{,}000$, respectively. For Reynolds numbers between 2,000 and 4,000, the flow is called transitional and can display characteristics of either laminar or turbulent flow, depending on upstream flow disturbances and pipe roughness.

Fig. 3.1 illustrates the range of Reynolds numbers commonly found in injection or production wells. Most water-injection wells will be in turbulent flow, at least in the upper part of the well. For example, the flow of water (1 cp [0.001 Pa·s]) in a 5-in. [12.7-cm] -ID pipe will be laminar at rates less than 100 B/D [16 m^3/d] and turbulent at rates greater than 220 B/D [35 m^3/d]. Of course, at some depth in virtually any well, enough flow will have been lost to the formation that the flow rate remaining in the wellbore is low enough for laminar flow to occur. When this transition from turbulent to laminar flow takes place, some logging responses are affected profoundly.

With high-viscosity fluids, such as polymer solutions, laminar flow will persist at much higher flow rates. Fig. 3.1 shows that for injection of a 10-cp [0.01-Pa·s] fluid in the same 5-in. [12.7-cm] -ID pipe, 2,200 B/D [350 m^3/d] is required to establish turbulent flow. This analysis assumes Newtonian viscosity behavior. The likelihood of laminar flow is high when logging in viscous fluids.

3.3. Velocity Profiles

In laminar flow, the velocity profile can be determined analytically (see Ref. 2). For flow in a circular pipe, such an analysis yields

$$v(r)=\frac{(\Phi_O-\Phi_L)r_i^2}{4\mu L}\left[1-\left(\frac{r}{r_i}\right)^2\right], \quad (3.2)$$

where $\Phi_O=p_O+\rho g z_O$; $\Phi_L=p_L+\rho g z_L$; p_O, p_L=pressures at longitudinal positions a distance L apart; z_O, z_L=heights above some datum at these axial positions; r_i=inside pipe radius; r=radial distance from the center of the pipe; and $v(r)$=velocity as a function of radial position. This equation shows that the velocity profile is parabolic in laminar flow, with the maximum velocity occurring at the center of the pipe (see Fig. 3.2).

In production logging, we are generally interested in determining the average velocity because it is simply related to the volumetric flow rate by

$$\bar{v}=\frac{q}{A}, \quad (3.3)$$

where q=volumetric flow rate and A=cross-sectional area. Measurements, however, are often made in the center of the pipe; hence, a value near the maximum flow velocity is actually measured. The relationship between the average and the maximum velocities in laminar flow can be readily determined from Eq. 3.2. The average velocity is found by integrating the velocity profile over the area of flow,

$$\bar{v}=\frac{\int_0^{2\pi}\int_0^{r_i}v(r)r\,dr\,d\theta}{\int_0^{2\pi}\int_0^{r_i}r\,dr\,d\theta}=\frac{(\Phi_O-\Phi_L)r_i^2}{8\mu L}, \quad (3.4)$$

while the maximum velocity is obtained by setting r in Eq. 3.2 to zero,

$$v_{max}=\frac{(\Phi_O-\Phi_L)r_i^2}{4\mu L}\left[1-\left(\frac{0}{r_i}\right)^2\right]=\frac{(\Phi_O-\Phi_L)r_i^2}{4\mu L}. \quad (3.5)$$

Dividing Eq. 3.4 by Eq. 3.5 shows that the relationship between average and maximum velocities in laminar flow is

$$\frac{\bar{v}}{v_{max}}=0.5. \quad (3.6)$$

Thus, a measurement of velocity in the center of the pipe could be twice as high as the average velocity.

Turbulent flow does not lend itself to simple analytical treatment as does laminar flow because of the random nature of turbulent flow. Empirical expressions have been developed from experiments to describe the velocity profile in turbulent flow. One such expression that is fairly accurate for $10^5 > N_{Re} > 10^4$ is the power-law model:

$$\frac{v(r)}{v_{max}} = \left(1 - \frac{r}{r_i}\right)^{1/7} \quad \text{(3.7)}$$

This expression can show that

$$\frac{\bar{v}}{v_{max}} \approx 0.8. \quad \text{(3.8)}$$

Thus, in turbulent flow, the velocity profile is much flatter than that found in laminar flow and the average velocity is closer to the maximum velocity (Fig. 3.2). The ratio $\bar{v}/v_{max}$ in turbulent flow varies with Reynolds number and pipe roughness, but is generally in the range of 0.75 to 0.86.[3,4] In interpretation of production logs, a value of $\bar{v}/v_{max}$ of 0.83[2,5] is typically used; the log interpreter should recognize that this value is not always correct and that there will often be some error in converting measured velocities to average velocities.

TABLE 3.1—CONSISTENT UNITS FOR CALCULATING N_{Re}

d_{ci}	$\bar{v}$	ρ	μ
ft	ft/sec	lbm/ft³	lbm/(ft-sec)
m	m/s	kg/m³	Pa·s (10 poise)
cm	cm/s	g/cm³	g/cm-s (poise)

3.4 Flow in Annuli

Often in production logging the flow of interest is in the annular space between the logging tool and the inner surface of the casing. Fig. 3.3 shows typical velocity profiles for flow in an annulus. As in the circular geometry, the velocity profile is parabolic when the flow is laminar and flatter in turbulent flow. Notice also that, in annular flow, the maximum velocity does not occur in the center of the annulus but is shifted toward the inner surface. For this geometry, in laminar flow, the average velocity is given by

$$\bar{v} = \frac{(\Phi_O - \Phi_L) r_i^2}{8\mu L}\left[\frac{1-F_d^4}{1-F_d^2} - \frac{1-F_d^2}{\ln(1/F_d)}\right], \quad \text{(3.9)}$$

where F_d is the ratio of tool diameter to casing ID, defined by

$$F_d = \frac{d_T}{d_{ci}}. \quad \text{(3.10)}$$

The maximum velocity is

$$v_{max} = \frac{(\Phi_O - \Phi_L) r_i^2}{4\mu L}\left(1 - \left[\frac{1-F_d^2}{2\ln(1/F_d)}\right] \times \left\{1 - \ln\left[\frac{1-F_d^2}{2\ln(1/F_d)}\right]\right\}\right). \quad \text{(3.11)}$$

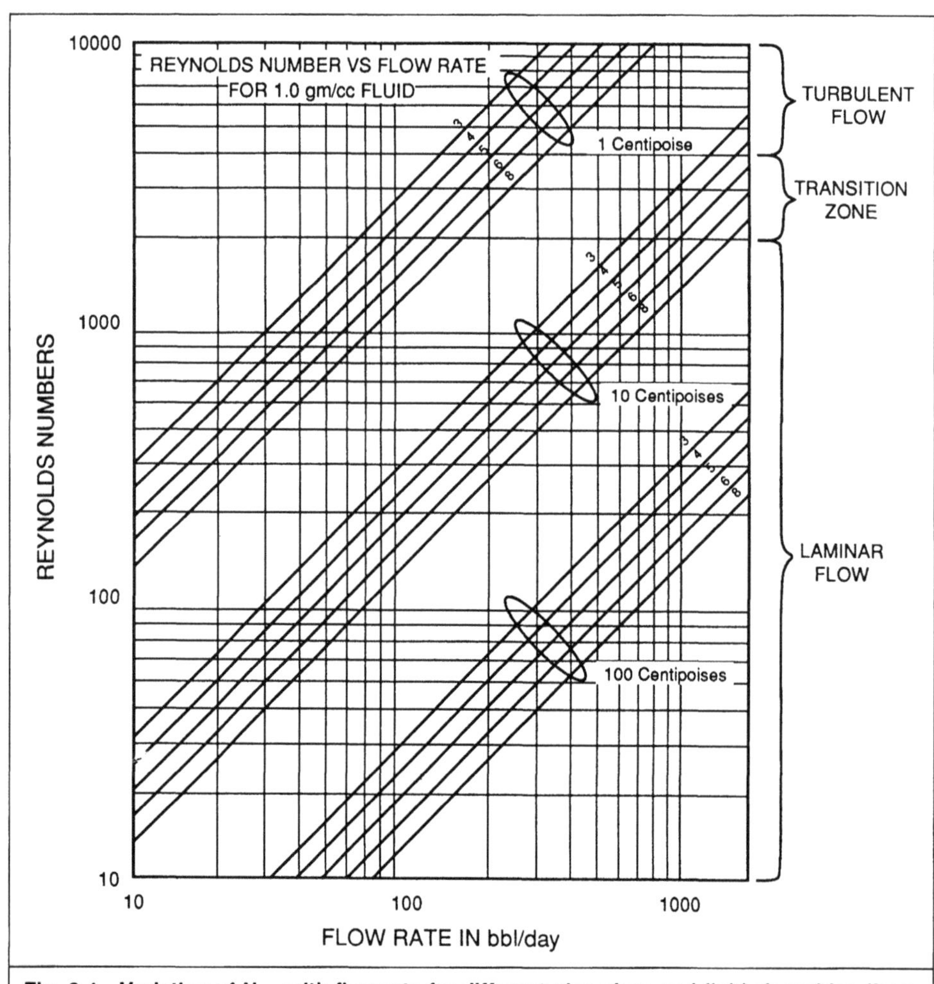

Fig. 3.1—Variation of N_{Re} with flow rate for different pipe sizes and fluid viscosities (from Ref. 1, courtesy Schlumberger).

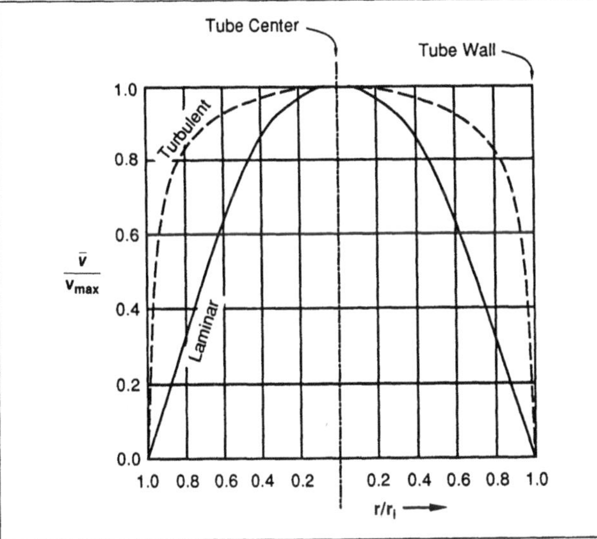

Fig. 3.2—Qualitative comparison of laminar and turbulent velocity distributions (from Ref. 2, *Transport Phenomena*, R.B. Bird *et al.*, ©1960, John Wiley & Sons Inc., reproduced with permission).

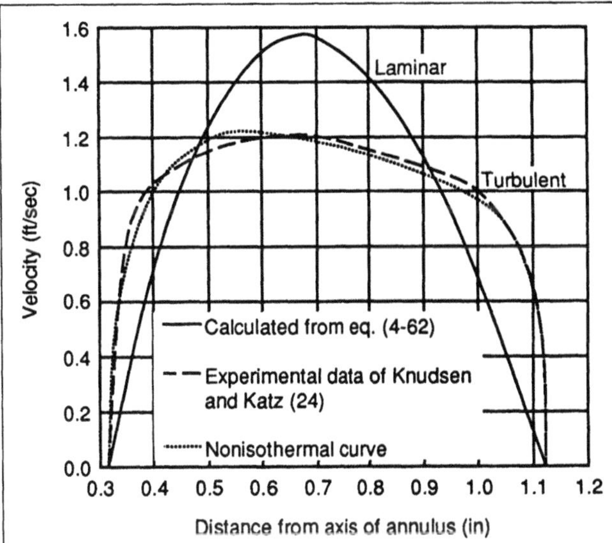

Fig. 3.3—Laminar and turbulent velocity profiles in an annulus (from Ref. 4, *Fluid Dynamics and Heat Transfer*, J.G. Knudsen and D.L. Katz, ©1958, McGraw-Hill Publishing Co., reproduced with permission).

Taking the ratio of these two quantities, it is obvious that the ratio is not a constant 0.5 as in a circular geometry, but depends on the relative dimensions of the tool and the pipe. Note that if the flow is laminar, $\bar{v}/v_{max}$ does not depend on Reynolds number, only the relative tool and pipe sizes. Fig. 3.4 shows the dependence of $\bar{v}/v_{max}$ on casing ID for the three most common sizes of production logging tools. In laminar flow in an annulus, the ratio of average to maximum velocity will range from 0.65 to 0.67 for typical sizes of pipe and production logging tools.

For turbulent flow in an annulus, approximate expressions have been derived for velocity profile.[4] It has been found experimentally that

$$\frac{\bar{v}}{v_{max}} \approx 0.88, \qquad (3.12)$$

though this ratio will vary with Reynolds number and pipe roughness as it does in a circular geometry.

In annular flow, the transition between laminar and turbulent flow also occurs for a Reynolds number of 2,000 to 4,000. The Reynolds number for annular flow, however, is defined slightly differently:

$$N_{Re,an} = \frac{d_{ci}(1-F_d)\bar{v}\rho}{\mu}. \qquad (3.13)$$

To compare the annular-flow Reynolds number with the Reynolds number for flow in the same pipe without a tool present, we can replace velocity with volumetric flow rate. Thus,

$$\bar{v} = \frac{q}{A}. \qquad (3.14)$$

For an annular geometry,

$$A = \frac{\pi}{4} d_{ci}^2(1-F_d^2), \qquad (3.15)$$

where d_{ci} is the pipe ID, so

$$\bar{v} = \frac{4q}{\pi d_{ci}^2(1-F_d^2)}. \qquad (3.16)$$

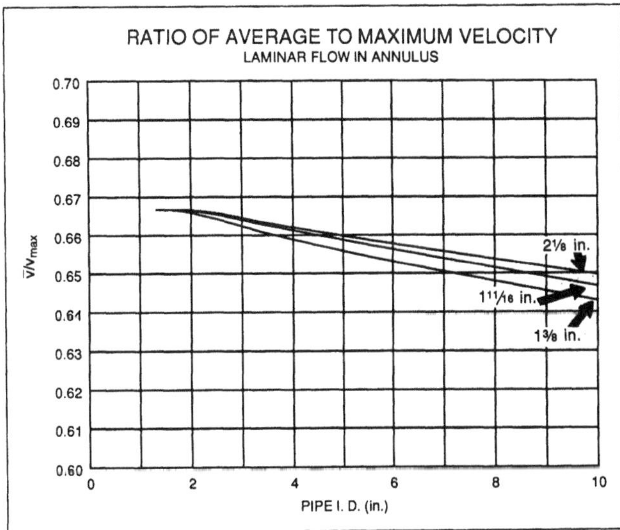

Fig. 3.4—$\bar{v}/v_{max}$ in annular flow between casing and a logging tool.

Substituting into Eq. 3.13 yields

$$N_{Re,an} = \frac{4q\rho}{\pi d_{ci}\mu(1+F_d)}. \qquad (3.17)$$

For the circular pipe (no tool present),

$$N_{Re,cir} = \frac{4q\rho}{\pi d_{ci}\mu}. \qquad (3.18)$$

Therefore,

$$N_{Re,an} = \frac{N_{Re,cir}}{1+F_d}. \qquad (3.19)$$

Eq. 3.19 shows that, for a given volumetric flow rate in a pipe, the Reynolds number will be lower when a tool is present than when no logging tool is in the pipe. Even though the velocity is higher in the annular geometry, the additional shear at the tool wall causes laminar flow to persist to higher flow rates than would occur without the tool present.

Nomenclature

A = cross-sectional area of flow, L^2
d_{ci} = casing ID, L
d_T = tool diameter, L
F_d = ratio of tool diameter to pipe ID
g = acceleration of gravity, L/t^2
L = length, L
N_{Re} = Reynolds number
$N_{Re,an}$ = Reynolds number, annular flow
$N_{Re,cir}$ = Reynolds number, pipe flow
p_L = pressure at Position L, m/Lt^2
p_O = pressure at Position O, m/Lt^2
q = volumetric flow rate, L^3/t
r = radial position, L
r_i = pipe inside radius, L
v = velocity, L/t
$\bar{v}$ = average velocity, L/t
v_{max} = maximum velocity, L/t
$v(r)$ = velocity as a function of radial position, L/t
z_L = height of Position L above a datum, L
z_O = height of Position O above a datum, L
θ = angle
μ = viscosity, m/Lt
ρ = density, m/L^3
Φ_L = potential at Position L, m/Lt^2
Φ_O = potential at Position O, m/Lt^2

References

1. *Production Log Interpretation,* Schlumberger (1973) 4.
2. Bird, R.B., Stewart, W.E., and Lightfoot, E.N.: *Transport Phenomena,* John Wiley & Sons Inc., New York City (1960) 42-46.
3. Nikuradse, J.: "Strömungsgesetze in rauhen Rohren," *VDI-Forschungsheft* (1933) 361.
4. Knudsen, J.G. and Katz, D.L.: *Fluid Dynamics and Heat Transfer,* McGraw-Hill Publishing Co., New York City (1958) 186.
5. *Interpretive Methods for Production Well Logs,* second edition, Atlas Wireline Services, Western Atlas Intl. Inc., Houston (1982) 24.

SI Metric Conversion Factors

ft	× 3.048*	E−01	=	m
in.	× 2.54*	E+00	=	cm
lbm/(ft-sec)	× 1.488 164	E+00	=	Pa·s
lbm/ft³	× 1.601 846	E+01	=	kg/m³
poise	× 1.0*	E−01	=	Pa·s

*Conversion factor is exact.

Chapter 4
Temperature Logging

4.1 Introduction

Production logging began in the late 1930's with the introduction of temperature measurements in oil and gas wells.[1,2] Spurred by the development of accurate, rapid-resolving resistance thermometers, temperature logging was originally investigated as a means to locate hydrocarbons. It soon became apparent, however, that the differences in thermal properties of hydrocarbons and water were too small to affect the overall thermal conductivity of reservoir rocks sufficiently to distinguish hydrocarbon-bearing zones from nonproductive intervals.[3-9]

Though the original application of temperature logs was not particularly successful, temperature logging was soon recognized as a means to evaluate production well characteristics by measuring and analyzing anomalous temperature behavior. Temperature logging applications described in early papers[1-12] included location of gas entries, detection of casing leaks and fluid movement behind casing, location of lost-circulation zones, and evaluation of cement placement. Temperature logs are still used for these applications and others, but the most common use now is qualitative identification of injection or production zones. Despite the many other logs that have been developed, the temperature log remains the workhorse of the production logging stable. This is primarily because of its reliability; no matter what the wellbore flow conditions, temperature can be measured accurately. Also, the temperature log tends to reflect the long-term behavior of a well, not just current conditions.

This chapter first describes temperature logging tools and operating procedures and presents the theory of temperature behavior in wells, laying the foundation for the following section on temperature log interpretation. All the common interpretation procedures are illustrated with examples. The chapter then discusses the use of temperature logs to locate hydraulic fractures, and finally presents guidelines for running and interpreting temperature logs.

4.2 Tools and Operations

Temperature instruments in use today are generally based on elements (typically wire coils) with resistances that vary with temperature, though in the past, instruments based on gas expansion were used.[13] The variable-resistance element is connected with bridge circuitry or a constant-current circuit so that a voltage response proportional to temperature is obtained. Fig. 4.1 shows a schematic of a typical temperature logging tool. The voltage signal from the temperature device is then usually converted to a frequency signal that is transmitted to the surface, where it is converted back to a voltage signal and recorded. The absolute accuracy of temperature logging instruments is not high (on the order of ±5°F [±2.5°C]), but resolution is good (generally 0.05°F [0.025°C] or better[14]), although this accuracy can be compromised by present-day digitization of the signal on the surface. The temperature instrument can usually be included in the string with other tools, such as radioactive-tracer tools or spinner flowmeters.

Running a temperature log consists of measuring the wellbore temperature as a function of depth. Often, temperature logs will be run while the well is flowing and while it is shut in, sometimes with a number of shut-in logs obtained at different times after shut-in. Temperature logs are run continuously, typically at cable speeds of 20 to 30 ft/min [6 to 9 m/min] because the temperature sonde responds rapidly. Temperature should be recorded while running in the hole. When suites of logs are run, the temperature log should be recorded first so that the wellbore temperature is not disturbed by tool motion. If the temperature log is run moving up the well, temperature anomalies will be smeared over greater distances, reducing the vertical resolution of the log.

Often, two presentations of the temperature measurement will be provided—a "gradient" curve, which is wellbore temperature as a function of depth, and a "differential" curve, which is a continuous plot of the gradient-curve slope. Temperature log interpretation is based almost entirely on the gradient curve, although the differential curve is sometimes useful in accentuation of slight temperature anomalies. Fig. 4.2 shows a standard temperature log display.

4.3 Theory of Temperature Behavior in Wellbores

To interpret a temperature log, the various factors that influence temperature behavior in a well must be understood. This section discusses these factors, which include the natural temperature of formations penetrated by a well, heat conduction between the well and surrounding formations, heat convection attending fluid flow, and thermal changes of fluids.

4.3.1 Geothermal Temperature Profile. Because of heat flux from the earth's interior to the atmosphere, temperatures in the earth's crust increase with depth. The temperature profile is called the geothermal temperature, and the temperatures in a wellbore in the absence of any thermal disturbance will correspond to the geothermal temperature. Of course, drilling a well without causing any thermal disturbance is impossible, so the geothermal temperature profile can be measured only in a well that has been inactive (shut in) for a long time period. The geothermal temperature profile will vary significantly from area to area and the slope of the geothermal temperature (geothermal gradient) will vary from formation to formation. The geothermal gradient depends on the rock thermal conductivity—the higher the thermal conductivity, the more easily heat is transported through the rock. Thus, a lower temperature gradient will exist. Quantitatively, this relationship is given by Fourier's law.

$$u=\lambda\frac{dT}{dD}, \qquad (4.1)$$

where u=heat flux, λ=thermal conductivity, and dT/dD=geothermal gradient (g_G). The heat flux will be approximately con-

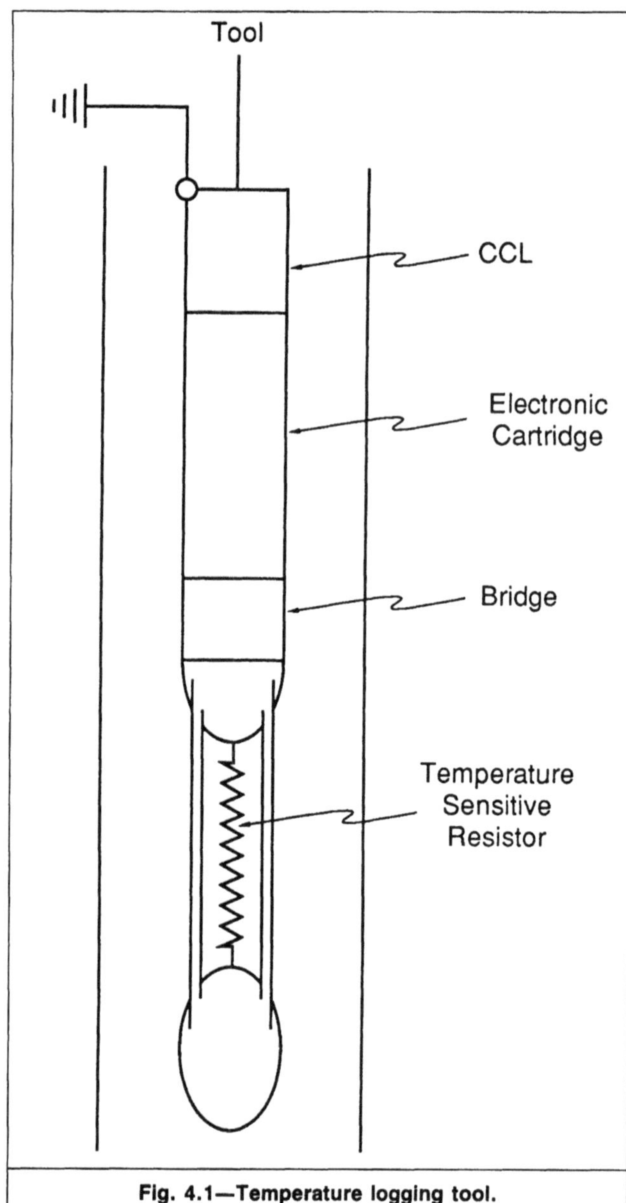

Fig. 4.1—Temperature logging tool.

Fig. 4.2—Typical gradient and differential temperature logs (from Ref. 15, courtesy Atlas Wireline Services, Western Atlas Intl.).

stant with depth, so the geothermal gradient will vary inversely with the thermal conductivity of the rock strata. Fig. 4.3 illustrates this hypothetical behavior.

A temperature log from a well in thermal equilibrium with the surrounding formations (Fig. 4.4) shows the actual variation of geothermal gradient with lithology. It is clear that the geothermal gradient varies significantly with stratigraphy. This is important to remember when interpreting temperature logs because virtually all quantitative interpretation techniques are based on the assumption that the geothermal gradient is constant with depth. Obviously, the more uniform the sediments penetrated by a well, the better this assumption is. Table 1 gives thermal properties of some common reservoir rocks. Prats[18] provides an excellent review of published data and correlations for determination of thermal properties of fluids and rocks around wellbores.

4.3.2 Wellbore Temperature in Regions of No Reservoir Flow. Wellbore temperature will be modified from geothermal temperature as fluids are injected or produced through a wellbore. In a producing well, the fluid will generally be warmer than the formations above the production zone, resulting in a wellbore temperature higher than the geothermal temperature. In an injection well, however, the injected fluid is usually cooler than most formations penetrated by the well; thus the wellbore temperature is lower than the geothermal temperature. Fig. 4.5 illustrates the expected temperature behavior in injection or production wells.

The temperature profile in a well changes with time and depends on many other factors, including flow rate, fluid and formation properties, and completion configuration. Ramey[19] developed equations that relate wellbore temperature to these parameters for the case of flow in tubing or casing with no flow in the formation (i.e., for wellbore locations not opposite production or injection intervals). Fig. 4.6 shows the heat-transfer problem posed by Ramey. For an incompressible fluid, a combination of a general energy balance and a mechanical energy balance shows that the rate of heat transfer, dQ, from a volume element of fluid of size $\pi r_1^2 dD$ is

$$dQ = wC_{pf}\,dT_w, \qquad (4.2)$$

where w=mass flow rate, C_{pf}=fluid heat capacity, and T_w=fluid wellbore temperature. Assuming that all the heat lost from the fluid is conducted radially to the outside of the casing at r_2,

$$dQ = -wC_{pf}\,dT_w = 2\pi r_1 U(T_w - T_{ce})\,dD, \qquad (4.3)$$

where U=overall heat-transfer coefficient and T_{ce}=temperature outside casing. Now, assuming that the heat will be conducted radially away from the casing through the formation, another

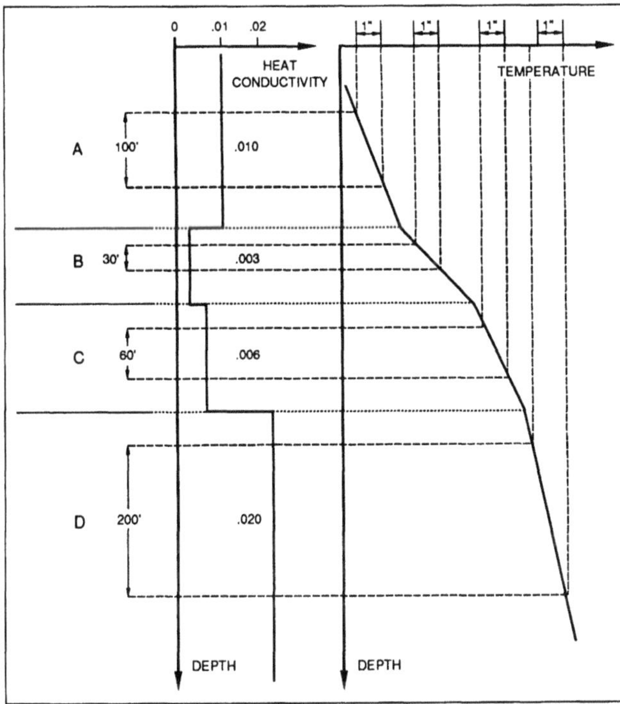

Fig. 4.3—Hypothetical geothermal temperature profile in horizontal beds with differing thermal conductivities (from Refs. 3 through 9, courtesy Gulf Publishing Co.).

equation for dQ is obtained as

$$dQ=-wC_{pf}dT_w=\frac{2\pi\lambda(T_{ce}-T_G)dD}{f(t)}, \qquad (4.4)$$

where T_G=geothermal temperature and $f(t)$=time-dependent function that depends on the boundary condition assumed for the heat conduction problem.

Combining Eqs. 4.3 and 4.4 to eliminate T_{ce} yields

$$\frac{dT_w}{dD}=-\frac{(T_w-T_G)}{Z}=-\frac{T_w}{Z}+\frac{T_G}{Z}, \qquad (4.5)$$

where the parameter Z is defined by

$$Z=\frac{wC_{pf}[\lambda+f(t)r_1U]}{2\pi\lambda r_1U}. \qquad (4.6)$$

If the geothermal temperature profile is known, Eq. 4.5 can be solved to yield the wellbore temperature, T_w, as a function of time

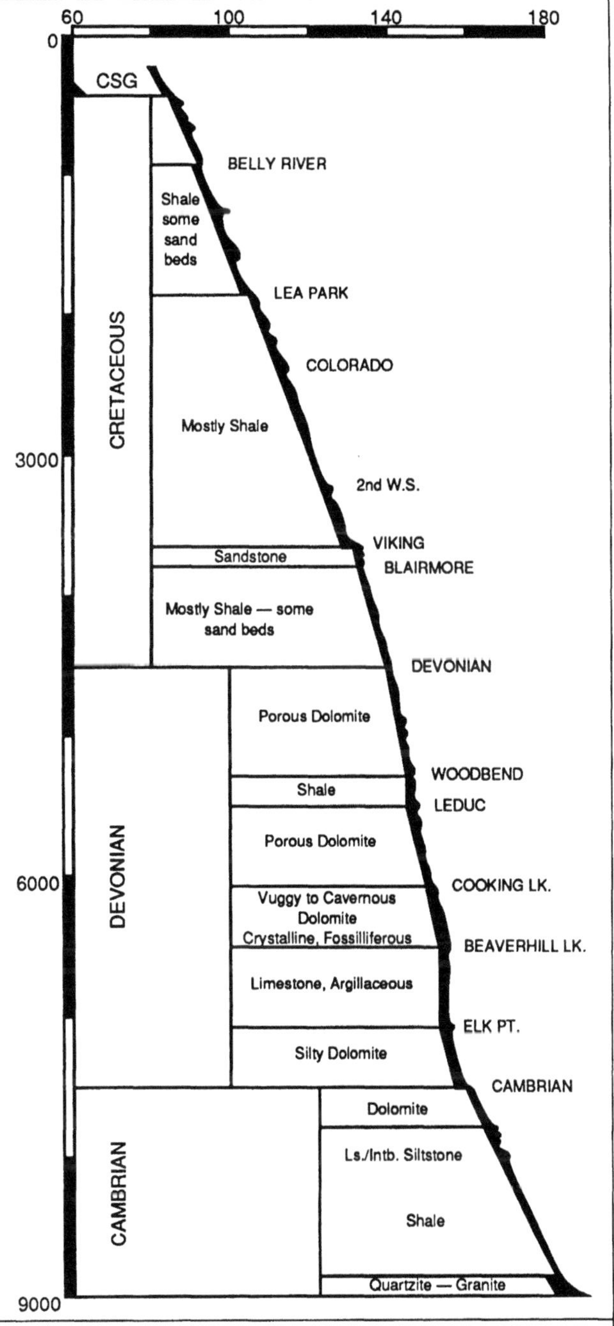

Fig. 4.4—Geothermal temperature profile, Leduc area, Canada (from Ref. 16, courtesy Society of Professional Well Log Analysts).

TABLE 4.1—THERMAL CHARACTERISTICS OF RESERVOIR ROCKS (from Ref. 17)

	Bulk Density (lbm/ft³)		Thermal Diffusivity Unsteady State (ft²/D at 200°F)		Thermal Diffusivity Steady State (ft²/D)		Thermal Conductivity Unsteady State (Btu/D-ft-°F at 200°F)	
Sample	Initial	Repeat	Initial	Repeat	90°F	275°F	Initial	Repeat
Bandera SS*	134.2	131.8	0.816	0.660	0.895	0.660	2.16	16.6
Berea SS	134.8	126.0	0.821	0.581	0.948	0.665	21.8	15.7
Boise SS	118.9	116.0	0.833	0.492	0.694	0.624	19.5	11.8
Limestone	140.2	78.5**	0.780	0.497	0.780	0.643	21.7	13.9
CaO**	78.5	—	0.924	—	—	—	—	—
Shale	137.1	128.2	0.936	0.516	0.950	0.698	26.2	13.9
Rock salt	135.0	128.0	2.95	—	1.7	1.5	8.30	8.30
Tuffaceous SS	115.3	107.2	0.444	—	0.504	—	9.55	9.55

*SS = sandstone.
**After reaction.

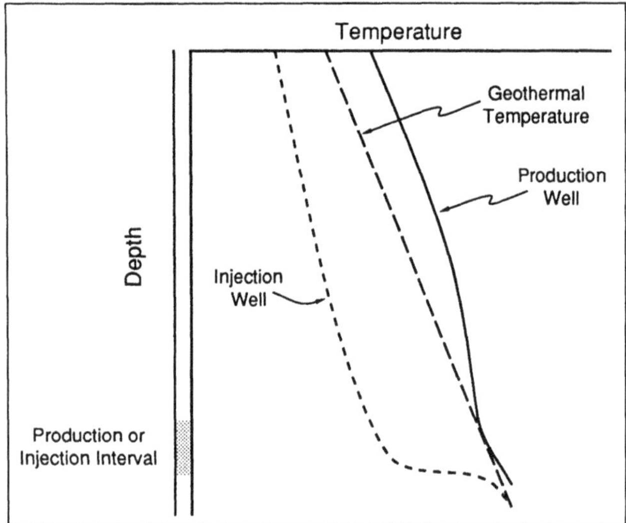

Fig. 4.5—Temperature behavior in injection or production wells.

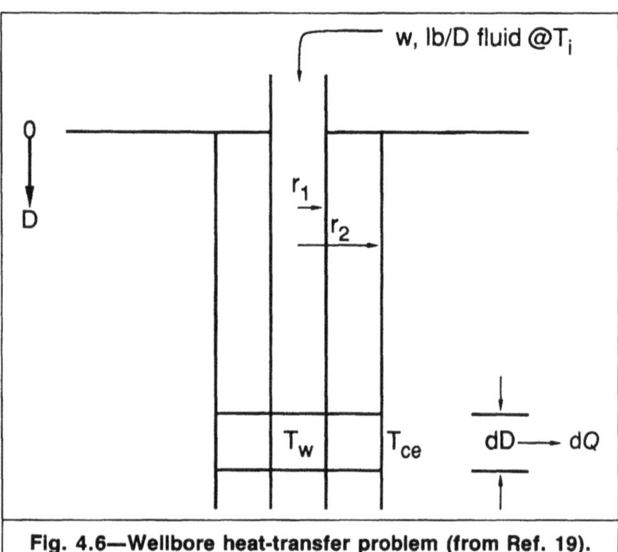

Fig. 4.6—Wellbore heat-transfer problem (from Ref. 19).

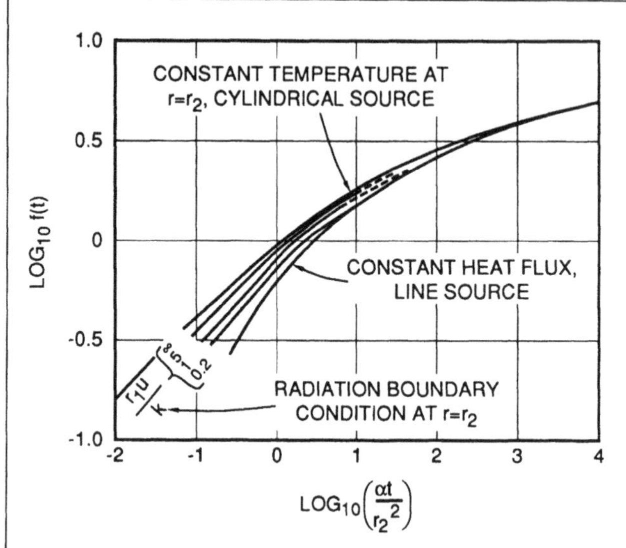

Fig. 4.7—Time function, $f(t)$, in the Ramey equation (from Ref. 19).

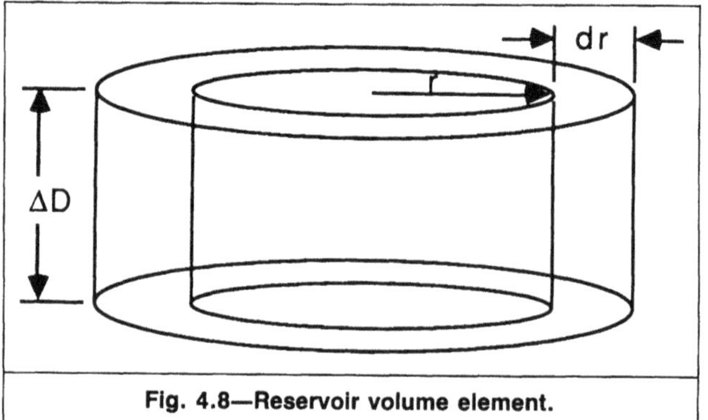

Fig. 4.8—Reservoir volume element.

and position (depth). If we assume a linear geothermal temperature profile, then

$$T_G = g_G D + T_b, \quad (4.7)$$

where g_G=geothermal gradient and T_b=temperature at $D=0$. Integration of Eq. 4.5 with the boundary condition that $T_w=T_i$ at $D=0$ yields

$$T_w(D,t) = g_G D + T_b - g_G Z + [T_i(t) + g_G Z - T_b]e^{-D/Z}. \quad (4.8)$$

To apply Ramey's equation (Eq. 4.8), the time function, $f(t)$, must be determined. The time function will depend on the boundary conditions assumed (Fig. 4.7), but for times greater than approximately 1 week, all solutions converge to the line-source solution, for which

$$f(t) = -\ln\frac{r_2}{2\sqrt{\alpha t}} - 0.290, \quad (4.9)$$

where α=rock thermal diffusivity and r_2=casing outside radius. In this equation, terms with a size on the order of $r_2^2/4\alpha t$ have been neglected. Thermal diffusivity is defined as

$$\alpha = \frac{\lambda}{\rho C_p}, \quad (4.10)$$

where λ=thermal conductivity, ρ=density, and C_p=heat capacity. Thermal diffusivity can be thought of as the ratio of a body's ability to conduct heat to its ability to store heat.

Examination of Eq. 4.8 shows that when D is large compared to Z, $e^{-D/Z}$ will approach zero and Eq. 4.8 will simplify to

$$T_w(D,t) = g_G D + T_b - g_G Z$$
$$= T_G - g_G Z. \quad (4.11)$$

Thus, for large D, the wellbore temperature profile will be a straight line parallel to the geothermal temperature.

In a more general case, the parameter Z must be calculated to construct the wellbore temperature profile. This requires information about the thermal properties of the rock surrounding the wellbore and of the overall heat-transfer coefficient for the region between the inner wall of the tubing and the outer wall of the casing. Heat-transfer coefficients can be calculated with correlations presented in the chemical engineering literature.[20,21]

One application of the Ramey equation is the determination of formation thermal conductivity from a temperature log, as shown by Romero-Juarez.[22] Rearrangement of Eq. 4.5 shows that

$$Z = \frac{T_G - T_w}{dT_w/dD}. \quad (4.12)$$

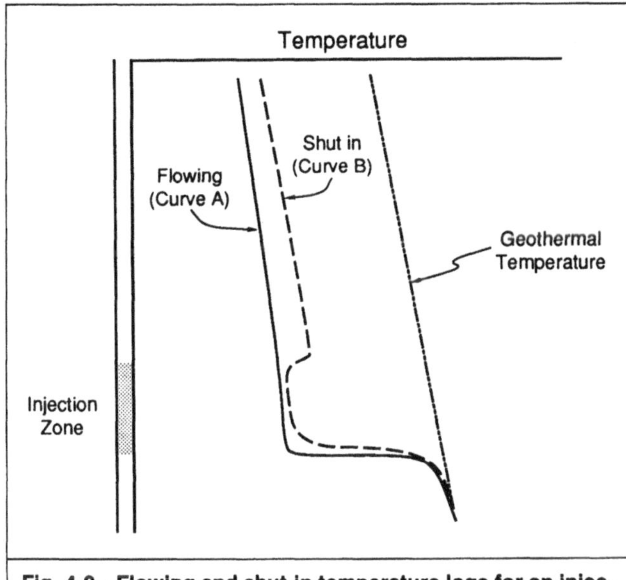

Fig. 4.9—Flowing and shut-in temperature logs for an injection well.

If $f(t)r_1U$ is large compared to the thermal conductivity, λ, then Z simplifies to

$$Z=\frac{wC_{pf}f(t)}{2\pi\lambda}, \quad (4.13)$$

and the thermal conductivity of the formation is then

$$\lambda=\frac{wC_{pf}f(t)}{2\pi Z}. \quad (4.14)$$

To use this method, the geothermal temperature profile must be known. From a temperature log obtained at a later time, the wellbore temperature, T_w, and the temperature gradient in the wellbore, $\mathrm{d}T_w/\mathrm{d}D$, can be determined for the zone of interest. From these data, the function Z and the thermal conductivity of the formation can be calculated. Another application of the Romero-Juarez method is to calculate mass flow rate from the obtained value of Z, assuming the thermal conductivity is known.

4.3.3 Wellbore Temperature Opposite Injection or Production Zones. The Ramey equation will apply for regions of the wellbore that are not near zones of injection or production. Where flow in the formation is taking place, heat transfer results from forced convection and conduction, and Joule-Thomson heating or cooling of the fluid can be significant.[23] The temperature distribution in a large region of the reservoir around the wellbore must be determined to calculate the temperature in the wellbore.

A general energy balance (the first law of thermodynamics) coupled with Fourier's law of heat transfer describes the temperature behavior in a reservoir with fluid movement. Consider a cylindrical element of reservoir with thickness ΔD and radial size $\mathrm{d}r$ (Fig. 4.8). Assuming that a single-phase, incompressible fluid is flowing radially and horizontally through this reservoir element at a steady mass flow rate, w, and that kinetic energy effects are negligible, an energy balance yields

$$(\overline{\rho C_p})\frac{\partial T}{\partial t}=\frac{wC_{pf}}{2\pi r\Delta D}\frac{\partial T}{\partial r}+\frac{q}{2\pi r\Delta D}\frac{\partial p}{\partial r}$$

$$+\frac{1}{r}\frac{\partial}{\partial r}\left(\lambda r\frac{\partial T}{\partial r}\right)+\frac{\partial}{\partial D}\left(\lambda\frac{\partial T}{\partial D}\right). \quad (4.15)$$

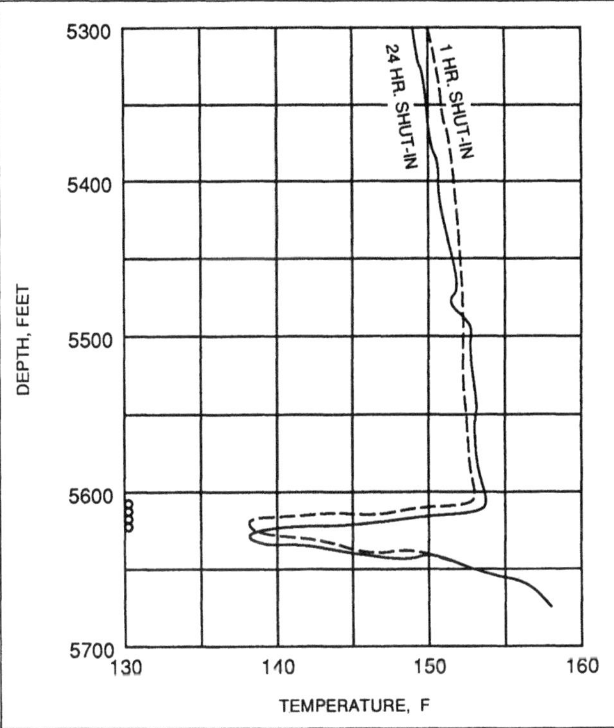

Fig. 4.10—Cool anomaly on temperature log caused by gas influx (from Ref. 14).

In this equation, $(\overline{\rho C_p})$=average density and heat capacity of all material in the reservoir element (rock and fluid), w=mass flow rate, q=volumetric flow rate, C_{pf}=fluid heat capacity, p=pressure, T=temperature, and λ=reservoir thermal conductivity. The term on the left side of the equation is the accumulation of energy in the reservoir element; on the right side of the equation, the first term is the sensible heat convected to the element by the fluid, the second term accounts for the heat generated by mechanical energy dissipation during flow (frictional heating), the third term is radial heat conduction, and the fourth term describes the vertical heat conduction.

To calculate the wellbore temperature opposite a zone of injection or production, Eq. 4.15 must be solved for a large region of the reservoir around the wellbore. This generally requires a numerical solution. The flow behavior is coupled with the energy balance through the frictional heating term so that a general solution would require a simultaneous solution of the diffusivity equation describing reservoir pressure behavior. For the simple case of single-phase radial flow of an incompressible fluid, for example, the pressure field is independent of temperature and the pressure gradient can be calculated directly from Darcy's law as

$$\frac{\mathrm{d}p}{\mathrm{d}r}=\frac{q\mu}{2\pi r\Delta Dk}, \quad (4.16)$$

where μ=viscosity and k=permeability. Substituting for $\partial p/\partial r$ in Eq. 4.15 yields

$$Q_{fr}=\frac{q}{2\pi r\Delta D}\left(\frac{q\mu}{2\pi r\Delta Dk}\right), \quad (4.17)$$

where Q_{fr}=rate of frictional heating per unit volume. To calculate the actual amount of heat generated by friction in a region of the reservoir, Eq. 4.17 can be integrated over a volume element of thickness ΔD from an inner radius r_1 to an outer radius r_2,

$$Q_{fr}=\Delta D\int_{r_1}^{r_2}\left(\frac{q}{2\pi r\Delta D}\right)\left(\frac{q\mu}{2\pi r\Delta Dk}\right)2\pi r\mathrm{d}r \quad (4.18)$$

or

$$Q_{fr}=q\int_{r_1}^{r_2}\frac{q\mu}{2\pi r\Delta Dk}dr. \quad (4.19)$$

Integration of Eq. 4.19 yields

$$Q_{fr}=q\left[\frac{q\mu\ \ln(r_2/r_1)}{2\pi\Delta Dk}\right]. \quad (4.20)$$

Because the expression in brackets in Eq. 4.20 is the Darcy's law expression for the pressure drop from r_1 to r_2, the frictional heating is

$$Q_{fr}=q(p_1-p_2), \quad (4.21)$$

where p_1=pressure at r_1 and p_2=pressure at r_2.

The amount of frictional heating will be greatest near the wellbore where the pressure gradient is the highest. To illustrate the amount of heat generated by frictional heating, consider the flow of 20 B/D [3.2 m^3/d] of water into a 1-ft [0.3-m] -thick and 1-ft [0.3-m] -deep element of a 10-md reservoir next to a wellbore with a 0.25-ft [0.08-m] radius. For this case, r_1=0.25 ft [0.08 m] and r_2=1.25 ft [0.4 m]. Solving for Q_{fr} from Eq. 4.20 or 4.21 yields a value of 0.109 Btu/sec [0.115 kJ/s] with appropriate unit conversions. If all this heat is absorbed by the fluid, the amount of heat absorbed per unit mass of fluid will be Q_{fr} divided by the mass flow rate, w, which in this case will be 1.34 Btu/lbm [3.12 kJ/kg]. For water, this would result in a temperature increase of 1.34°F [0.74°C]. Of course, some heat will be conducted away according to Eq. 4.15, yielding a smaller temperature rise in the fluid.

The equations developed here are strictly applicable only to horizontal, single-phase, radial flow of an incompressible fluid under steady-state conditions. They will adequately describe the temperature behavior in many water-injection cases, but will be inadequate for most production well cases. Prats[18] presents equations that describe reservoir temperature behavior for more complex flow fields.

Computer solutions of the energy balance equation in an injection zone show that the wellbore temperature will be nearly constant across the flow zone. Thus, a typical flowing temperature log for an injection well with one injection zone will look like Curve A in Fig. 4.9. The temperature profile flattens across the injection interval, then increases rapidly to the geothermal temperature below the injection zone. In a long injection interval, the temperature opposite the zone will increase slightly with depth. When the well is shut in, all temperatures will tend toward the geothermal temperature. However, because of the large volume of cool fluid that has been placed in the injection zone, temperatures in this region will return more slowly to the geothermal temperature, resulting in a cool temperature anomaly in the region of injection (Curve B).

In an oil production well, the flowing temperature log is more complex. Fluids will enter the wellbore at the geothermal temperature unless affected significantly by Joule-Thomson heating or cooling. They will then be mixing with warmer fluids flowing up the wellbore from greater depths. This will result in a break to lower temperatures at fluid entry locations. Above the production zones, the wellbore temperature will be higher than the geothermal temperature and the Ramey equation can be applied by setting D=0 at the production zone depth and using the production zone temperature as T_i.

4.3.4 Wellbore Temperature Behavior Opposite Gas Zones. In a gas production well, the expansion of gas as it enters the wellbore often results in a temperature decrease because of the Joule-Thomson effect. The gas flow through the porous formation near the wellbore can be considered a throttling process, one for which enthalpy is constant. For such a process, the temperature change of the gas can be determined from the Joule-Thomson coefficient, K_{JT}, defined as

$$K_{JT}=\left(\frac{\partial T}{\partial p}\right)_H. \quad (4.22)$$

For real gases, K_{JT} is usually positive at low-to-moderate pressures; thus, temperature and pressure decrease as gas flows to the wellbore. Thus, gas entry points can often be located by cool anomalies on temperature logs, as seen in Fig. 4.10 for a very low-permeability formation.

The Joule-Thomson coefficient for methane over a wide range of temperatures and pressures is given in Table 2. Notice that, for a given temperature, the Joule-Thomson coefficient is positive (indicating that the gas will cool as pressure drops) at low pressures, but that K_{JT} decreases as pressure increases, eventually reaching zero. Above this pressure, called the inversion point, K_{JT} is negative and the gas will be heated in a throttling process. For methane at typical reservoir temperatures, the inversion point occurs at about 500 atm or about 7,300 psi [50 MPa]; above this pressure, the gas will be heated as it flows into the wellbore and no cool anomaly will be observed on a temperature log.

The Joule-Thomson coefficient will vary significantly with gas composition, temperature, and pressure. The Joule-Thomson coefficient for any gas can be calculated from commonly available thermodynamic data or equations of state with the equation

$$K_{JT}=\frac{1}{C_p}\left[T\left(\frac{\partial V}{\partial T}\right)_p-V\right], \quad (4.23)$$

where C_p=heat capacity, T=temperature, and V=specific volume. Sources of data for calculating the Joule-Thomson coefficient of hydrocarbon gases are Refs. 26 and 27.

4.4 Temperature Log Interpretation

Temperature log interpretation techniques can be placed in two broad categories—quantitative techniques, in which flow rates at various positions in the wellbore are calculated from the temperature log, and qualitative methods, in which the general characteristics of the well profile are inferred from the shape of the temperature log.

4.4.1 Quantitative Analysis of Temperature Logs. Some attempts have been made to interpret temperature logs quantitatively—i.e., to calculate the flow rates to or from reservoir zones from a temperature log analysis. These methods can be placed in two categories—those based on analytical solutions of energy balance equations, such as the Ramey equation, and those based on computer solutions of energy balance equations.

Analytical Methods. Ramey Equation Methods. It was shown that for distances sufficiently great from the injection point or production zone ($D \gg Z$), the Ramey equation simplifies to an asymptotic solution (Eq. 4.11). Rearranging this equation to solve for Z yields

$$Z=\frac{T_G-T_w}{g_G}, \quad (4.24)$$

and, if the heat transfer coefficient is large, Z is given by

$$Z=\frac{q\rho C_{pf}f(t)}{2\pi\lambda}, \quad (4.25)$$

where $q\rho=w$.

TABLE 4.2—METHANE JOULE-THOMSON COEFFICIENTS, K/atm (from Ref. 25, courtesy British Petroleum Co. Ltd.)

Pressure (atm)	Temperature (K) 140	150	160	170	180	190	200	210	220	230	240	250	260	270	280	290	300
1	1.970	1.905	1.825	1.642	1.347	1.122	0.988	0.900	0.808	0.728	0.664	0.605	0.552	0.512	0.478	0.446	0.417
2.5		1.720	1.664	1.510	1.280	1.112	0.976	0.899	0.807	0.727	0.663	0.604	0.551	0.510	0.477	0.445	0.416
5			1.642	1.477	1.230	1.060	0.960	0.898	0.806	0.726	0.662	0.603	0.550	0.509	0.476	0.444	0.415
10				1.265	1.176	1.035	0.944	0.896	0.805	0.724	0.661	0.602	0.548	0.508	0.474	0.443	0.414
15					1.147	1.057	0.970	0.895	0.804	0.723	0.660	0.601	0.547	0.507	0.473	0.442	0.413
20							1.000	0.893	0.803	0.722	0.659	0.600	0.546	0.506	0.472	0.441	0.412
30							1.003	0.885	0.801	0.721	0.657	0.598	0.545	0.505	0.471	0.439	0.410
40							0.944	0.836	0.770	0.710	0.654	0.597	0.543	0.503	0.469	0.436	0.406
50							0.817	0.783	0.742	0.697	0.646	0.594	0.540	0.494	0.453	0.422	0.396
60							0.718	0.704	0.688	0.663	0.627	0.584	0.533	0.477	0.438	0.410	0.386
80							0.450	0.522	0.557	0.564	0.554	0.530	0.485	0.444	0.413	0.387	0.365
100							0.302	0.330	0.354	0.374	0.384	0.390	0.391	0.384	0.374	0.357	0.337
120							0.186	0.209	0.230	0.251	0.270	0.287	0.298	0.304	0.309	0.305	0.297
140							0.103	0.133	0.158	0.182	0.203	0.224	0.238	0.250	0.261	0.265	0.262
160							0.057	0.085	0.111	0.137	0.159	0.180	0.199	0.212	0.220	0.224	0.227
180							0.030	0.057	0.085	0.106	0.128	0.146	0.158	0.170	0.179	0.184	0.189
200							0.018	0.042	0.064	0.083	0.100	0.115	0.128	0.140	0.150	0.158	0.163
250							-0.0055	0.0075	0.0215	0.0346	0.0450	0.0560	0.0652	0.0735	0.0819	0.0884	0.0941
300							-0.0290	-0.0175	-0.0065	0.0050	0.0145	0.0238	0.0321	0.0392	0.0460	0.0521	0.0576
350							-0.0455	-0.0342	-0.0226	-0.0130	-0.0025	0.0051	0.0110	0.0168	0.0225	0.0285	0.0326
400							-0.0585	-0.0472	-0.0364	-0.0267	-0.0180	-0.0104	-0.0040	0.0015	0.0060	0.0098	0.0136
450							-0.0662	-0.0556	-0.0455	-0.0360	-0.0282	-0.0215	-0.0154	-0.0100	-0.0042	-0.0008	0.0020
500							-0.0721	-0.0625	-0.0528	-0.0437	-0.0359	-0.0290	-0.0223	-0.0170	-0.0129	-0.0097	-0.0068
550							-0.0768	-0.0672	-0.0578	-0.0485	-0.0411	-0.0346	-0.0286	-0.0245	-0.0200	-0.0170	-0.0146
600							-0.0805	-0.0712	-0.0618	-0.0529	-0.0450	-0.0385	-0.0333	-0.0295	-0.0265	-0.0236	-0.0212
650							-0.0846	-0.0755	-0.0652	-0.0566	-0.0484	-0.0421	-0.0375	-0.0338	-0.0317	-0.0282	-0.0264
700							-0.0875	-0.0790	-0.0686	-0.0593	-0.0514	-0.0455	-0.0412	-0.0376	-0.0354	-0.0327	-0.0310
750							-0.0902	-0.0818	-0.0713	-0.0610	-0.0538	-0.0484	-0.0443	-0.0411	-0.0392	-0.0369	-0.0352
800							-0.0922	-0.0840	-0.0733	-0.0624	-0.0553	-0.0507	-0.0470	-0.0442	-0.0420	-0.0400	-0.0383
850							-0.0927	-0.0847	-0.0747	-0.0642	-0.0574	-0.0525	-0.0490	-0.0464	-0.0449	-0.0427	-0.0410
900							-0.0931	-0.0851	-0.0756	-0.0655	-0.0588	-0.0542	-0.0507	-0.0485	-0.0472	-0.0449	-0.0434
950							-0.0933	-0.0853	-0.0761	-0.0665	-0.0601	-0.0555	-0.0524	-0.0502	-0.0486	-0.0465	-0.0453
1,000							-0.0935	-0.0855	-0.0765	-0.0675	-0.0611	-0.0563	-0.0536	-0.0512	-0.0492	-0.0475	-0.0464

Pressure (atm)	Temperature (K) 310	320	330	340	350	360	370	380	390	400	410	420	430	440	450	460	470
1	0.393	0.368	0.347	0.326	0.308	0.291	0.276	0.258	0.241	0.227	0.216	0.204	0.193	0.183	0.173	0.162	0.153
2.5	0.390	0.366	0.346	0.324	0.307	0.290	0.274	0.257	0.240	0.226	0.214	0.203	0.192	0.182	0.172	0.161	0.151
5	0.388	0.365	0.344	0.323	0.305	0.288	0.272	0.255	0.238	0.224	0.212	0.201	0.190	0.180	0.170	0.159	0.149
10	0.387	0.364	0.343	0.322	0.304	0.287	0.270	0.253	0.236	0.222	0.210	0.199	0.188	0.178	0.168	0.157	0.147
15	0.386	0.363	0.342	0.321	0.303	0.285	0.267	0.251	0.234	0.220	0.208	0.197	0.187	0.176	0.166	0.155	0.145
20	0.385	0.362	0.340	0.319	0.302	0.284	0.265	0.248	0.232	0.218	0.207	0.195	0.185	0.174	0.164	0.153	0.143
30	0.384	0.360	0.337	0.317	0.298	0.280	0.262	0.243	0.226	0.212	0.198	0.187	0.177	0.166	0.156	0.147	0.137
40	0.380	0.354	0.331	0.309	0.290	0.271	0.253	0.234	0.217	0.204	0.192	0.180	0.171	0.162	0.152	0.144	0.134
50	0.370	0.348	0.325	0.304	0.283	0.264	0.247	0.231	0.215	0.203	0.190	0.176	0.165	0.154	0.144	0.136	0.128
60	0.364	0.342	0.318	0.298	0.277	0.257	0.240	0.225	0.212	0.199	0.187	0.172	0.162	0.151	0.142	0.133	0.125
80	0.342	0.317	0.296	0.275	0.255	0.237	0.222	0.205	0.196	0.184	0.172	0.160	0.149	0.140	0.131	0.123	0.119
100	0.316	0.295	0.276	0.257	0.240	0.223	0.208	0.197	0.186	0.175	0.164	0.153	0.142	0.133	0.124	0.117	0.112
120	0.285	0.270	0.252	0.236	0.220	0.205	0.192	0.180	0.170	0.160	0.151	0.142	0.133	0.124	0.115	0.110	0.106
140	0.257	0.247	0.232	0.218	0.205	0.192	0.178	0.167	0.158	0.150	0.141	0.132	0.123	0.114	0.108	0.103	0.098
160	0.225	0.218	0.206	0.195	0.183	0.172	0.161	0.152	0.142	0.134	0.127	0.118	0.111	0.104	0.098	0.094	0.090
180	0.190	0.187	0.182	0.175	0.165	0.155	0.148	0.140	0.132	0.125	0.117	0.110	0.104	0.097	0.092	0.087	0.084
200	0.166	0.165	0.162	0.158	0.151	0.143	0.133	0.125	0.117	0.112	0.105	0.100	0.094	0.089	0.084	0.080	0.077
250	0.0976	0.100	0.109	0.108	0.103	0.0995	0.0984	0.0960	0.0925	0.0880	0.0826	0.0772	0.0724	0.0685	0.0654	0.0628	0.0604
300	0.0625	0.0667	0.0692	0.0705	0.0714	0.0720	0.0718	0.0706	0.0680	0.0639	0.0603	0.0576	0.0538	0.0512	0.0493	0.0472	0.0454
350	0.0368	0.0400	0.0420	0.0445	0.0457	0.0472	0.0468	0.0466	0.0460	0.0438	0.0432	0.0410	0.0387	0.0372	0.0357	0.0338	0.0324
400	0.0170	0.0198	0.0219	0.0235	0.0247	0.0258	0.0266	0.0270	0.0270	0.0267	0.0263	0.0257	0.0250	0.0243	0.0233	0.0220	0.0207
450	0.0052	0.0075	0.0107	0.0095	0.0109	0.0130	0.0127	0.0137	0.0138	0.0140	0.0142	0.0136	0.0133	0.0132	0.0125	0.0117	0.0112
500	-0.0045	-0.0026	-0.0008	-0.0005	0.0017	0.0029	0.0036	0.0042	0.0047	0.0050	0.0052	0.0054	0.0051	0.0048	0.0044	0.0037	0.0030
550	-0.0126	-0.0107	-0.0097	-0.0086	-0.0075	-0.0060	-0.0055	-0.0050	-0.0045	-0.0042	-0.0041	-0.0040	-0.0039	-0.0038	-0.0038	-0.0037	-0.0036
600	-0.0197	-0.0182	-0.0169	-0.0158	-0.0147	-0.0139	-0.0130	-0.0123	-0.0118	-0.0115	-0.0112	-0.0110	-0.0108	-0.0106	-0.0104	-0.0102	-0.0100
650	-0.0246	-0.0232	-0.0220	-0.0215	-0.0202	-0.0192	-0.0188	-0.0180	-0.0173	-0.0170	-0.0167	-0.0165	-0.0163	-0.0162	-0.0160	-0.0154	-0.0151
700	-0.0295	-0.0282	-0.0273	-0.0266	-0.0254	-0.0243	-0.0240	-0.0230	-0.0225	-0.0222	-0.0218	-0.0215	-0.0210	-0.0209	-0.0207	-0.0201	-0.0198
750	-0.0337	-0.0325	-0.0316	-0.0311	-0.0300	-0.0287	-0.0285	-0.0277	-0.0271	-0.0267	-0.0263	-0.0259	-0.0254	-0.0252	-0.0248	-0.0244	-0.0240
800	-0.0373	-0.0364	-0.0354	-0.0344	-0.0335	-0.0326	-0.0320	-0.0314	-0.0308	-0.0303	-0.0298	-0.0293	-0.0289	-0.0284	-0.0280	-0.0277	-0.0274
850	-0.0400	-0.0390	-0.0380	-0.0372	-0.0364	-0.0355	-0.0350	-0.0347	-0.0340	-0.0335	-0.0330	-0.0326	-0.0320	-0.0317	-0.0313	-0.0308	-0.0304
900	-0.0423	-0.0415	-0.0407	-0.0398	-0.0394	-0.0384	-0.0379	-0.0375	-0.0370	-0.0363	-0.0359	-0.0355	-0.0350	-0.0346	-0.0340	-0.0336	-0.0332
950	-0.0444	-0.0435	-0.0428	-0.0421	-0.0417	-0.0409	-0.0405	-0.0399	-0.0395	-0.0387	-0.0384	-0.0380	-0.0375	-0.0372	-0.0366	-0.0363	-0.0359
1,000	-0.0457	-0.0450	-0.0445	-0.0438	-0.0432	-0.0427	-0.0422	-0.0417	-0.0412	-0.0408	-0.0405	-0.0401	-0.0397	-0.0394	-0.0390	-0.0387	-0.0384

Thus, knowing thermal properties of the formation and the geothermal temperature profile, the asymptotic portions of a temperature log can be analyzed to yield the flow rate from a zone. Fig. 4.11 presents an example of this method. The analysis of this log is simplified if it can be assumed that the formation properties (e.g., density and thermal conductivity) are uniform throughout the region logged. Then, designating 5,820 and 5,890 ft [1774 and 1795 m] as Positions 1 and 2, respectively,

$$\frac{Z_2}{Z_1}=\frac{(T_w-T_G)_2}{(T_w-T_G)_1}=\frac{q_2}{q_1} \quad (4.26)$$

The difference between the asymptotic portion of the temperature log and the geothermal temperature is proportional to flow rate. For the log shown in Fig. 4.11, Zone A receives 25% of the injected fluid and the remainder enters Zone B.

The applicability of this interpretation scheme depends on the asymptotic portion of the temperature profile being reached between reservoir zones. This requires that D be substantially greater than Z for interpretation to be possible. Furthermore, D below a fluid-entry zone must be measured from that fluid-entry location. Thus, for the analysis method to hold, reservoir zones must be far enough apart for the asymptote to be reached. Also, Z increases with time because $f(t)$ is an increasing function with time; therefore, this procedure will be feasible primarily early in the well life.

To extend the applicability of the Ramey solution for quantitative temperature log interpretation, Curtis and Witterholt[29] use the exponential portion of the temperature profile to interpret temperature logs in injection wells. In this procedure, Z is adjusted to obtain the best least-squares fit of the Ramey equation (Eq. 4.8) to the

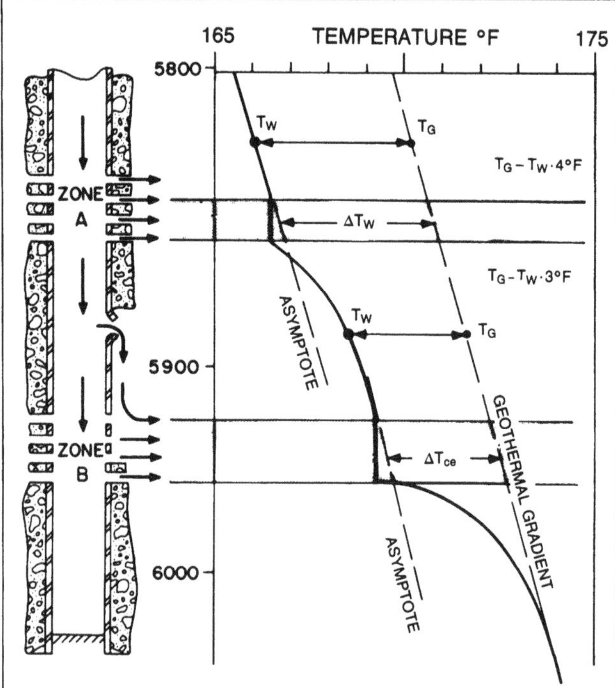

Fig. 4.11—Witterholt and Tixier temperature log interpretation (from Ref. 28).

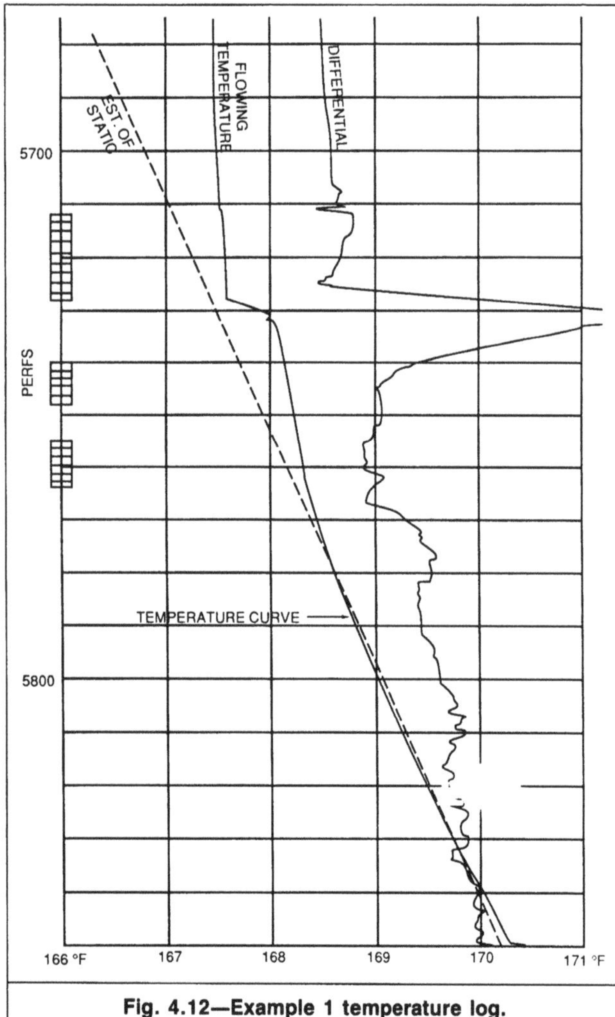

Fig. 4.12—Example 1 temperature log.

TABLE 4.3—ROMERO-JUAREZ TEMPERATURE LOG INTERPRETATION, EXAMPLE 1

Station	Depth (ft)	dT_w/dD (°F/ft)	$T_w - T_G$ (°F)	Z (ft)	Total Flow (%)
A	5,690	0.0033	1.1	333	100
B	5,740	0.008	0.4	50	15

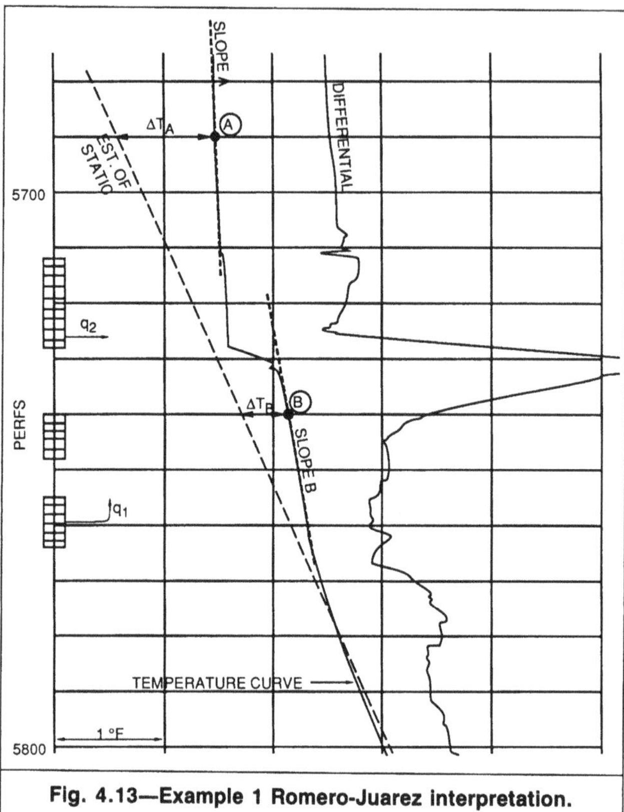

Fig. 4.13—Example 1 Romero-Juarez interpretation.

temperature log. The flow rate is then estimated from the determined value of Z.

A simpler method with similar accuracy is proposed by Romero-Juarez.[22] As shown in Eq. 4.12, if the Ramey solution holds, Z can be calculated at any depth from

$$Z=\frac{T_G-T_w}{dT_w/dD}.$$

Again, if thermal properties of the formation and the time function, $f(t)$, are known, the flow rate can be calculated from Z. Thus, if the geothermal profile is known, the flow rate at any depth (away from a production or injection zone) can be determined from the difference between the temperature profile in the wellbore and the geothermal profile (T_G-T_w) and the derivative of the temperature log (dT_w/dD).

All log interpretation methods that require use of the Ramey equation are precisely applicable only to flow of a single, incompressible phase. These methods can be applied to two-phase flow, however, if the two-phase mixture can be assumed to have uniform thermal properties. The validity of this assumption will likely vary from case to case.

Example 1—Temperature Log Interpretation by the Romero-Juarez Method. Fig. 4.12 shows a temperature log from a well producing 78% water and 22% oil from three sets of perforations between 5,712 and 5,764 ft [1741 and 1757 m]. An approximate flow profile for this well can be obtained by applying the Romero-Juarez method at locations between the perforations, assuming that the perforations are far enough apart for the Ramey equation to apply. The analysis consists of the following steps—(1) choose

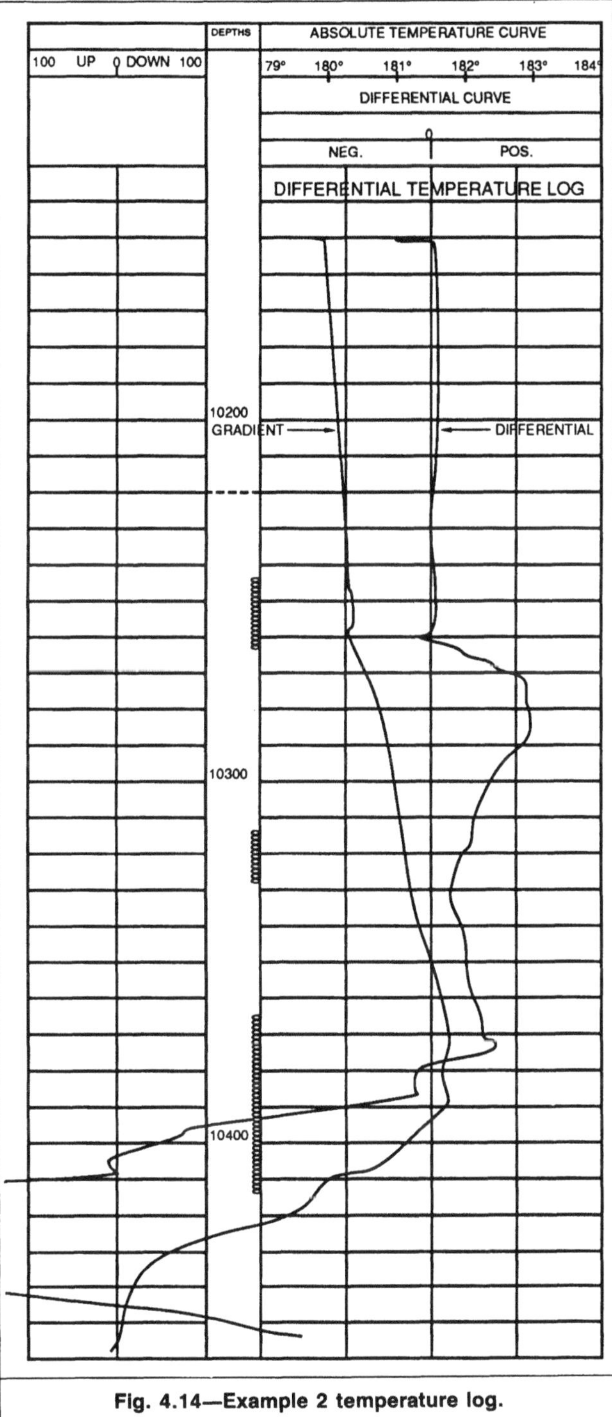

Fig. 4.14—Example 2 temperature log.

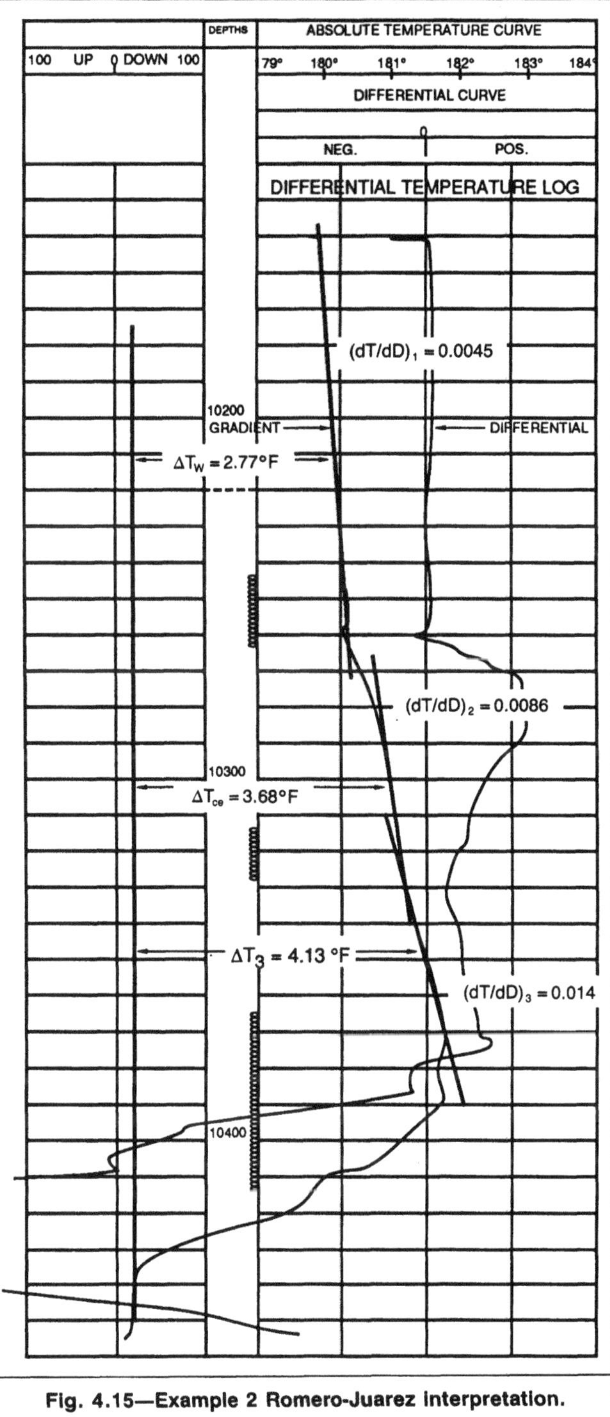

Fig. 4.15—Example 2 Romero-Juarez interpretation.

stations approximately midway between perforations for interpretation; (2) measure the slope of the temperature log, dT_w/dD, at each station; (3) measure the difference between the wellbore temperature and the geothermal temperature, $T_w - T_G$, at each station; and (4) calculate Z for each station (use Eq. 4.12). If the rock and fluid properties are assumed the same at all stations, the volumetric flow rate at each station is proportional to Z, so the fraction of total flow at each station can be calculated from

$$\frac{q_i}{q_{100}} = \frac{Z_i}{Z_{100}}, \quad \text{(4.27)}$$

where q_{100} and Z_{100} are for the station above all perforations and q_i and Z_i are values for any other station.

For the example log, the middle set of perforations did not appear to be producing any fluids, so stations were chosen at 5,690 ft [1734 m], above the top perforations, and at 5,740 ft [1750 m], approximately midway between the top and bottom perforations. In this well, the geothermal temperature could be extrapolated from the temperature log response in the rathole below the lowest perforations. Table 4.3 summarizes the values of dT_w/dD and $T_w - T_G$ for each station that were obtained as shown in Fig. 4.13. An additional section of the log above that shown in Fig. 4.13 was used to determine dT_w/dD at Station A.

The interpretation shows about 85% of the total flow coming from the top zone and about 15% from the bottom zone. It should be remembered that many assumptions are made in this interpretation; one of the most critical is the assumed geothermal temperature profile, which is seldom known with any accuracy. The interpretation is also very sensitive to the measured slope, dT_w/dD, because it is typically a small number and is in the denominator when Z is calculated. Thus the interpreted flow profile should be considered approximate.

TABLE 4.4—ROMERO-JUAREZ TEMPERATURE LOG INTERPRETATION, EXAMPLE 2

Station	Depth (ft)	dT_w/dD (°F/ft)	$T_w - T_G$ (°F)	Z (ft)	Total Flow (%)
A	10,210	0.0045	2.77	615	100
B	10,300	0.0086	3.68	428	70
C	10,350	0.014	4.13	295	50

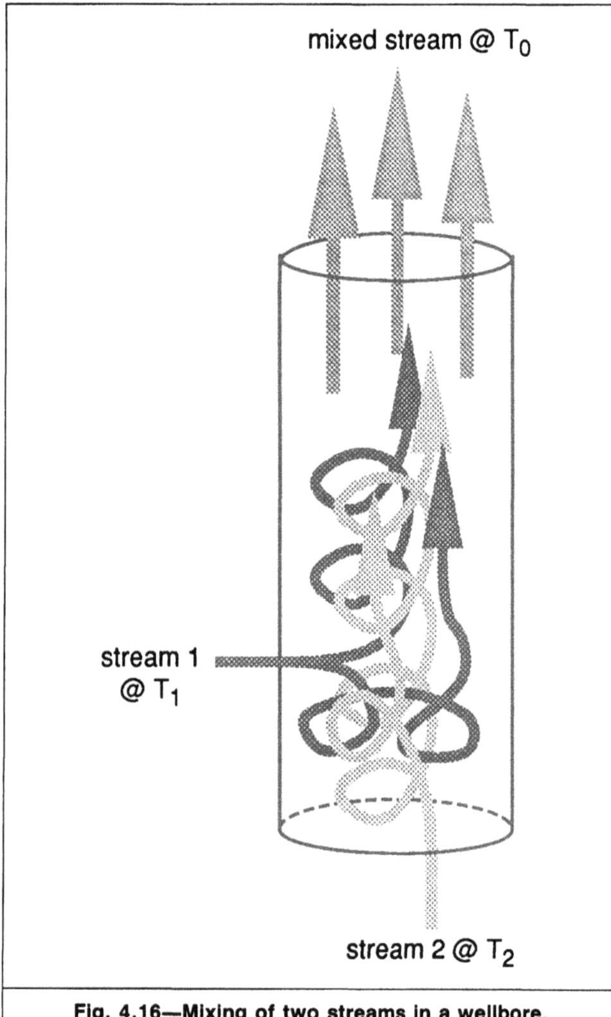

Fig. 4.16—Mixing of two streams in a wellbore.

Example 2—Romero-Juarez Method in New Production Well. When the Ramey equation is applied, it is assumed that the wellbore exchanges heat with a thermally undisturbed reservoir at its geothermal temperature. Mud circulation during drilling, however, cools the formation around the wellbore, creating an altered heat-transfer environment. During the early stages of production, the reference temperature for use with the Ramey equation should reflect this altered state. Fig. 4.14 is a temperature log from a well that produced 985 B/D [157 m^3/d] oil and 1.6 MMscf/D [46×10^3 std m^3/d] gas from three sets of perforations between 10,245 and 10,415 ft [3123 and 3175 m] for a short time after drilling. Notice that the temperature is virtually constant below all production (10,440 to 10,450 ft [3182 to 3185 m]). This flat profile is typical shortly after drilling. To apply the Romero-Juarez method in this case, the rathole temperature should be extrapolated for use as T_G (see Fig. 4.15). With this assumption, values of dT_w/dD and $T_w - T_G$ were obtained for three stations, as shown in Fig. 4.15 and Table 4.4. Note that Stations B and C, which are between perforated zones, are about midway between the major fluid entries. The interpretation shows about 30% of the flow coming from the top zone, 20% from the middle zone, and 50% from the bottom zone.

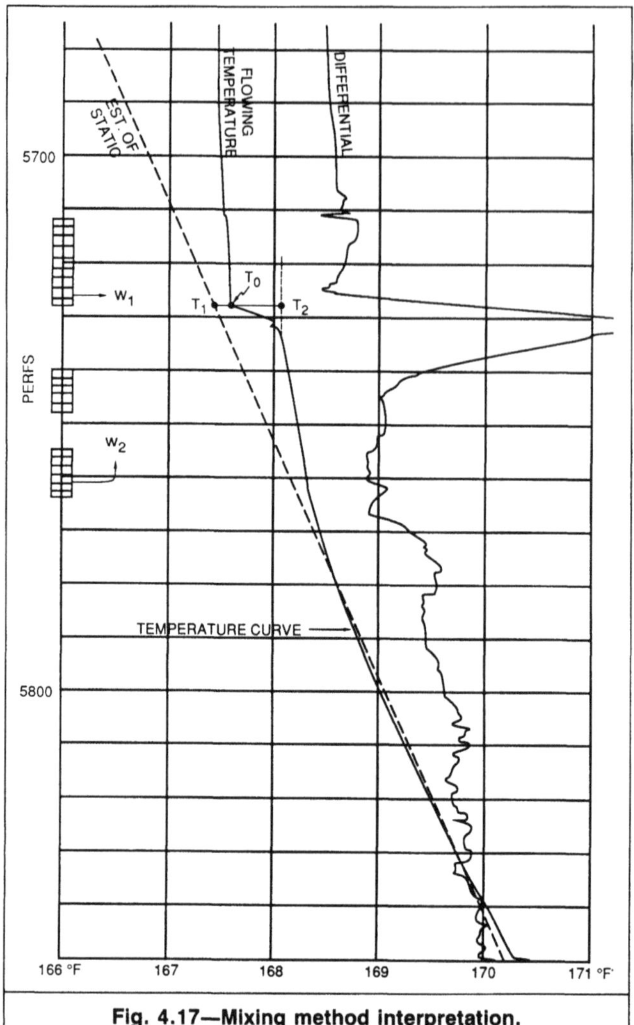

Fig. 4.17—Mixing method interpretation.

Mixing Method. Another analytical method that can be used for production wells is McKinley's* mixing method. In this method, an enthalpy balance applied to the mixing of two streams of fluid at different temperatures into one combined stream is used to determine the relative flow rates of the streams. Consider a region of a wellbore where Stream 1 enters the wellbore from the reservoir and mixes with Stream 2, which is flowing from a lower part of the wellbore (see Fig. 4.16). Stream 2 will be hotter than Stream 1 because Stream 2 has entered the wellbore at a lower depth and because Stream 1 will be close to the geothermal temperature for that depth. An energy balance applied to this system yields

$$w_1 C_{p1}(T_0 - T_1) + w_2 C_{p2}(T_0 - T_2) = 0, \quad (4.28)$$

where w_i = mass flow rates of the streams, C_{pi} = heat capacities of the streams, T_1 and T_2 = temperatures of the two streams, and T_0 = the mixed-stream temperature. Assuming that the fluids have approximately the same heat capacities,

$$\frac{w_1}{w_2} = -\frac{T_0 - T_2}{T_0 - T_1}. \quad (4.29)$$

Because

$$w_1 + w_2 = w_0, \quad (4.30)$$

*From personal communication with R.M. McKinley, Exxon Production Research Co., Houston (1987).

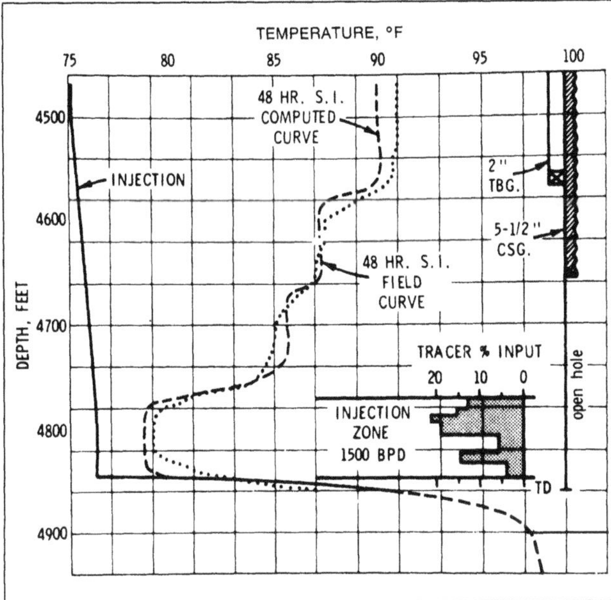

Fig. 4.18—Effect of wellbore geometry and thermal properties of rock and cement on temperature log (from Ref. 24).

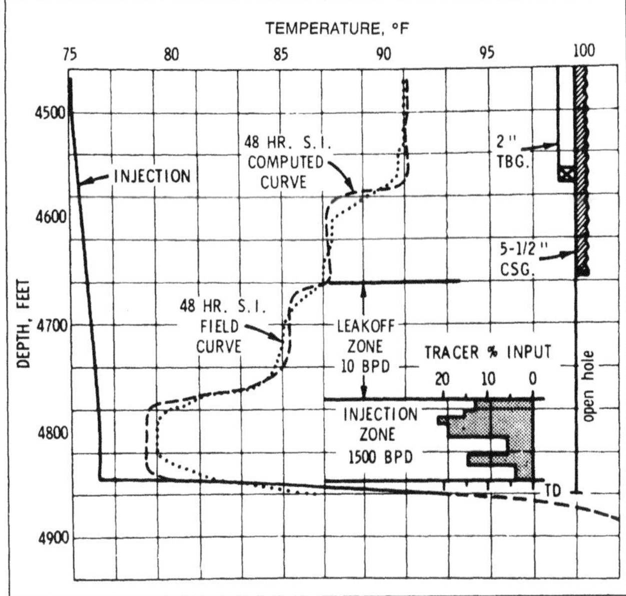

Fig. 4.19—Influence of low-rate leakoff on temperature log (from Ref. 24).

where w_0=mass flow rate above Producing Zone 1, the fraction of flow from Zone 1 is

$$\frac{w_1}{w_0}=\frac{T_0-T_2}{T_1-T_2}. \quad (4.31)$$

Example 3—Mixing Method for Production Wells. The log in Fig. 4.12 can be interpreted by the mixing method. When T_0-T_2 and T_1-T_2 are measured from the log as shown in Fig. 4.17, Zone 1 is calculated to contribute 80% of the total flow. This compares with 85% determined with the Romero-Juarez method.

Computer Simulation of Temperature Logs. The analytical log-interpretation methods based on the Ramey equation are all somewhat limited in application because they cannot be applied within production or injection zones. Some workers use computer solutions of the energy-balance equations to obtain more detailed flow profile information from temperature logs. In some cases, a better estimate of the flow profile can be obtained by matching the temperature logs with computer solutions. Two major difficulties of this approach are the nonuniqueness of the solutions obtained, as illustrated by Smith and Steffenson,[24] and sensitivities to the wellbore completion and to changes in injection water temperature.

Smith and Steffenson point out the difficulties of estimating flow profiles from computer analysis of temperature logs. They analyzed in detail a shut-in log from an injection well, varying parameters that influence temperature behavior. One result of this study is the illustration of a temperature log's sensitivity to rock and cement thermal properties and to wellbore geometry. In Fig. 4.18, the cooler temperatures in the openhole interval above the major injection zone were matched by including the different heat-transfer characteristics of this section as compared with the upper part of the well, where flow is in tubing.

Another well condition that can markedly affect a temperature log is a low-rate fluid-leakoff zone. Fig. 4.19 shows that another good match of the actual temperature log was obtained by assuming a flow rate of 10 B/D [1.6 m^3/d] into the 100 ft [30 m] of openhole interval above the pay zone. This minor amount of fluid loss is sufficient to cause a 2°F [1°C] shift in the wellbore temperature across this interval. This effect will be more pronounced the longer a well is on injection.

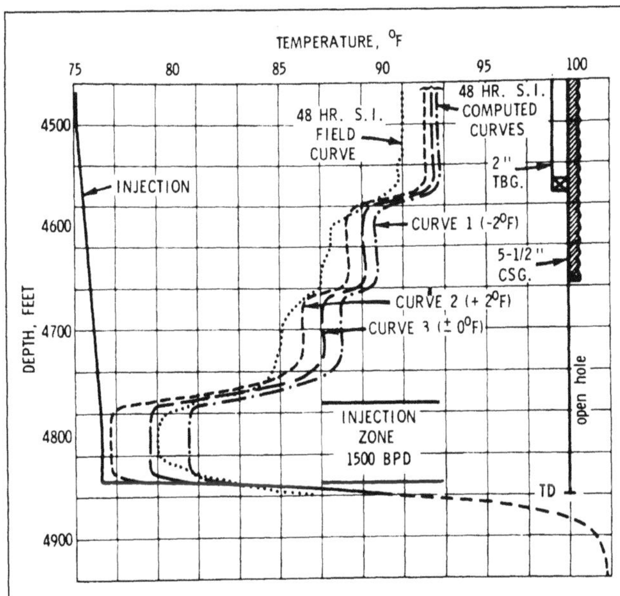

Fig. 4.20—Sensitivity of temperature log to injection water temperature (from Ref. 24).

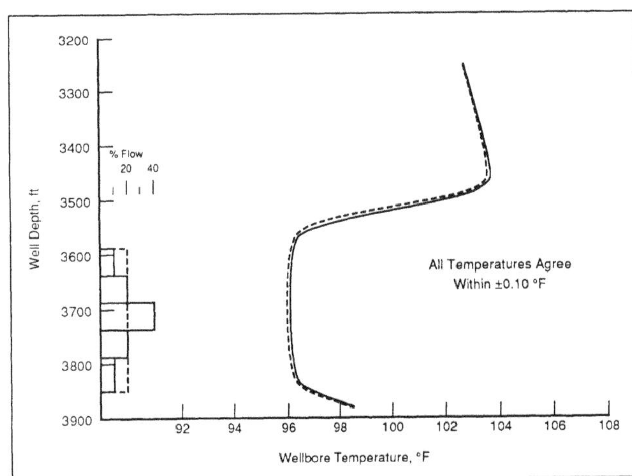

Fig. 4.21—Computer simulation of temperature logs for dissimilar injection profiles, mature injection well (from Ref. 30).

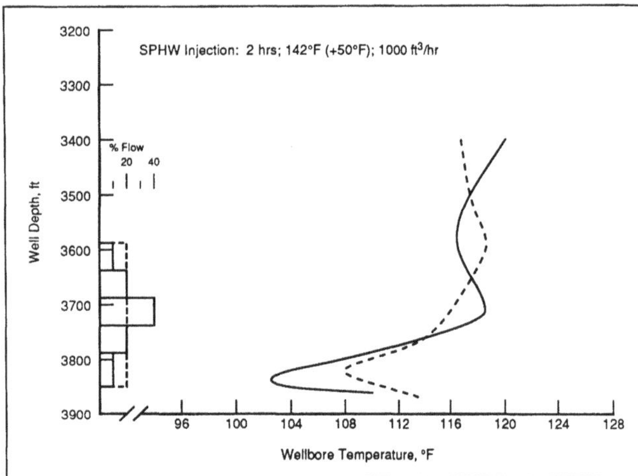

Fig. 4.22—Sensitivity to injection profile after a short period of hot-water injection (from Ref. 30).

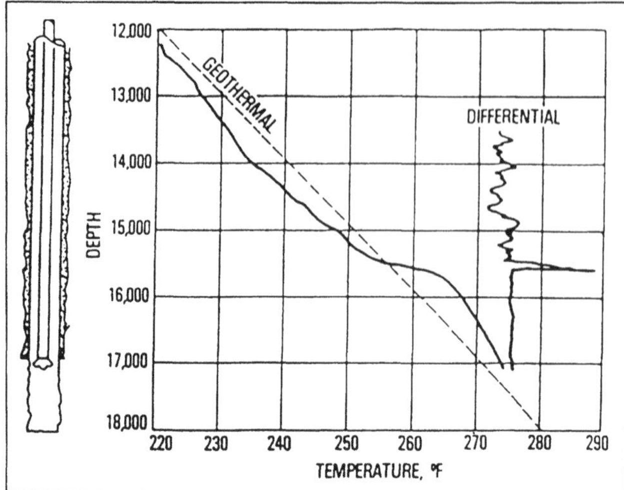

Fig. 4.23—Temperature log from well with underground blowout (from Ref. 14).

Computer matching of temperature logs can also be complicated by small changes in injection-water temperature. Fig. 4.20 shows that an injection-water temperature change of 2°F [1°C] for 6 hours before shut-in results in a shift in the wellbore temperature of almost 2°F [1°C] opposite the injection zone. In this simulation, the injection-water temperature at the injection zone was changed by 2°F [1°C]; the surface temperature would likely have to change by a greater amount to result in this change at the bottomhole location. Significant daily fluctuations in injection-water temperature are common, however, and can be large enough to effect a change in bottomhole injection temperature of this magnitude. Thus, to simulate an injection-well-temperature log properly, a careful record of the injection-water temperature is needed.

Quantitative temperature-log analysis can be enhanced by changing injected fluid temperature. In particular, a short period of hot-water injection has been shown to enhance temperature anomalies in an injection zone.[30] In a mature injection well, the injection-profile details cannot be determined from a temperature log unless an unreasonably long shut-in time is used. For example, Fig. 4.21 shows two dissimilar injection profiles that yield essentially the same wellbore temperature profile. In this comparison, temperature profiles are obtained from a computer model. If, however, a small amount of water heated to 50°F [28°C] higher than the usual injection-water temperature is injected for a short period before shut-in, the simulated temperature logs are different for the two injection profiles (Fig. 4.22). In this case, the hot water was injected for 2 hours before shut-in at a rate of 6 bbl/min [0.02 m^3/d]; the simulated shut-in time was 24 hours. Matching curves in which the effect of the injection profile on the temperature log is amplified enhances the chances of determining an injection profile with a computer simulation. About 2 bbl/ft [1.0 m^3/m] of formation is generally needed to create significant temperature anomalies during injection of a fluid with a different temperature. The larger the temperature change of the injection water, the smaller the volume required.

4.4.2 Qualitative Temperature Log Interpretation. When the uncertainties illustrated in the previous section prevent quantitative interpretation, temperature logs are used as qualitative tools to determine gross characteristics of injection or production wells. In an injection well, a flowing temperature log can positively identify the lowest point of injection as the depth at which the temperature log increases markedly toward the geothermal temperature. Logging deep enough in a well to locate this lower injection boundary, however, is sometimes impossible. A sequence of shut-in temperature logs in an injection well can be used to identify the gross interval of injection qualitatively.

Temperature logs in injection wells can, in some instances, positively identify channeling. Downward channeling is indicated when both the flowing and shut-in logs do not return to the geothermal temperature until a depth well below the bottom of the perforations is reached. Injected fluid is moving downward from the perforations, through a channel either in the cement or in the formation, to cause this temperature behavior. Similarly, in the case of upward channeling, the shut-in log shows a cool anomaly extending a significant distance above the perforations.

Except for the case of a very young well, an injection-well temperature log is useful primarily to identify the gross injection interval and should not be relied on for detailed information about the injection profile. In many injection wells, however, the injection-water temperature varies enough through the year that temperature logs will show more detail about the profile than would be expected for a mature injection well. This effect is similar to what is obtained by a short period of hot-water injection.

Temperature logs are sometimes the best means of determining the flow profile in production wells, as shown in the previous example, which made use of the Romero-Juarez method. Though the temperature log interpretation in such cases is approximate, other logging methods may be even more inaccurate because of the effects of multiphase flow on tool responses. In some cases, a temperature log can be used to locate gas-entry points because of the Joule-Thomson cooling effect. If significant Joule-Thomson cooling occurs, a gas-entry location will be characterized by a cool anomaly on a temperature log.

During drilling and completion operations, temperature logs can identify zones of fluid movement in much the same manner as in production or injection wells. Thus, temperature logs can locate sources of well kicks, underground blowouts, or lost-circulation zones. Locating the cement top is another straightforward application of a temperature log.

The following examples illustrate some of the qualitative information that can be obtained from temperature logs.

Example 4—Location of Underground Blowout. An underground blowout occurs during drilling when a fluid influx from a high-pressure zone causes a lower pressure zone to break down, taking the fluid from the high-pressure zone. A temperature log can often identify the fluid influx source and the underground blowout location. Fig. 4.23 shows a shut-in temperature log run on a well experiencing an underground blowout. The elevated temperature from 15,600 ft [4755 m] to the lowest depth logged indicates that fluid is being produced from the lower part of the well. The sharp decrease in temperature at 15,600 ft [4755 m] back to the cool side of geothermal temperature, particularly noticeable in this case on the differential temperature plot, means that the produced fluid is not traveling up the wellbore past this location. This locates the underground blowout at 15,600 ft [4755 m]. A noise log subsequently run on the well confirmed this interpretation of the temperature log.

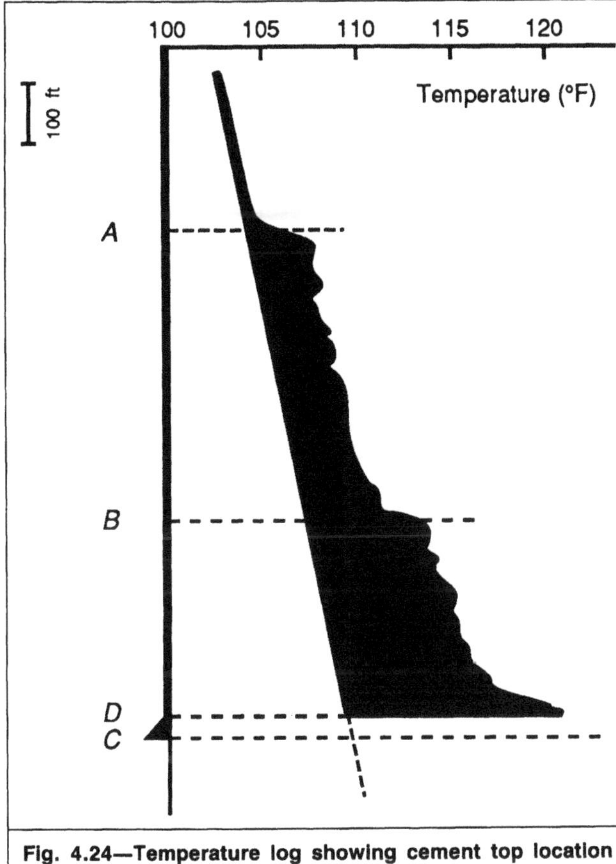

Fig. 4.24—Temperature log showing cement top location (from Ref. 1, courtesy The Inst. of Petroleum).

Example 5—Cement Top Location. The curing of cement is an exothermic reaction and will cause a rise in wellbore temperature that can be readily detected with a temperature log. Fig. 4.24 is a temperature log run about 24 hours after cementing. The top of cement is identified as Location A by the rise in temperature that occurs at this depth. The temperature increase at Location B results from an increased heat generation rate, caused by an accelerator that was added to the last sacks of cement. The rise in temperature at Location D was caused by the larger volume of cement in the wellbore because the cement plug had not been drilled.

Example 6—Identification of Bottom of Injection Interval With Flowing Temperature Log. Fig. 4.25 shows the temperature log measured with this well on injection and the flow profile determined with a velocity-shot log. The temperature log on this mature well gives no information about the details of the profile; the temperature is almost constant across the entire injection interval. The bottom of injection is, however, clearly indicated on the temperature log as the depth (4,698 ft [1432 m]) where the temperature breaks markedly toward the geothermal temperature. The lowest velocity-shot measurement showed about 15% of the total flow moving down from 4,689 ft [1429 m], but a no-flow condition was not detected with the velocity-shot log. The temperature log clearly shows that the fluid moving down at 4,689 ft [1429 m] enters the formation by 4,698 ft [1432 m], below the indicated perforations.

Example 7—Flowing and Shut-In Logs, Mature Injection Well. Fig. 4.26 shows a flowing and a 3-hour shut-in temperature log and the velocity-shot-determined flow profile for a mature injection well. The flowing temperature log yields little information in this case; the dynamic temperature is flat across the entire injection interval, and it was impossible to log to the lowest injection depth. The flowing and shut-in logs show no break to higher temperatures at the bottom of the well, confirming that fluid is moving downward beyond the lowest depth logged. The shut-in log shows the broad cool anomaly (from about 5,010 ft [1527 m] to total depth) resulting from water injection, but little detail about the current injection profile can be obtained from the temperature log in this mature well. It appears that, during the life of the well, most of the water has

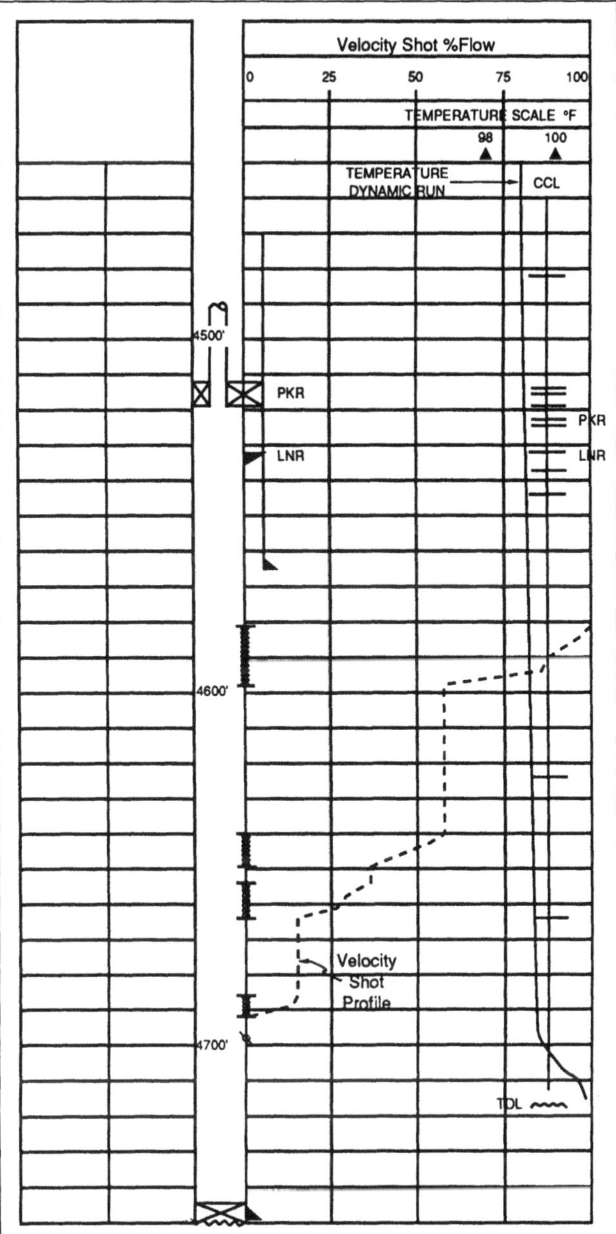

Fig. 4.25—Flowing temperature log showing bottom of injection.

entered the formation below 5,100 ft [1555 m] because there is little separation between the flowing and shut-in logs over this interval. It is interesting that the radioactive-tracer log indicates that about 30% of the flow is now entering the top set of perforations (5,080 to 5,100 ft [1548 to 1555 m]). A comparison of the temperature and tracer results suggests that the well profile has changed recently, with the top zone now receiving more injection.

Example 8—Flowing and Shut-In Temperature Logs, Young Well. More definition about the profile can be obtained from the temperature logs, particularly shut-in logs, on a well that has not been on injection for as long a time. Fig. 4.27 shows the flowing and shut-in logs from a well that had been receiving 900 B/D [140 m^3/d] of water injection for 3 years. The log obtained after a 3-day shut-in period shows two distinct injection intervals, as indicated by the cool anomalies on the log. The flowing log confirms the two main injection intervals by the flattening of the temperature across these zones and suggests that the upper injection interval (5,900 to 5,970 ft [1798 to 1820 m]) also is receiving some injection. The definition obtained on this log results from the relatively short injection period and the lengthy shut-in period. The more typical 3- to 24-hour shut-in period would not allow as much

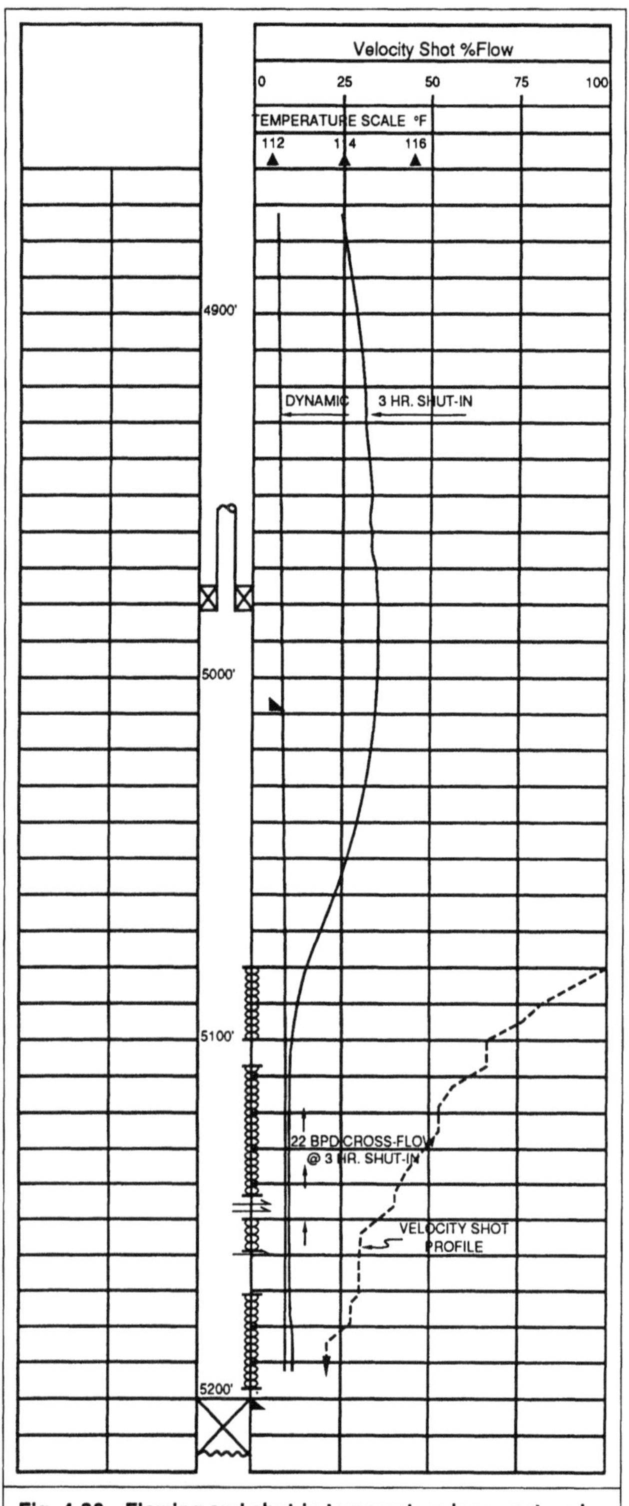

Fig. 4.26—Flowing and shut-in temperature logs, mature injection well.

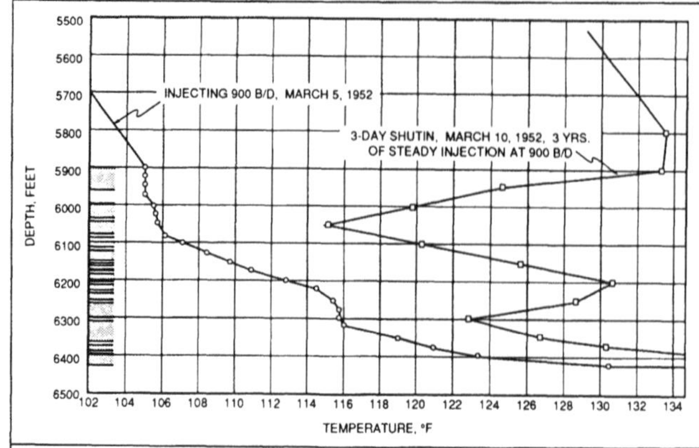

Fig. 4.27—Flowing and shut-in temperature logs, young injection well (from Ref. 31).

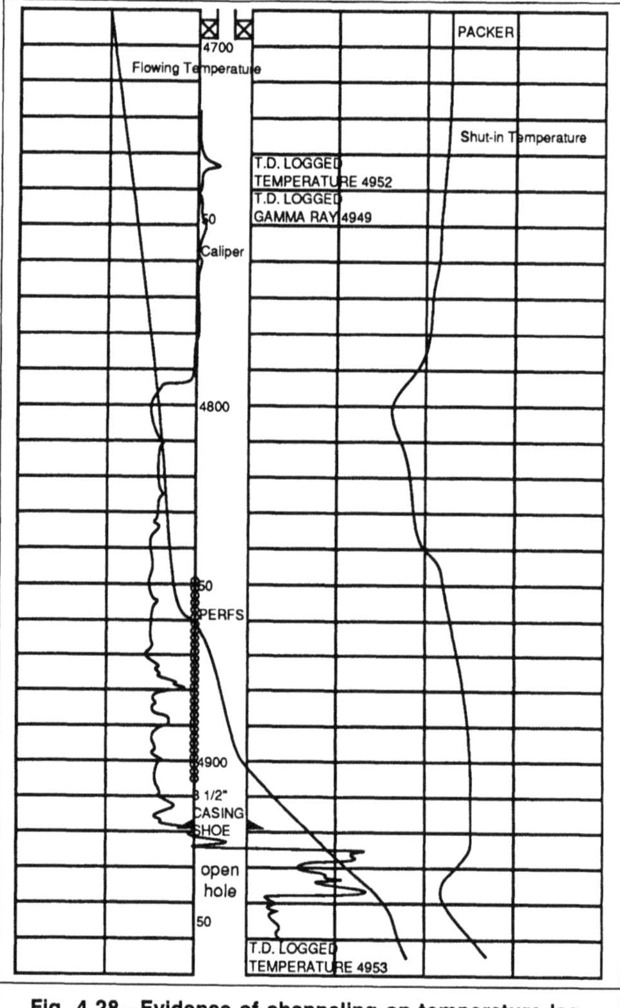

Fig. 4.28—Evidence of channeling on temperature log.

distinction between the two injection intervals to develop. This log was recorded by temperature measurement with the tool stationary at discrete locations indicated by the points on the plots. A continuous recording might have provided even more definition about the flow profile. A final point is that the relative flow rates into each zone cannot be estimated from the size of the temperature anomalies on the shut-in log because the rate of warming of the zones after shut-in will not depend directly on the injection rate into the zones.

Example 9—Channel Identification With a Shut-in Temperature Log. Fig. 4.28 is an 18-hour shut-in temperature log from a well with a suspected channel. Look first at the perforated interval. No cool anomaly has developed in this region, so it appears that little of the injected fluid enters the formation opposite the perforations. The cool anomaly from 4,790 to 4,844 ft [1460 to 1476 m] suggests that fluid enters the formation over this interval after channeling up from the perforations beginning at 4,848 ft [1478 m]. The cool anomaly at 4,924 to 4,950 ft [1500 to 1510 m] is likely caused by injection into the openhole section of the well at this depth.

Example 10—A Succession of Shut-in Logs. Fig. 4.29 shows a series of shut-in logs from a young water-injection well that had two perforated intervals. The logs were run ½, 1, 2, and 4 hours after shut-in. As shut-in time increases, an increasingly distinct cool anomaly develops, locating the two zones of injection. It is also

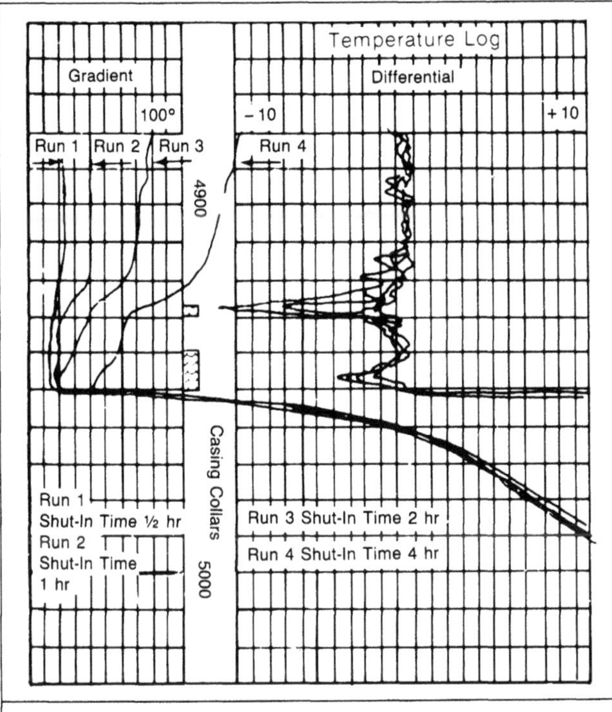

Fig. 4.29—Succession of shut-in logs (from Ref. 15, courtesy Atlas Wireline Services, Western Atlas Intl.).

clear that the temperature opposite the upper zone is increasing toward the geothermal temperature more rapidly than the temperature opposite the lower zone, indicating that the lower zone has received relatively more injection than the upper zone, assuming that the thermal properties of the two zones are similar. Notice also that on these logs, the differential temperature logs clearly mark the top and bottom of injection, because distinct changes in the slopes of the temperature curves occur at these locations.

Example 11—Location of Gas Entries From a Temperature Log. Fig. 4.30 shows a temperature log run on a flowing San Juan basin gas well producing 412 Mcf/D [11.7×10^3 m^3/d]. This log shows four distinct cooling anomalies that correspond to four major gas-entry locations. Gas entries will not always appear as distinctly as the ones shown here. Factors that contribute to the pronounced anomalies on this log are the short production time—the well had only been on production for 8 hours at the time the log was run—and the gas-filled wellbore with no liquid present.

4.4.3 Temperature Log Limitations. Though the temperature log can sometimes be extremely valuable, its utility can be limited at other times, as is true with any individual production log. One caution is that the completion and the lithology affect the temperature log. The computer simulations of Smith and Steffenson[24] clearly point out the importance of the well completion to temperature log behavior.

The length of time a well has been on injection, or more precisely, the cumulative volume injected, has a large bearing on the utility of a temperature log. In general, the older the well, the less distinct the temperature anomalies used to identify injection intervals will be. Similarly, temperature effects in wells can persist for a long time, so what appears as an injection zone on a temperature log can instead be a location where injection occurred in the past. This point is important to remember when trying to evaluate workovers with production logs—the temperature log may be misleading for a long time after workover because of the persistence of thermal effects caused by the old flow profile.

4.5 Detection of Hydraulic Fractures With Temperature Logs

Temperature logs are commonly used to evaluate the height of hydraulic fractures. As first shown by Agnew,[32] the vertical extent

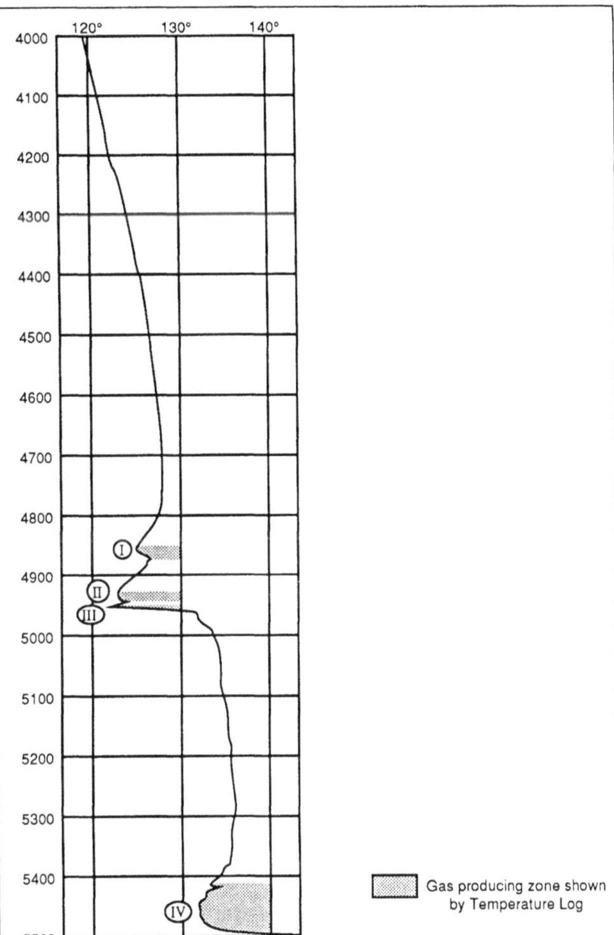

Fig. 4.30—Gas-entry locations indicated by temperature log (from Ref. 12).

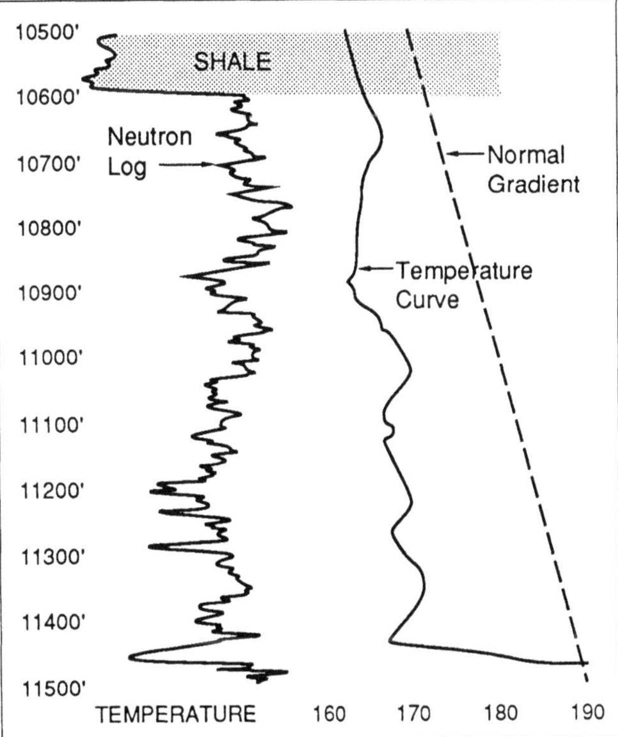

Fig. 4.31—Cool anomaly on temperature log indicates fracture height (from Ref. 32).

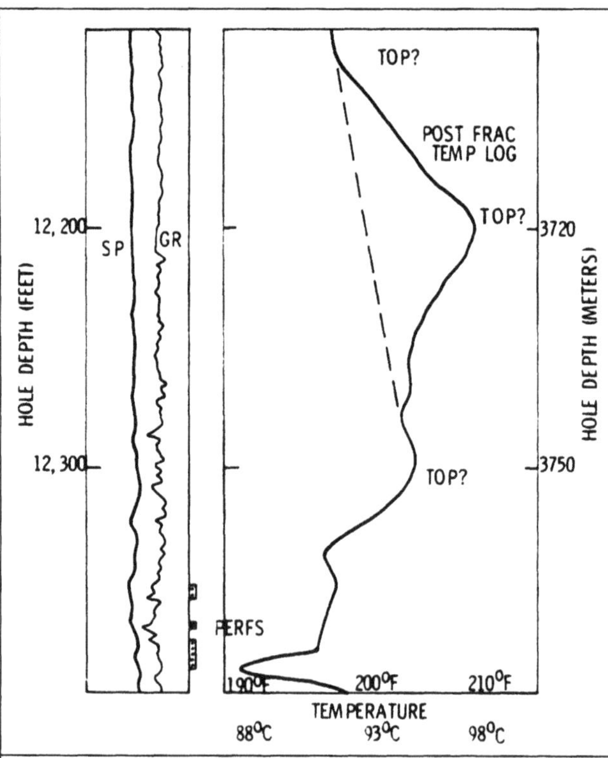

Fig. 4.32—Temperature log with warm anomaly above treatment zone (from Ref. 33).

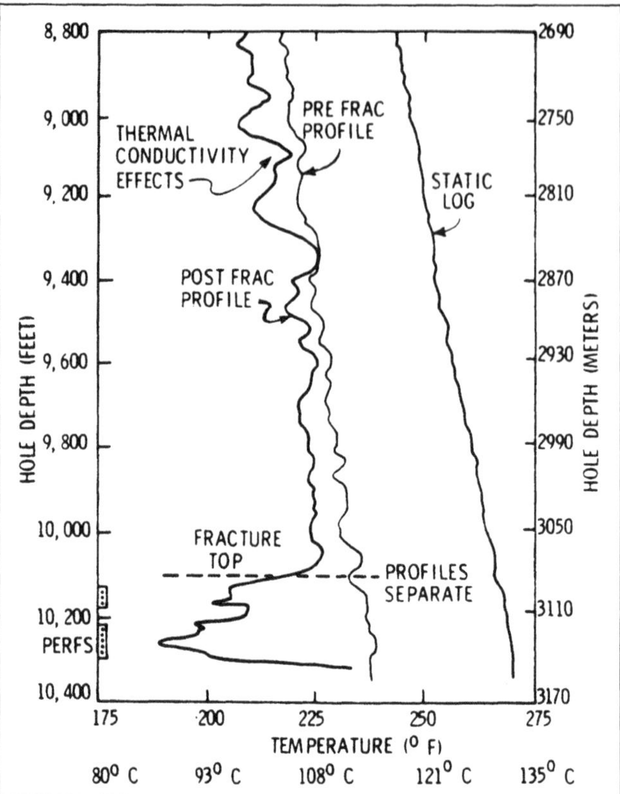

Fig. 4.33—Pre- and postfracture temperature profiles show thermal conductivity effects (from Ref. 33).

of a hydraulic fracture is often indicated by a warm or cool anomaly on a shut-in temperature log run shortly after completion of the fracture treatment. The fracturing fluid injected will generally be warmer or cooler than the formation being fractured; with the types of fluids used today, however, the fracture fluid will usually be cooler than the formation. As the treatment proceeds, the unfractured formation around the wellbore is cooled by radial heat conduction as in any injection well, while cool fluid is placed in the fracture. When the well is shut in, the wellbore opposite the unfractured formation begins to return to the geothermal temperature by unsteady radial heat conduction, while in the fractured region, the wellbore is warmed by linear heat conduction from the formation to the fracture. Because the radial heat transfer is more rapid than linear heat conduction, the fractured region will warm more slowly, yielding a cool anomaly on a shut-in temperature log. Fig. 4.31 illustrates a straightforward application of a temperature log for fracture detection. This well was fractured with 110,000 gal [416 m^3] of ambient-temperature, gelled brine in two stages through 17 perforations. The cool anomaly from 10,650 to 11,450 ft [3246 to 3490 m] indicates a vertical extent of 800 ft [245 m] for the fracture, with the top of the fracture about 200 ft [60 m] above the highest perforation.

Location of a hydraulic fracture from a temperature log is not always as clear-cut as the example just shown. In particular, warm anomalies are often observed when wells are fractured with cool fracture fluids.[33,34] Fig. 4.32 illustrates this complication. This shut-in log shows the cool anomaly expected opposite the perforations after fracturing, but there is also a pronounced warm anomaly above the perforations. The question to answer is which anomaly corresponds to the top of the fracture. Dobkins explains that this effect results from differences in the thermal conductivity of the formations.

> The temperatures measured by postfracture temperature surveys are related inversely to the rock thermal conductivities. Rocks with high thermal conductivity tend to change temperature more slowly than those with low thermal conductivity. Therefore, when a cool fluid is pumped down a hot wellbore, the high thermal conductivity zones cool less than the low thermal conductivity zones. After pumping has stopped, the high thermal conductivity zones exhibit relatively high temperature. Therefore, these high thermal conductivity zones show up as warm anomalies on temperature surveys.[33]

Note that the thermal diffusivity, rather than the thermal conductivity, will control the rate of temperature change during the transient period. Generally, those formations with high thermal conductivities also have high thermal diffusivities, so Dobkins' statements about thermal conductivity dependence are correct.

Warm anomalies on postfracture temperature logs can also result from frictional heating of the fracture fluid as it is injected at a high rate through perforations and in the fracture[33] or from fluid movement within the fracture. Dobkins speculated that fluid moving from one wing of a fracture to another after shut-in or vertical fluid movement in the fracture after shut-in could cause the "warm nose" often observed on postfracture temperature logs. Dobkins shows that warm anomalies caused by formation thermal-property variations can be distinguished from those caused by fluid movement effects by running a prefracture shut-in temperature log after circulating cool fluid in the wellbore, as illustrated in Fig. 4.33. The warm anomalies on the postfracture temperature log correspond to warm anomalies on the prefracture survey, demonstrating that these anomalies result from thermal property variations. When a warm anomaly appears on a postfracture temperature log and no corresponding anomaly exists on the prefracture log, the warm anomaly is apparently caused by fluid movement in the fracture after shut-in. The warm anomaly region would thus be included as part of the interpreted fractured zone.

4.6 Guidelines for Running and Interpreting Temperature Logs

Smith and Steffenson[35] suggest several guidelines for the running and interpretation of temperature logs in injection wells. Some of these recommendations for injection wells and a few recommendations for production wells are presented below as general guidelines to conducting and interpreting temperature surveys.

4.6.1 Recommendations for Running Temperature Logs.

1. For routine surveys, stabilize injection or production conditions (rate and temperature) for 48 hours before surveying.

2. More information about the flow profile can usually be obtained from a shut-in log than from a flowing temperature log. Whenever possible, run a shut-in log. In many instances, running a succession of shut-in logs will aid in flow profile definition.

3. Permit little or no surface leakoff during an injection survey. Permit no injection or leakoff during a shut-in survey. Even a few barrels of backflow can impair the survey.

4. Check the lubricator grease head to ensure proper pressure balance to prevent injection of large amounts of grease down the tubing. Be sure the temperature instrument is responding properly (is free of grease, etc.).

5. Log when entering the hole to record undisturbed temperatures if possible. If forced to log up, use as slow a cable speed as possible.

6. Logging speeds should not exceed about 20 ft/min [6 m/min] with current temperature-sensing instruments. After the continuous survey is finished, make a few stationary measurements to check the response rate.

7. Allow sufficient time between runs on decay-time surveys (successive shut-in logs) for temperature equilibrium to be restored within the wellbore. Typically, 1 to 1½ hours should be allowed between surveys.

4.6.2 Guidelines for Interpreting Temperature Logs.

1. A suite of flowing and shut-in logs is generally easier to interpret and more diagnostic than a single log.

2. Long cumulative injection times result in slower temperature recovery after shut-in and more vertical smearing of temperature profiles. Injection times exceeding roughly 2 years decrease identification of multiple injection intervals (within a gross injection zone) on shut-in temperature curves.

3. Previous injection intervals significantly affect the temperature shut-in curve for up to 6 months (or even longer) after injection has been terminated in a particular interval; as wells become more mature, distinction between previous and current injection intervals is increasingly difficult without varying the injection water temperature.

4. Identification of current injection intervals in mature wells can be improved by injection of warmer or cooler water a few hours before shut-in.

5. Injection zones as thin as 6 ft [2 m] or less can be identified on the temperature profile.

6. Thief-zone losses of 5 B/D-ft [0.3 $m^3/d \cdot m$] or more cause anomalies on the shut-in temperature curve of the same magnitude as major injection intervals. Even thief losses as low as 0.5 B/D-ft [0.03 $m^3/d \cdot m$] cause sizable anomalies on shut-in curves after sufficient injection time.

7. Wellbore arrangement of tubing, casing, and open hole significantly affects the shut-in temperature curve in noninjection intervals. Added insulation at any depth generally speeds the return to geothermal temperature.

8. Wellbore arrangements, except for shot holes, have little effect on shut-in temperature curves in the injection intervals themselves.

9. Shot and enlarged holes cause anomalies that can be mistaken for injection intervals on the shut-in temperature curves.

10. Large masses of cement in enlarged holes, whether in open hole or behind casing, will cause warm anomalies on late-shut-in-time (24-hour) curves in injection wells and cool anomalies on late shut-in logs from production wells.

11. Backflow at the surface during shut-in will decrease temperature anomalies.

12. To apply quantitative interpretation methods based on the Ramey equation, reservoir zones must be far enough apart that vertical heat conduction is not significant between zones.

13. Though temperature logs have limited depth resolution in defining a flow profile, they are sometimes the most accurate logs available in multiphase production wells. For this reason, they should be routinely included in the suite of logs run in a multiphase-flow well.

14. A sharp break toward geothermal temperature positively identifies the bottom of the fluid injection interval.

Nomenclature

C_p = specific heat capacity, Btu/lbm-°F [MJ/kg-°C], L^2/t^2-T
C_{pf} = specific fluid heat capacity, Btu/lbm-°F [MJ/kg-°C], L^2/t^2-T
d_{ce} = casing OD, ft [m], L
d_{ci} = casing ID, ft [m], L
D = depth, ft [m], L
ΔD = depth increment, ft [m], L
$f(t)$ = Ramey time function, dimensionless
g_G = geothermal gradient, °F/ft [°C/m], T/L
H = enthalpy, Btu [kJ], m-L^2/t^2
k = permeability, md, L^2
K_{JT} = Joule-Thomson coefficient, °F/psi [°C/kPa], T-t^2-L/m
p = pressure, psi [kPa], m/L-t^2
q = volumetric flow rate, ft^3/sec [m^3/s], L^3/t
Q = heat transfer rate, Btu/sec [kJ/s], m-L^2/t^3
Q_{fr} = frictional heating rate, Btu/ft^3-sec [kW/m^3], m/L-t^3
r = radial position, ft [m], L
r_1 = Radial Position 1
r_2 = Radial Position 2
t = time, seconds, t
T = temperature, °F [°C], T
T_b = temperature at D=0, °F [°C], T
T_{ce} = outside casing temperature, °F [°C], T
T_G = geothermal temperature, °F [°C], T
T_i = injection temperature, °F [°C], T
T_w = wellbore temperature, °F [°C], T
T_0 = temperature of Stream 0, °F [°C], T
T_1 = temperature of Stream 1, °F [°C], T
T_2 = temperature of Stream 2, °F [°C], T
u = heat flux, Btu/ft^2-sec [kW/m^2], m/t^3
U = overall heat transfer coefficient, Btu/ft^2-sec-°F [kW/$m^2 \cdot$K], m/T-t^3
V = specific volume, L^3/m
w = mass flow rate, lbm/sec [kg/s], m/t
Z = Ramey equation parameter, ft [m], L
α = thermal diffusivity, ft^2/sec [m^2/s], L^2/t
λ = thermal conductivity, Btu/°F-sec-ft [kW/m·K], m-L/t^3-T
μ = viscosity, cp [Pa·s], m/L-t
ρ = density, lbm/ft^3 [kg/m^3], m/L^3
ρ_f = fluid density, lbm/ft^3 [kg/m^3], m/L^3
$\overline{\rho C_p}$ = average density and specific heat capacity, Btu/ft^3-°F [kW/m·K], m/t^2-L-T

References

1. Schlumberger, M., Doll, H.G., and Perebinossoff, A.A.: "Temperature Measurements in Oil Wells," *J. Inst. Pet. Technologists* (Jan. 1937) **23,** No. 159, 1–25.
2. Millikan, C.V.: "Temperature Surveys in Oil Wells," *Trans.*, AIME (1941) **142,** 15–23.
3. Guyod, H.: "Temperature Well Logging, Part 1," *Oil Weekly* (Oct. 21, 1946) **123,** No. 8, 35.
4. Guyod, H.: "Temperature Well Logging, Part 2," *Oil Weekly* (Oct. 28, 1946) **123,** No. 9, 33.
5. Guyod, H.: "Temperature Well Logging, Part 3," *Oil Weekly* (Nov. 4, 1946) **123,** No. 10, 32.
6. Guyod, H.: "Temperature Well Logging, Part 4," *Oil Weekly* (Nov. 11, 1946) **123,** No. 11, 50.
7. Guyod, H.: "Temperature Well Logging, Part 5," *Oil Weekly* (Dec. 2, 1946) **124,** No. 1, 26.
8. Guyod, H.: "Temperature Well Logging, Part 6," *Oil Weekly* (Dec. 9, 1946) **124,** No. 2, 36.
9. Guyod, H.: "Temperature Well Logging, Part 7," *Oil Weekly* (Dec. 16, 1946) **124,** No. 3, 38.

10. Halbouty, M.T.: "Temperature Effects on Oil Well Drilling," *Oil Weekly* (Dec. 18, 1939) **96,** No. 2, 10-16.
11. Goins, W.C. Jr. and Dawson, D.D. Jr.: "Temperature Surveys to Locate Zone of Lost Circulation," *Oil & Gas J.* (June 22, 1953) **52,** No. 7, 170-71.
12. Kunz, K.S. and Tixier, M.P.: "Temperature Surveys in Gas Producing Wells," *JPT* (July 1955) **7,** 111-19; *Trans.*, AIME, **204.**
13. Briggs, F.: "The Temperature Bomb," *Oil Weekly* (June 28, 1943) **110,** No. 4, 15-17.
14. McKinley, R.M.: "Production Logging," paper SPE 10035 presented at the 1982 SPE Intl. Petroleum Exhibition and Technical Symposium, Beijing, March 18-26.
15. *Interpretive Methods for Production Well Logs,* second edition, Atlas Wireline Services, Western Atlas Intl. Inc., Houston (1982) 85-98.
16. Connolly, E.T.: "Resume and Current Status of the Use of Logs in Production," paper presented at the 1965 SPWLA Annual Logging Symposium, Dallas, May 4-7.
17. Somerton, W.H. and Boozer, G.D.: "Thermal Characteristics of Porous Rocks at Elevated Temperatures," *JPT* (June 1960) 77-81; *Trans.*, AIME, **219.**
18. Prats, M.: *Thermal Recovery,* Monograph Series, SPE, Richardson, TX (1982) **7,** 16-29, 201-38.
19. Ramey, H.J. Jr.: "Wellbore Heat Transmission," *JPT* (April 1962) 425-35: *Trans.*. AIME. **225.**
20. McCabe, W.L. and Smith, J.C.: *Unit Operations of Chemical Engineering,* second edition, McGraw-Hill Publishing Co., New York City (1967) 279-321.
21. McAdams, W.H.: *Heat Transmission,* second edition, McGraw-Hill Publishing Co., New York City (1942) 133-53.
22. Romero-Juarez, A.: "A Note on the Theory of Temperature Logging," *SPEJ* (Dec. 1969) 375-77.
23. Steffenson, R.J. and Smith, R.C.: "The Importance of Joule-Thomson Heating (or Cooling) in Temperature Log Interpretation," paper SPE 4636 presented at the 1973 SPE Annual Meeting, Las Vegas, Sept. 30-Oct. 3.
24. Smith, R.C. and Steffenson, R.J.: "Computer Study of Factors Affecting Temperature Profiles in Injection Wells," *JPT* (Nov. 1970) 1447-58; *Trans.*, AIME, **249.**
25. Tester, H.E.: *Thermodynamic Properties of Methane,* British Petroleum Co. Ltd., Sunbury-on-Thames, England (1959) 57-58.
26. Katz, D.L. *et al.*: *Handbook of Natural Gas Engineering,* first edition, McGraw-Hill Publishing Co., New York City (1959) 69-188.
27. Starling, K.E.: *Fluid Thermodynamic Properties for Light Hydrocarbon Systems,* Gulf Publishing Co., Houston (1973).
28. Witterholt, E.J. and Tixier, M.P.: "Temperature Logging in Injection Wells," paper SPE 4022 presented at the 1972 SPE Annual Meeting, San Antonio, Oct. 8-11.
29. Curtis, M.R. and Witterholt, E.J.: "Use of the Temperature Log for Determining Flow Rates in Producing Wells," paper SPE 4637 presented at the 1973 SPE Annual Meeting, Las Vegas, Sept. 30-Oct. 3.
30. Fagley, J. *et al.*: "An Improved Simulation for Interpreting Temperature Logs in Water Injection Wells," *SPEJ* (Oct. 1982) 709-18.
31. Nowak, T.J.: "The Estimation of Water Injection Profiles from Temperature Surveys," *Trans.*, AIME (1953) **198,** 203-12.
32. Agnew, B.G.: "Evaluation of Fracture Treatments with Temperature Surveys," *JPT* (July 1966) 892-98; *Trans.*, AIME, **237.**
33. Dobkins, T.A.: "Improved Methods to Determine Hydraulic Fracture Height," *JPT* (April 1981) 719-26.
34. Wages, P.E.: "Interpretation of Postfracture Temperature Surveys," paper SPE 11189 presented at the 1982 SPE Annual Technical Conference and Exhibition, New Orleans, Sept. 26-29.
35. Smith, R.C. and Steffenson, R.J.: "Interpretation of Temperature Profiles in Water-Injection Wells," *JPT* (June 1975) 777-84; *Trans.*, AIME, **259.**

SI Metric Conversion Factors

°API	141.5/(131.5+°API)		=	g/cm^3
atm	× 1.013 250*	E+05	=	Pa
bbl	× 1.589 873	E−01	=	m^3
Btu	× 1.055 056	E+00	=	kJ
ft	× 3.048*	E−01	=	m
ft^2	× 9.290 304*	E−02	=	m^2
ft^3	× 2.831 685	E−02	=	m^3
°F	(°F−32)/1.8		=	°C
in.	× 2.054*	E+00	=	cm
lbm	× 4.535 924	E−01	=	kg

*Conversion factor is exact.

Chapter 5
Radioactive-Tracer Logging

5.1 Introduction

One of the most common logging methods being used today for quantitatively evaluating injection profiles is radioactive-tracer logging. Radioactive-tracer logging methods can be placed in two broad categories: (1) a tracer material is injected from the surface and (2) a radioactive tracer is ejected from a logging tool in the wellbore. The first category includes such techniques as injecting radioactive proppant during a fracture treatment. Running a gamma ray detector after treatment gives some indication of the fracture location. Another application in the first category is tagging cement with a radioactive tracer. Again, a subsequent gamma ray survey locates the tagged cement. Because logs in the first category are generally self-explanatory and are discussed in detail elsewhere,[1] we will focus on logs from the second category. These logs are generally run to evaluate injection profiles.

Obtaining injection profiles from radioactive-tracer logs is based on the ability of the tracer, which is miscible with the wellbore fluids, to disperse rapidly and then to travel with the wellbore fluids. If the tracer moves with the wellbore fluid, monitoring the velocity or loss of tracer from the wellbore should mirror the distribution of velocity or fluid loss of the injected fluid. In addition, because the gamma radiation emitted by the radioactive tracer can penetrate through casing and cement, a radioactive-tracer log can, in some instances, be used to detect channels behind casing, though other logs, such as temperature or noise logs, often identify channels more conclusively.

In this chapter, the tools and operational procedures used in radioactive-tracer logging are described first. Next, the two most common types of tracer logs, tracer-loss and velocity-shot, are described. A section on tracer placement describes the effect of tracer distribution on log quality. Special considerations for tracer logging in laminar flow are then presented, followed by a description of a new tracer-logging method, the two-pulse method. Finally, general guidelines for running and interpreting radioactive-tracer logs are presented. This chapter deals with the use of radioactive-tracer logs in single-phase flow. Radioactive tracers are also applied in multiphase production wells, and these applications will be described in Chap. 9.

5.2 Tools and Operations

Fig. 5.1 shows a typical instrument used for radioactive-tracer logging. A solution of radioactive tracer material is loaded on the surface into a chamber in the tool. A motor then ejects tracer whenever it is activated by an electrical signal from the surface. One or, more commonly, two gamma ray detectors—either scintillation counters or, more often, Geiger-Müller tubes—are incorporated in the tool.

The tracers used are radioactive isotopes that emit gamma rays as they decay. The most commonly used isotope in tracer logging is iodine 131 (^{131}I), which decays to stable xenon (^{131}Xe) through the emission of five beta particles and six gamma ray photons of various energies. Iodine 131 has a half-life of 8.1 days. In a typical water-injection well application, the radioactive tracer will be contained in a dilute aqueous solution to ensure miscibility with the wellbore fluids. Oil- or gas-soluble radioactive tracers are also available for logging in those media. Generally, about 5×10^{-6} Ci [185×10^{3} Bq] of radioactive material is placed in the tool.

In any radioactive-tracer logging, centralizing the logging tool is always a good practice if the well completion will allow. With the tool decentralized, tracer may be ejected against the casing wall, leading to poor mixing of the tracer in the injection stream. Because of small tubing size, to allow more rapid tool movement, or simply for ease of operation, radioactive-tracer tools are often run without centralizers. In this case, the operator should be aware of the possible poor mixing of the tracer with the wellbore fluids and should rerun the log when an unusually dispersed tracer profile indicates poor tracer distribution.

Two radioactive-tracer logging methods—tracer-loss (timed-slug) and velocity-shot—are commonly used today. In the tracer-loss method, a single slug of tracer material is ejected into the wellbore above all zones of fluid loss. The tracer concentration is then measured as a function of depth by passing a gamma ray detector repeatedly through the tracer as the tracer slug moves down the wellbore. The tracer-loss method was developed for openhole completions with irregular wellbore diameters. The velocity-shot method consists of measuring the transit time of a slug of tracer between two points (usually between two gamma ray detectors, but sometimes between the ejector and one gamma ray detector). Velocity-shot measurements are repeated at various locations in the wellbore. We will consider these two methods of injection-profile logging in detail in the next two sections.

5.3 Tracer-Loss Log

5.3.1 Running a Tracer-Loss Log. Running a tracer-loss log begins when the baseline gamma ray intensity is recorded throughout the well before any tracer is ejected. The baseline gamma ray log is preferably run with high sensitivity to help judge tracer movement behind pipe. Though its gamma ray levels are usually quite low compared with the levels measured in the tracer slug, this baseline log must be subtracted from the subsequent measurements of the tracer slug. Next, a fairly large slug of radioactive tracer is ejected into the wellbore above all zones of fluid exit. The tracer will ideally be ejected well below (20 to 30 ft [6 to 9 m]) the tubing tail and a similar distance above the first perforated interval. After ejecting the tracer, the operator will often move the tool rapidly up and down through the tracer slug, attempting to mix the tracer thoroughly with the injection water. The logging tool is then dropped below the tracer slug, and gamma ray intensity is measured with a gamma ray detector as the tool passes back up through the tracer. It is important that this first measurement of the tracer slug be made before any of the tracer reaches a fluid exit location. If there is sufficient distance between the tubing tail and the top of the perforations, making several passes through the tracer before tracer reaches the perforations is helpful. This provides a measure of the accuracy of the tracer response in the maximum flow-rate portion of the well,

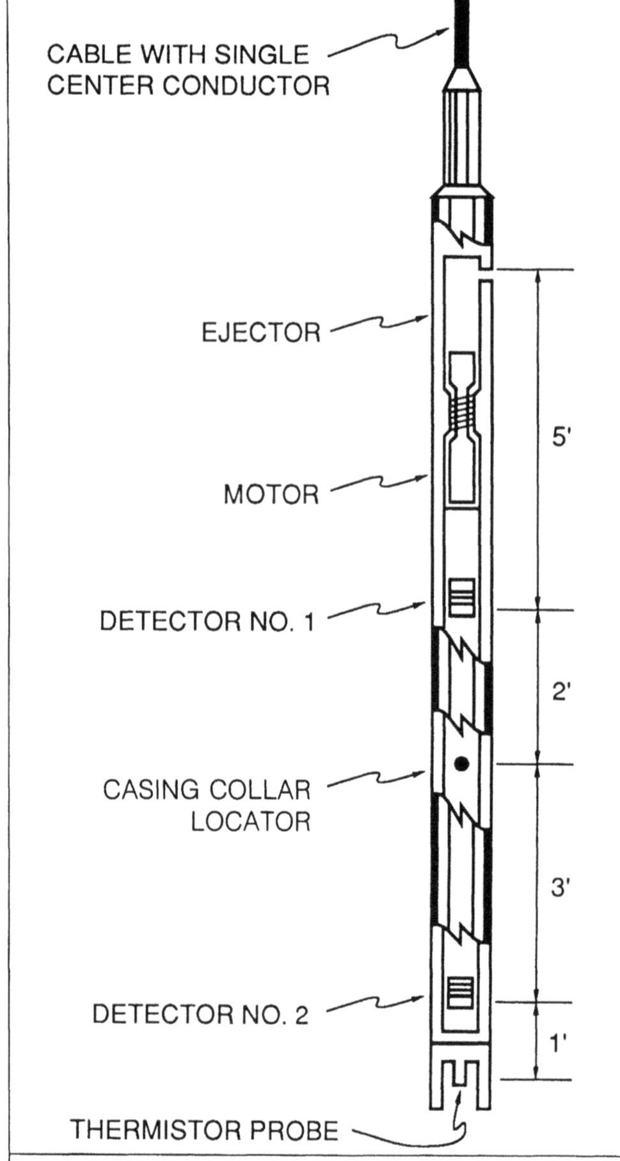

Fig. 5.1—Radioactive-tracer logging tool (from Ref. 2, courtesy Southwestern Petroleum Short Course).

where an accurate measurement is crucial because this first response is used as a reference point in the log analysis procedure.

The rest of the logging procedure consists of repeatedly lowering the tool through the tracer slug and then measuring gamma ray intensity while logging back up through the tracer slug. These measurements are made as rapidly as possible to enhance depth resolution as the tracer slug moves down the wellbore. Logging continues until the tracer slug has ceased moving appreciably or is no longer detectable. The resultant log will consist of a series of plots of gamma ray intensity vs. depth and the time each peak concentration was recorded (see Fig. 5.2). As seen in the figure, the tracer slug diminishes in intensity, spreads in extent, and slows as it moves down the wellbore because of turbulent dispersion and fluid exiting.

Because the movement of the tool through the tracer slug tends to spread the tracer longitudinally, the number of passes that can be made is limited, with 15 being a typical maximum. Thus, in a well with a long injection interval, the tool may be moved at a slower rate to obtain data over the entire injection zone before the tracer has become too dispersed. In any case, the logging passes made up through the tracer slug should be made at a high cable speed to minimize slug distortion caused by tool motion (see Sec. 5.3.4 for a discussion of this effect).

5.3.2 Tracer-Loss Log Interpretation. To determine an injection profile from a tracer-loss log, we must first relate the measured gamma ray intensity to tracer concentration. The gamma rays emitted by the disintegrating isotopes travel in random directions and can travel 1 to 2 ft [0.3 to 0.1 m] before being absorbed. The gamma rays are scattered and absorbed by all the materials in the tracer surroundings—e.g., the tool, water, casing, and formation. Thus, much of the radiation emitted by the tracer will never reach the detector. Furthermore, the number of gamma rays incident on the detector will vary inversely with the square of the distance of the source from the detector. The overall effect is that the gamma ray detector is influenced primarily by tracer that is near the detector—i.e., the tracer inside the casing. Because the number of disintegrations is proportional to the number of radioactive atoms, the gamma ray intensity measured by the detector should reflect the mass of tracer in the wellbore. Bearden *et al.*[3] show that in full-scale tests, the gamma ray intensity measured does "map" the amount of tracer present. Thus, the first fundamental basis of the tracer-loss method is that gamma ray intensity measured by the detector is proportional to tracer mass. It is also convenient to assume that gamma ray intensity is proportional to tracer concentration, which implies uniform mixing of the tracer across the wellbore cross section.

As the tracer slug moves down the wellbore, some of the tracer will be carried into the formation with the wellbore fluid in regions

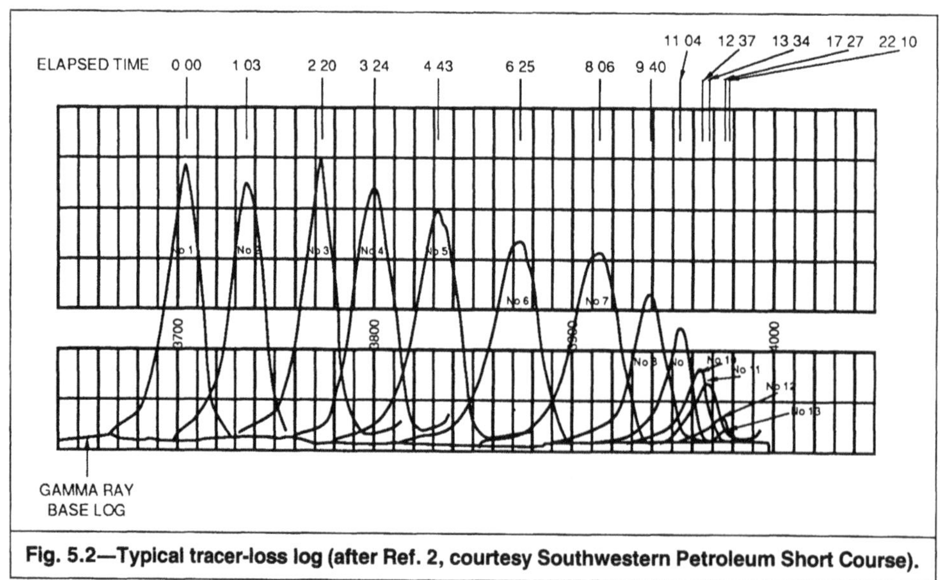

Fig. 5.2—Typical tracer-loss log (after Ref. 2, courtesy Southwestern Petroleum Short Course).

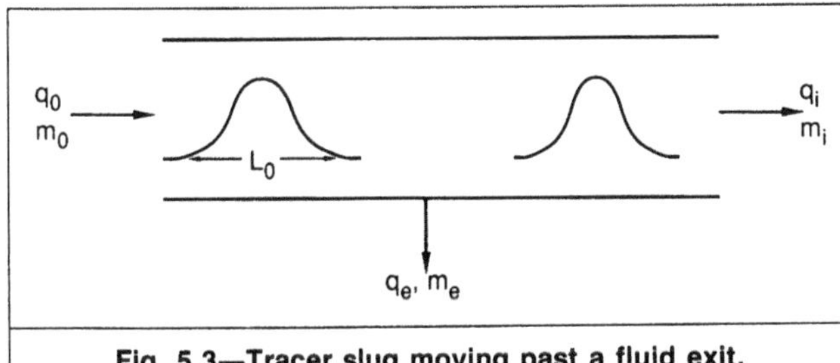

Fig. 5.3—Tracer slug moving past a fluid exit.

of fluid exit. If the tracer is uniformly mixed with the wellbore fluid, tracer should be lost to the formation in the same proportion as fluid is lost to the formation. If, for example, the tracer slug passes a zone taking 30% of the total injection, 30% of the tracer should exit the wellbore. A logging pass through the slug after it has passed the fluid exit would show 30% less tracer in the slug than was measured with a logging pass through the tracer slug before it reached the fluid-loss zone. This proportionality between tracer loss and fluid loss is the second fundamental basis for the tracer-loss log.

Tracer-Loss Interpretation—The Area Method. Tracer-loss logs are interpreted by calculating the areas under the curves of gamma ray intensity vs. depth, with each area assumed proportional to the volumetric flow rate at that slug location. To derive the area method, consider a slug of radioactive tracer as it moves past a fluid exit location (see Fig. 5.3). Upstream of the fluid exit, a total mass of tracer, m_0, moves with a volumetric flow rate q_0. The volumetric flow rate and mass of tracer downstream of the exit point are denoted q_i and m_i, respectively. From a mass balance for an incompressible fluid,

$$q_0=q_e+q_i \quad (5.1)$$

and

$$m_0=m_e+m_i, \quad (5.2)$$

where q_e and m_e=volumetric flow rate and mass of tracer exiting, respectively. The amount of tracer in the slug upstream from the exit, m_0, and the amount of tracer exiting, m_e, can be related to the volumetric flow rates by integrating with time. For a point upstream of the fluid exit, the time required for the entire tracer slug to pass this point, assuming that the tracer is moving at the average flow velocity (q/A_w), is

$$\Delta t=\frac{L_0 A_w}{q_0}, \quad (5.3)$$

where Δt=time required for passage of the tracer slug, L_0=length of the slug, and A_w=cross-sectional area of the pipe. Integrating on time, the amount of tracer passing the point upstream of the fluid exit is

$$m_0=\int_0^{L_0 A_w/q_0} q_0 c\,dt, \quad (5.4)$$

where c=tracer concentration. Because q_0 is constant,

$$m_0=q_0\int_0^{L_0 A_w/q_0} c\,dt. \quad (5.5)$$

In a similar fashion, the length of time tracer will exit is also $L_0 A_w/q_0$. The amount of tracer exiting is then

$$m_e=q_e\int_0^{L_0 A_w/q_0} c\,dt. \quad (5.6)$$

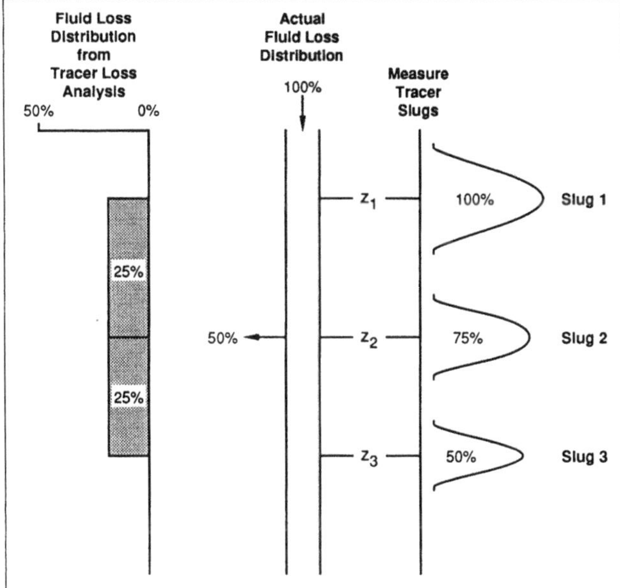

Fig. 5.4—Lack of depth resolution in tracer-loss method (from Ref. 5).

Dividing Eq. 5.6 by Eq. 5.5 yields

$$\frac{m_e}{m_0}=\frac{q_e}{q_0} \quad (5.7)$$

because the integrals are identical. Finally, q_i and m_i are substituted for q_e and m_e by using Eqs. 5.1 and 5.2, and

$$\frac{m_0-m_i}{m_0}=\frac{q_0-q_i}{q_0} \quad (5.8)$$

or

$$\frac{m_i}{m_0}=\frac{q_i}{q_0}. \quad (5.9)$$

Thus, the flow rate at any depth, q_i, is proportional to the amount of tracer present. The amount of tracer is obtained by integrating the curves for gamma ray intensity vs. depth, assuming the gamma ray intensity is proportional to tracer concentration.

$$m_i=\int A_w c\,d\ell-\int A_w c_b\,d\ell. \quad (5.10)$$

The second integral in Eq. 5.10 accounts for the background radiation that must be subtracted so that the calculations account only for radiation from the tracer. It is generally assumed that the gamma ray intensity recorded is proportional to the total mass of tracer in the wellbore, though there is no clear justification for this assumption. Thus, if a tracer slug moved past a point where the cross-sectional area changed, the gamma ray intensity would not change. If this is true, the quantity $A_w c$ is not affected by changes in A_w and the mass of tracer is proportional to the area under the gamma ray intensity curve. Thus,

$$m_i=\int C\gamma_i\,d\ell \quad (5.11)$$

or

$$m_i=CA_{\gamma i}, \quad (5.12)$$

where C=a constant, γ_i=gamma ray intensity measured from Slug i, and $A_{\gamma i}$=area under the gamma ray intensity curve. The volu-

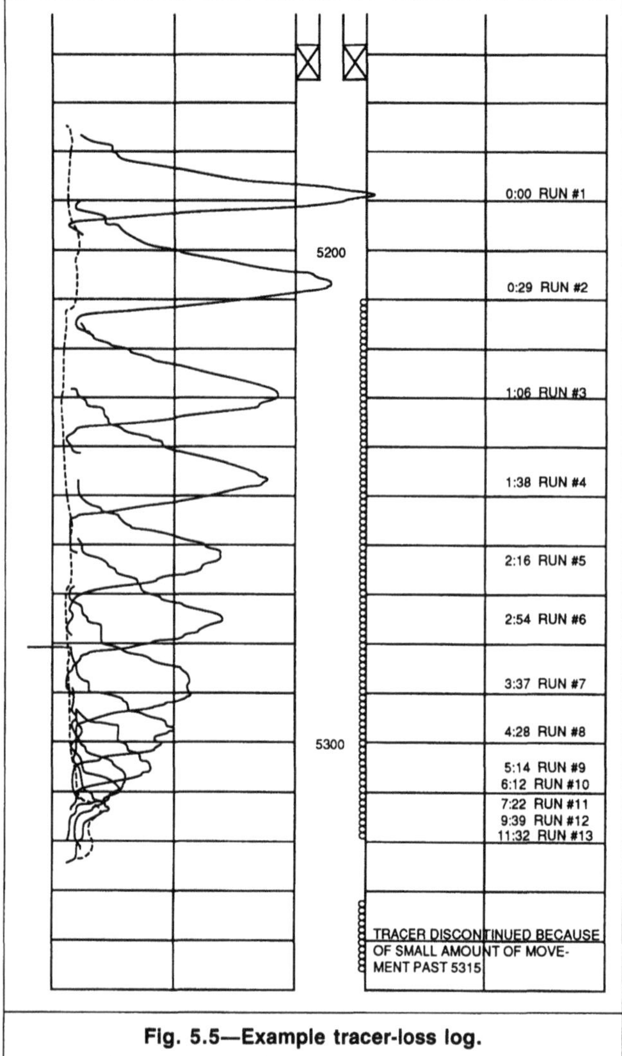

Fig. 5.5—Example tracer-loss log.

metric flow rate is then related to the volumetric flow rate above all fluid exits by

$$\frac{q_i}{q_{100}}=\frac{A_{\gamma i}}{A_{\gamma 100}}, \qquad (5.13)$$

where q_{100} and $A_{\gamma 100}$ refer to the flow rate and area under the gamma ray intensity curve above any fluid exits. It is important to recognize that the area method of tracer-loss analysis is based on the assumptions that the tracer material is uniformly mixed across the wellbore cross section, that the recorded gamma ray intensity is proportional to the amount of tracer in the wellbore, and that the tracer slugs measured are not opposite fluid-exit zones. We will discuss the last point in more detail.

Wiley and Cocanower[4] present a slightly modified method for calculating the amount of tracer present from the gamma ray intensity curve that accounts for system-resolving time. In this method, the gamma ray response, γ_i, is replaced in the integrations by the true count rate, γ_{it}, given by

$$\gamma_{it}=\frac{\gamma_i}{1-t_\gamma \gamma_i}, \qquad (5.14)$$

where t_γ=system response time, which is on the order of 2.5×10^{-4} seconds. It can be seen that for high count rates (500 counts/sec), this correction can be significant.

The flow rates calculated by the area method are assigned to the depths of the respective gamma ray intensity peaks. A significant error is introduced in tracer-loss interpretation because of the often large distances between recorded tracer slugs and because some tracer slugs are opposite fluid-exit zones. As pointed out by Lichtenberger,[2] these effects tend to "smear" the actual fluid losses over greater distances than are actually occurring.

Fig. 5.4 illustrates this problem. Fifty percent of the flow stream exits at a depth z_2. Measurements are made when the tracer slug is above the fluid exit (z_1), when exactly half the tracer slug has passed the fluid exit (z_2), and when the entire slug has passed the fluid exit (z_3). Because only half the tracer slug has passed the fluid exit, the amount of tracer in Slug 2 will be 75% of the amount in Slug 1. Thus, at z_2, the location of the peak of Slug 2, the volumetric flow rate will be interpreted to be 75% of the rate at z_1. This results in a 25% fluid loss being assigned to the interval z_1 to z_2. Slug 3 will have an area that is 50% of Slug 1, so another 25% fluid loss will be assigned to the interval z_2 to z_3. Thus, the actual point of fluid exit has been distributed over the distance between the first and third measurements, which in practice can be a considerable distance. Obviously, if Slug 2 had not been centered about the fluid exit, the interpreted injection profile would have been skewed above or below the actual fluid-loss location.

The tracer-loss method, then, is useful primarily for locating the general area of major fluid loss. Other logging techniques with better depth resolution, such as velocity-shot logs, should be used for more precise definition of the injection profile.

Example—Tracer-Loss Log Interpretation With Use of the Area Method. A tracer-loss log is shown on the left side of Fig. 5.5. The tracer slug was first logged through when the tracer peak was located at 5,189 ft [1582 m]; this slug is considered to represent 100% flow. The baseline gamma ray log showed a background gamma ray reading between 0.5 and 1.2 chart divisions throughout the logged interval, as shown by the dashed line. The region below the background gamma ray log should not be used in the calculation of the tracer-slug areas.

TABLE 5.1—TRACER-LOSS ANALYSIS RESULTS

Tracer Slug	Depth Assignment (ft)	Area	Flow Using Slug 1 (1.82) (%)	Flow Using Average of First Four Slugs (1.91) (%)	Area Triangles	Flow Triangles (%)
1	5,189	1.82	100	100	1.66	100
2	5,207	2.00	110	100	1.61	100
3	5,229	1.88	103	100	1.66	100
4	5,246	1.94	107	100	1.62	100
5	5,262	1.31	72	69	1.04	63
6	5,275	1.38	76	72	0.90	55
7	5,288	1.27	70	66	0.90	55
8	5,297	0.87	48	46	0.65	40
9	5,305	0.55	30	29	0.35	21
10	5,307	0.35	19	18	0.27	16
11	5,309	0.21	12	11	0.14	8
12	5,313	0.15	8	8	0.10	6

To calculate each tracer-slug area, any slug that does not extend to the baseline gamma ray reading (e.g., Slug 1) must first be extrapolated to the baseline. The area of each slug is then determined planimetrically or by dividing the slugs into small trapezoidal pieces. Table 5.1 shows the area obtained from the log in Fig. 5.5 by planimetry. The fraction of total flow at each location is then calculated from

$$f_i=\frac{q_i}{q_{100}}=\frac{A_{\gamma i}}{A_{\gamma 100}}, \quad \text{(5.15)}$$

where f_i = fraction of total flow at Location i, q_i and $A_{\gamma i}$ = flow rate and area under the gamma ray curve for any Slug i, respectively, and q_{100} and $A_{\gamma 100}$ = flow rate and area under the gamma ray curve at the uppermost slug location, respectively. A problem is immediately noticed in applying the analysis procedure. Note that the area under Slug 2 is greater than the area under Slug 1, suggesting that more tracer is in the wellbore at the time Slug 2 was logged than was initially present. This is, of course, not possible. The apparent increase in tracer between Slugs 1 and 2 likely results from incomplete mixing of the tracer with the wellbore fluids, a common problem with tracer-loss logs. Applying the interpretation method as given by Eq. 5.15 yields the profile shown in Column 3 of Table 1, showing an apparent increase in wellbore flow rate at some positions in the wellbore.

A technique commonly used to circumvent the problem of subsequent slug areas being greater than the initial slug area is to use an average of the areas of all slugs logged before significant fluid loss occurs as the area of the initial slug. In the example log, the areas of Slugs 1 through 4 would yield 1.91 as the initial slug area to be used in calculating the fractional flow throughout the rest of the well. Comparing the interpreted profile obtained by this technique with that obtained with use of the Slug 1 area as the initial slug size shows that the interpreted profile is not greatly affected and that the unrealistic interpretation in the upper part of the well is eliminated. Remember, however, that averaging is simply a way to smooth errors in the log; the variation in areas of the first four slugs in this log indicates inaccuracy in the log.

A technique that is sometimes used to give a quick approximation of the area method is to draw tangents on each side of the tracer slugs and a straight baseline at the bottom of the slug (see Fig. 5.6). The areas of the triangles created are then used to approximate the areas under the tracer slugs. If this method is applied, the areas and flow profile shown in Cols. 5 and 6 of Table 5.1 are obtained. Comparing these results with the profile obtained planimetrically clearly shows that significant errors have been introduced by the triangle approximations. This technique should not be used other than to give a rough idea of the flow profile.

As can be seen from the tracer-loss log, the tracer slugs are up to 25 ft [7.6 m] long and as much as 22 ft [6.7 m] apart. Thus, the interpreted flow profile has poor depth resolution. The results of the log interpretation are presented in Fig. 5.7, a plot of the percent of total flow vs. depth and a bar graph of the percent of flow entering different intervals. Considering the errors caused by incomplete mixing of the tracer and limited depth resolution, the log should be used only as a first estimate of the flow profile. This log shows significant fluid entries occurring in the 5,246- to 5,262-ft and 5,288- to 5,313-ft [1599- to 1604-m and 1612- to 1619-m] intervals. Other logging methods, such as velocity-shot logs, should be applied to assess the flow profile more accurately.

Usually, the time at which each tracer peak is logged through is recorded. Fig. 5.5 shows these times in minutes and seconds for this example. From these times and the distance between tracer peaks, fluid velocity in the interval between tracer peaks can be estimated, as in a velocity-shot analysis (see Sec. 5.4). This procedure is referred to as a timed-slug analysis. For this example log, such an analysis yields the flow profile shown by the line with squares in Fig. 5.7. A comparison of this result with the profile obtained by the area method shows a discrepancy with the depth assignments—the fluid entries appear at higher elevations with the timed-slug approach. This discrepancy again illustrates the limited depth resolution of the tracer-loss log.

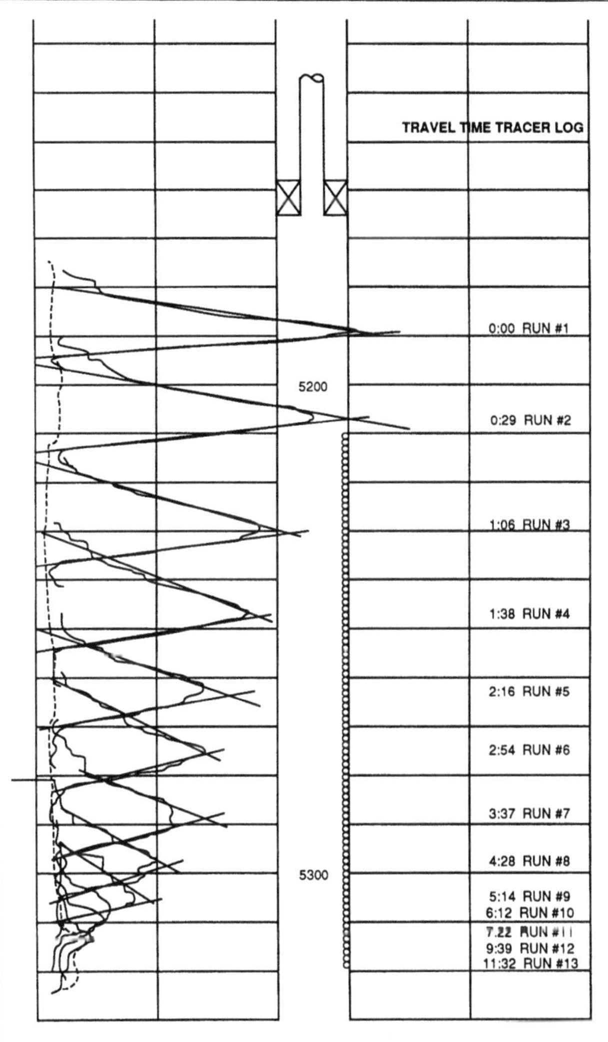

Fig. 5.6—Approximation of tracer slugs with triangles.

Tracer-Loss Interpretation—Self Method. Another interpretation method that has been applied to tracer-loss logs is Self's geometric method.[6] In this technique, triangles are constructed from the gamma ray intensity profiles by constructing a baseline and two tangents to the tracer curve. The base and height of each triangle are added, and these sums are assumed proportional to volumetric flow rate. The Self method is inconsistent with a tracer material balance (the basis of the area method) because the sum of the base plus the height of such a triangle is not proportional to the area under the tracer-concentration curve. Furthermore, the method is dimensionally inconsistent because the results obtained depend on the scale used to plot the gamma ray response. For example, if a tracer-loss log is plotted as gamma ray intensity (in arbitrary units) vs. depth (in feet), the Self method yields a certain flow profile. If the depth scale for the plots were changed to inches, the base of each triangle constructed would be increased by a factor of twelve. Clearly, the result obtained by the Self method would be different for this different depth scale. Because of these inconsistencies, the Self method is not recommended for tracer-loss log interpretation.

5.3.3 Channel Identification—Tracer-Loss Logging. Because gamma rays generated by radioactive tracers can penetrate 1 to 2 ft [0.3 to 0.6 m], observing fluid movement outside of the casing is possible in tracer-loss logs. If the fluid movement can be clearly identified as being outside the casing, the tracer-loss log can give a positive indication of channeling.

Channeling is identified on a tracer-loss log by the development of a secondary peak of tracer concentration (gamma ray intensity). Fig. 5.8 illustrates this technique. After tracer enters the perforations

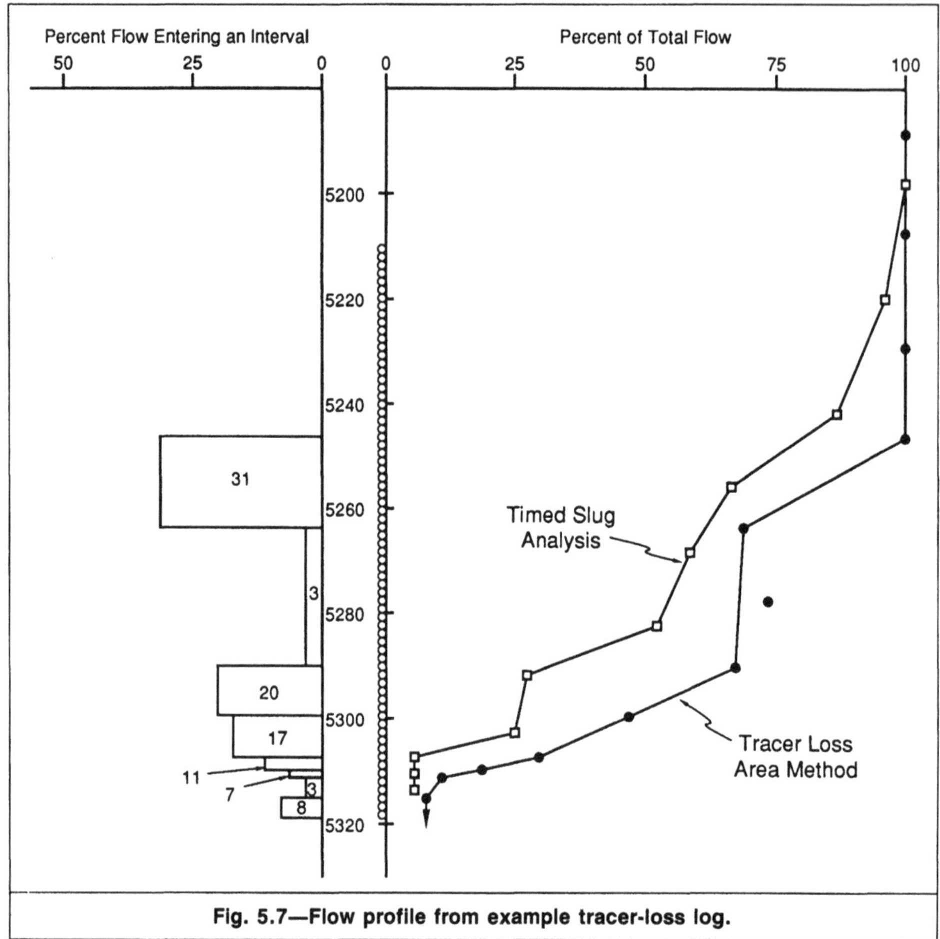

Fig. 5.7—Flow profile from example tracer-loss log.

at Sand 3, a secondary peak that moves back up the well (e.g., Peaks f, j, n, v) develops. This movement indicates fluid channeling up the casing/formation annulus to Sand 4. Also, tracer is detected moving below the lowest perforated interval at Sand 2 (Sand 1, Peak p). Presumably, this movement results from channeling down to Sand 1. Finally, the secondary tracer peak that remains stationary at the packer is attributed to tracer caught in the hardware and in turbulent eddies and does not indicate channeling.

The flow rate in a channel identified by the movement of a secondary peak cannot be determined with any accuracy from a tracer-loss log. Because the gamma rays are penetrating a variable and unknown amount of solids (e.g., casing wall and cement), the area under the secondary tracer peak cannot be compared with peaks in the wellbore to determine the mass of tracer in the channel. Similarly, because of the unknown cross-sectional area of the channel, the velocity of tracer movement in the channel cannot be used to calculate the volumetric flow rate in the channel.

Another method of determining channeling from tracer-loss logs is by comparison with velocity-shot logs. Because the tracer-loss log responds to flow inside or outside the casing and because the velocity-shot log determines flow inside the casing only, a discrepancy between the two can result from channeling. Considering the errors that can occur in both methods, however, this analysis should be applied with caution. The only case in which the difference between a velocity-shot profile and a tracer-loss profile clearly indicates channeling is one in which the tracer-loss log shows tracer movement in a region where the velocity-shot log indicates no flow in the wellbore.

Example—Channel Identification With Tracer-Loss Log. Figs. 5.9a through 5.9d show the passes through a tracer slug from a tracer-loss log of an injection well in which most of the injected fluid was channeling upward. The first four passes (Fig. 5.9a) indicate a single slug of tracer moving down the wellbore, as would normally be expected. By the fifth pass, however, after the main tracer slug has reached the casing shoe, a secondary peak of gamma ray intensity developed above the main slug. Subsequent passes through the tracer (Fig. 5.9b) show this secondary peak moving steadily upward while the main slug progresses slowly down the wellbore. Though the flow rate in the channel cannot be quantitatively determined, the channel is clearly taking a large portion of the flow because the area of the secondary peak is almost as large as the main peak in the wellbore. Because the radiation from the secondary peak must penetrate through the casing and perhaps some cement, the actual amount of tracer in the channel is more than would be calculated by comparing the area of the secondary peak with the area of the main slug.

The behavior of the secondary peak later in the log (Figs. 5.9c and 5.9d) indicates the ultimate destination of the channel. Passes 11 through 13 in Fig. 5.9c show a rapid decrease in the gamma ray intensity of the secondary slug, which might suggest that tracer is entering the formation in the 4,080- to 4,090-ft [1244- to 1247-m] region. Note, however, that this depth corresponds to the location of the packer and tubing tail, so it is likely that the diminished response is caused by increased gamma ray absorption by the tubing and packer. At later times, Passes 15 through 17 (Fig. 5.9d) show secondary peaks that are almost stationary and gradually diminishing in strength. This behavior indicates that the tracer is likely entering the formation along the interval from 3,980 to 4,030 ft [1213 to 1228 m].

5.3.4 Limitations of the Tracer-Loss Log. For a variety of reasons, the tracer-loss log is an inaccurate log for injection profiling purposes and should be used only to obtain a rough estimate of the profile. Its primary utility should be as a means to determine the approximate locations of major fluid exits so that more precise logs, such as velocity-shot logs, can be concentrated in these locations. The causes of the inaccuracy of the tracer-loss log are summarized in this section.

The tracer-loss log depends on the proportionality of the measured gamma ray intensity to the average tracer concentration in the pipe.

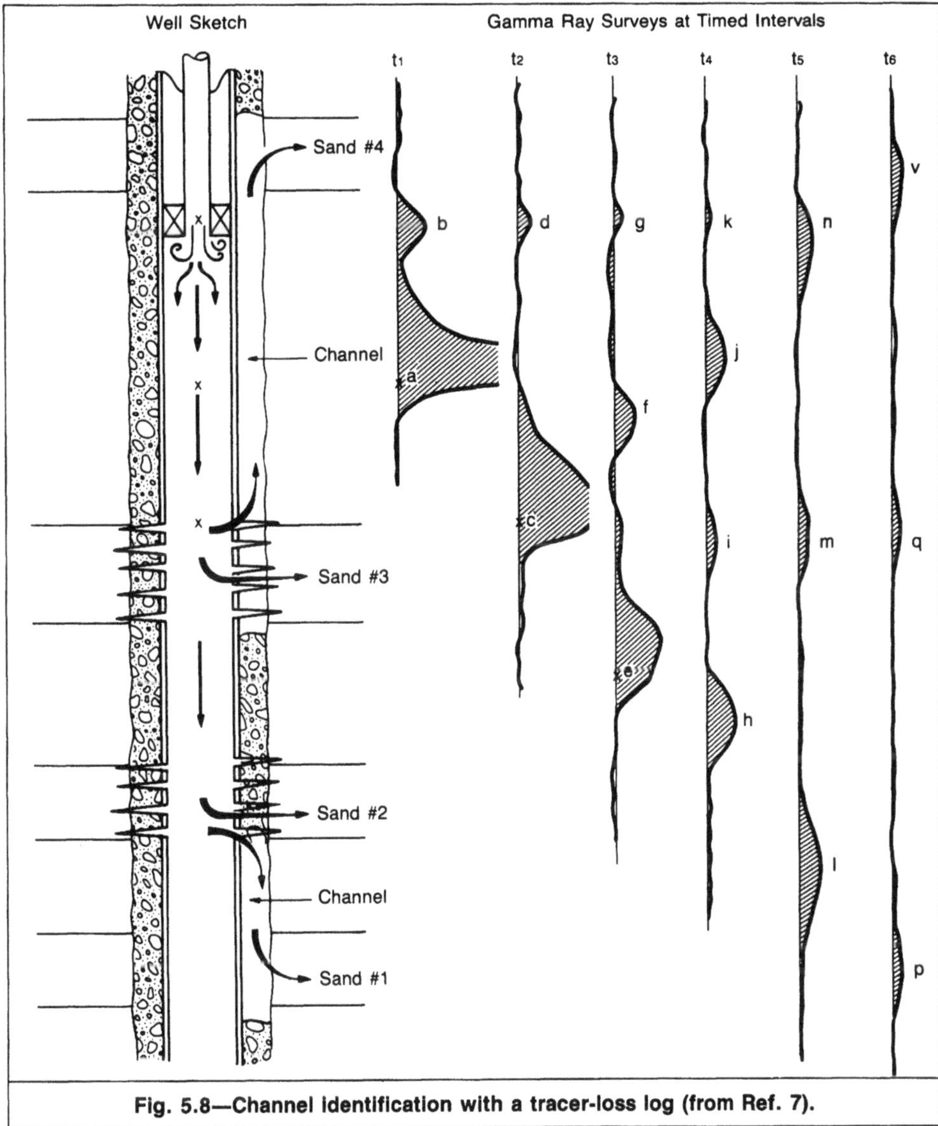

Fig. 5.8—Channel identification with a tracer-loss log (from Ref. 7).

A finite and sometimes long mixing region, however, is required for the tracer to be distributed uniformly in the wellbore fluid after it is ejected from the tool. Tracer slugs measured before the tracer is well mixed may be measured as higher or lower in gamma ray intensity than a well-mixed slug would be, depending on whether the bulk of the tracer was near the tool or near the wall of the wellbore when the detector passed through the slug. Errors caused by poor mixing will be most serious in the first few slugs logged, but because the flow profile is determined by dividing the areas of all subsequent slugs by the area of the first slug, the error in the measurement of the first slug will propagate through the entire log interpretation.

Poor depth resolution is another factor that limits the accuracy of the tracer-loss log. Depth resolution is poor because the tracer slug may move 20 to 40 ft [6 to 12 m] between logging passes. In addition, depth resolution is impaired whenever part of the tracer slug is opposite fluid-entry zones, as discussed in Sec. 5.3.2.

A third factor that contributes to poor accuracy of tracer-loss logs is the distortion of the slug caused by the finite velocity of the tool as it passes through the tracer slug. Tracer-loss-log interpretation implicitly assumes that we take a series of "photographs" of the tracer slug at different times as it moves down the wellbore. In practice, the slugs are measured by moving the tool up through the tracer slug as the slug moves down the wellbore. This results in the measured slug being shorter in length than the actual tracer slug. Fig. 5.10 illustrates this effect. As the detector is pulled upward, it encounters the leading edge of the tracer slug at Position 2 (z_2), and the tool begins recording increasing gamma radiation. While the tool moves upward, the tracer slug continues its downward travel so that the detector senses the back edge of the slug at Position 1 (z_1). This results in an apparent slug length, l_m, measured by the tool, while the actual slug length is l_s. Neglecting dispersion effects, it is easily shown that these are related by

$$\frac{l_m}{l_s}=\frac{1}{1+v_s/v_T}, \quad (5.16)$$

where v_s and v_T=slug and tool velocities, respectively. Thus, the more slowly the tool moves relative to the fluid velocity, the more compressed the measured tracer slug is compared with its real shape. Because this distortion depends on the fluid velocity, it will be a changing and unknown factor as the slug moves down the wellbore, adding to the inaccuracy of the tracer-loss log. A timed-slug analysis is not affected by the distortion of the tracer slug.

A technique that can be used to eliminate the distortion of the slug by the tool movement is to measure the tracer slug as it moves past a stationary tool. With this method, the tool is dropped below the tracer slug, then held stationary with the recorder on time drive as the slug passes. When the slug is past, the tool is again dropped below the slug and held stationary, and gamma ray intensity is recorded until the slug is past. The process is repeated until the slug is no longer distinguishable. The series of plots of gamma ray intensity vs. time obtained in this manner can then be analyzed with the area method. Though this technique eliminates the distortion

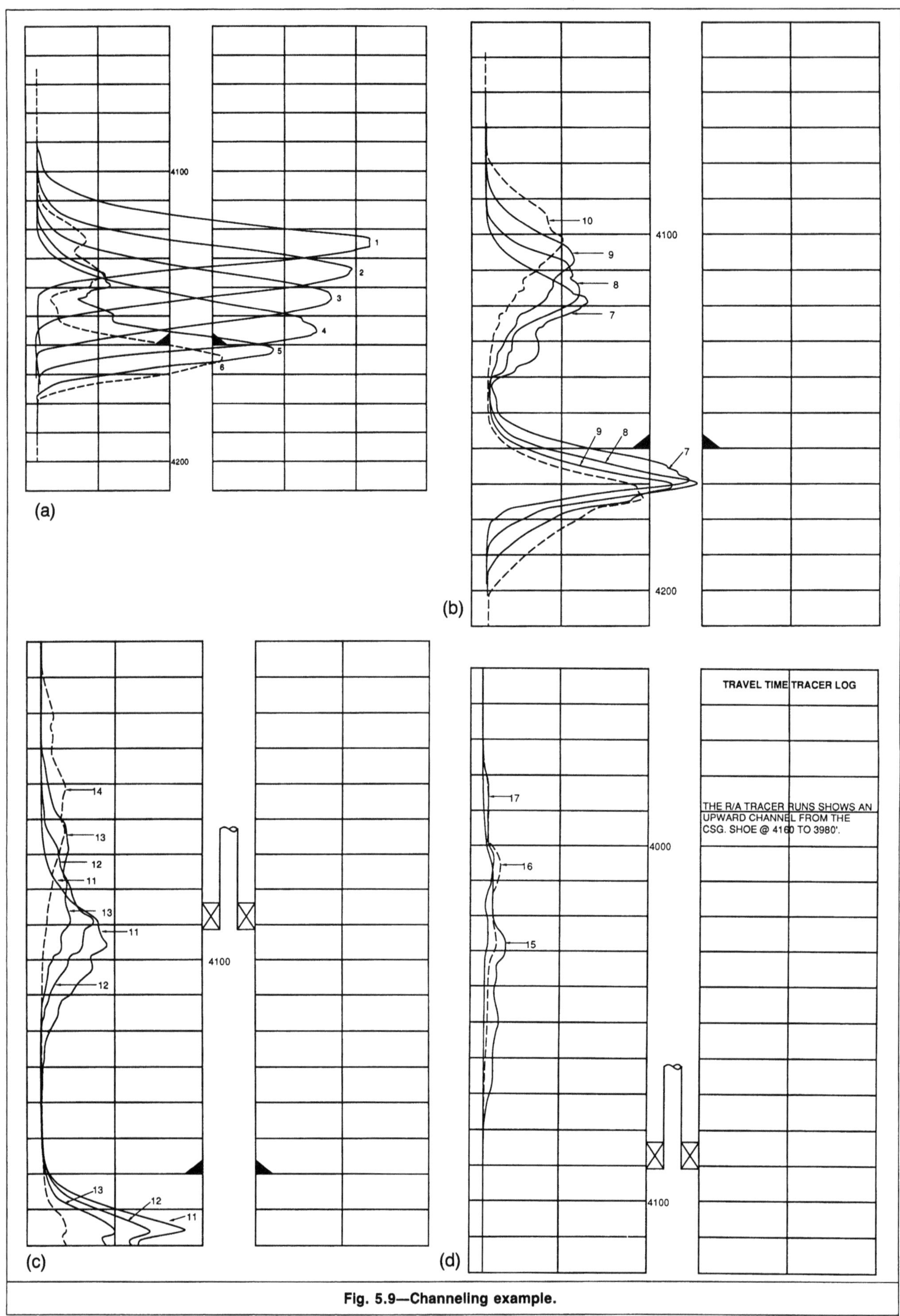

Fig. 5.9—Channeling example.

of the tracer slug by tool movement, it also reduces the number of tracer-slug recordings that can be obtained.

In summary, the tracer-loss log is a means of obtaining an approximate injection profile to be used primarily to guide other logging methods. Its main utility may be in its ability to locate channels positively in certain instances.

5.4 Velocity-Shot Log

5.4.1 Running a Velocity-Shot Log. In velocity-shot logging, a small quantity (shot) of tracer is ejected from the logging tool, and the time required for the tracer to travel between detection points is measured. In most cases, the measured transit time is the travel time between two gamma ray sensors that send signals of gamma ray intensity as the tracer moves by them. Fig. 5.11 shows a typical velocity-shot response by two gamma ray detectors. Some operators have run velocity-shot logs using only one gamma ray detector and measuring the time between tracer ejection and arrival at the gamma ray detector. Because of the variable time required to eject the tracer, however, this technique is less accurate than the two-gamma-ray-detector approach. Thus, most operators now use two detectors to measure travel time. Another variation of velocity-shot logging is timed-slug analysis, the Ford method.[8,9] In this technique, a single tracer slug is ejected in the upper part of the wellbore. The tracer slug is repeatedly logged as it moves down the wellbore, with the travel time between each measurement of the tracer slug noted. Fluid velocities are then calculated from these travel times. The timed-slug analysis is usually performed as part of a tracer-loss log by measuring the time at which each tracer peak is logged while the log is run.

With most tracer-logging tools, the detector spacing can be varied by adding or removing blank tool sections. It is generally desirable to use the smallest detector spacing possible while allowing a sufficient travel time between detectors for an accurate transit-time measurement in the highest-flow-rate portion of the well. The smaller the detector spacing, the better the depth resolution of the log will be. Detector spacings from 1 to 20 ft [0.3 to 6.1 m] are used in practice.

5.4.2 Velocity-Shot Log Analysis. In velocity-shot log analysis, a fluid velocity is calculated from the time required for a tracer slug to move between two detectors according to

$$v_f=\frac{L}{\Delta t}, \qquad (5.17)$$

where v_f=fluid velocity, L=spacing between detectors, and Δt=transit time. If the calculated velocity is assumed to be the average velocity, the volumetric flow rate is given by

$$q=\frac{\pi(d_{ci}^2-d_{Te}^2)L}{4\Delta t}, \qquad (5.18)$$

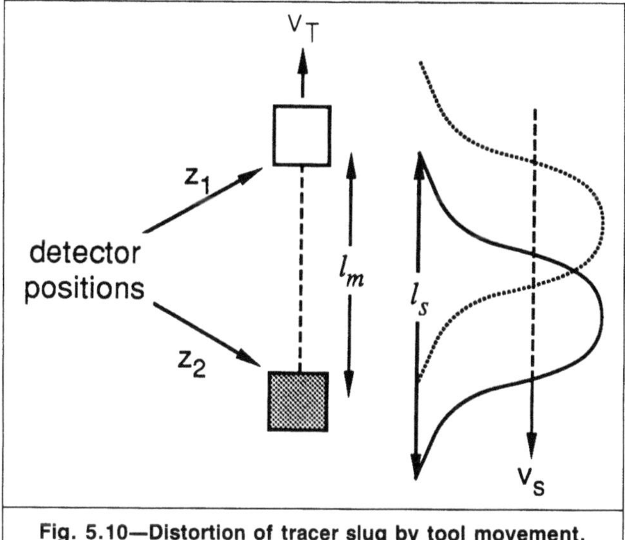

Fig. 5.10—Distortion of tracer slug by tool movement.

where q=volumetric flow rate, d_{ci}=casing ID, and d_{Te}=logging tool OD.

Taylor[10] showed that the peak concentration of a tracer in turbulent flow in circular pipes moves at the average flow velocity (q/A_w). Aris[11] extended this result to other flow geometries. Thus, if the Δt measured from the velocity-shot log is a peak-to-peak transit time (Δt_{pp}), this transit time can be used directly in Eq. 5.18 to calculate volumetric flow rate. Often, however, the transit time measured from a velocity-shot log is the time between the arrivals of the leading edge of the tracer slug at the detectors (Δt_{le}). This arrival time is measured between the points where gamma ray intensity first departs from baseline (see Fig. 5.11). Because the leading edge of the tracer slug moves at the maximum flow velocity, a volumetric flow rate calculated from Δt_{le} must be corrected for the difference between maximum and average velocity. Thus,

$$q=\frac{\pi(d_{ci}^2-d_{Te}^2)L}{4\Delta t_{le}}B, \qquad (5.19)$$

where

$$B=\frac{\bar{v}}{v_{max}}. \qquad (5.20)$$

Knudsen and Katz[12] showed that the correction factor, B, is approximately 0.88 in turbulent, annular flow. This factor will vary somewhat with Reynolds number and flow geometry (i.e., the deviation of the tool position from the pipe centerline). B can be esti-

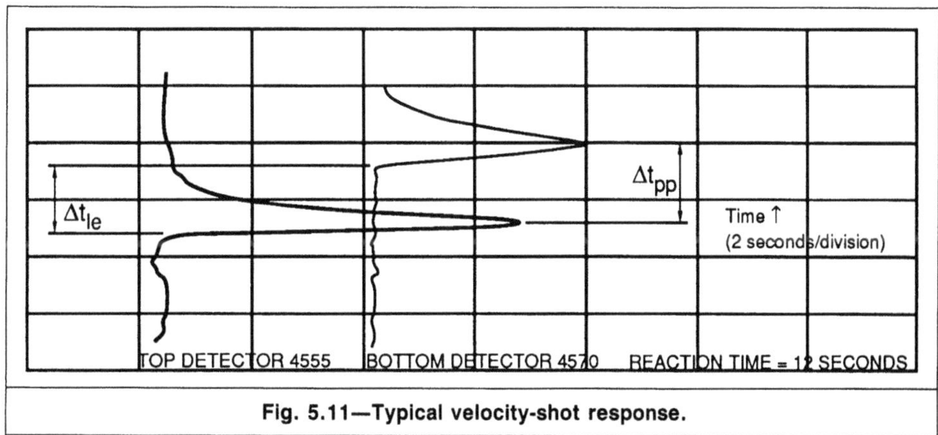

Fig. 5.11—Typical velocity-shot response.

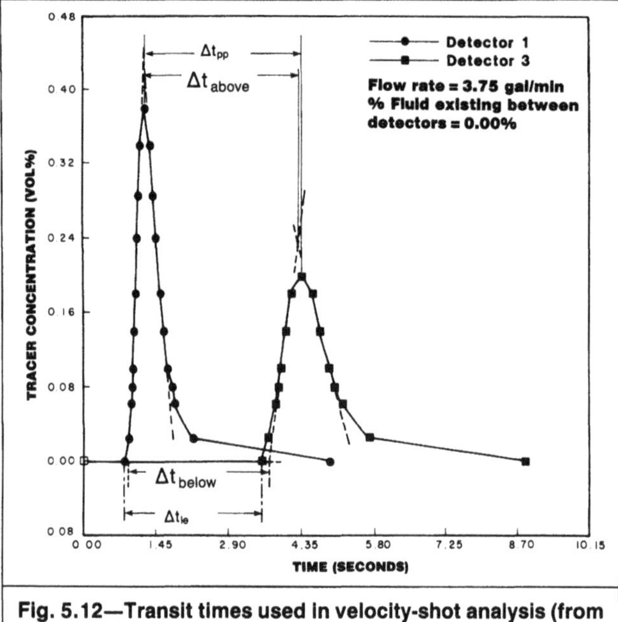

Fig. 5.12—Transit times used in velocity-shot analysis (from Ref. 5).

mated from a velocity-shot log by comparing Δt_{pp} with Δt_{le} because

$$B=\frac{\Delta t_{le}}{\Delta t_{pp}}. \quad (5.21)$$

In the example velocity shot shown in Fig. 5.11,

$$B=\frac{\Delta t_{le}}{\Delta t_{pp}}=\frac{12 \text{ seconds}}{14 \text{ seconds}}=0.86.$$

If the wellbore cross-sectional area is constant and the velocity profile correction factor, B, is assumed constant throughout the well, then all quantities except Δt on the right sides of Eqs. 5.18 and 5.19 are constant, and the flow rate at any depth is inversely proportional to transit time. Thus, the fraction of total flow at any depth can be calculated by

$$\frac{q_i}{q_{100}}=\frac{\Delta t_{100}}{\Delta t_i}, \quad (5.22)$$

where q_i and Δt_i=flow rate and transit time for Depth i, respectively, and q_{100} and Δt_{100}=flow rate and transit time measured above any fluid exit, respectively.

Hill and Solares[5] investigated two other transit-time measurements that are sometimes used: Δt_{above}, the time difference between points formed by the intersection of two tangents to the tracer gamma ray intensity profiles at each detector, and Δt_{below}, the time difference between points of intersection of the baseline gamma ray intensity with tangents on the upstream side of the gamma ray intensity profiles. These transit times are illustrated in Fig. 5.12. It was found that Δt_{above} corresponded very closely with Δt_{pp} and that Δt_{below} was essentially identical to Δt_{le}. Thus it appears that no accuracy is lost in using these transit times, and they may be advantageous when the tracer peak is indistinct or when the signal is noisy.

Other measures of transit time are also used, including the centroid transit time of the tracer slug and one-half peak-height transit time. The one-half peak-height transit time is measured by locating the one-half peak-concentration points on the leading edge of the tracer slug at each detector location. The tracer slug centroid will move at the average velocity in both laminar and turbulent flow; thus, no correction factor is needed when this transit time is used.

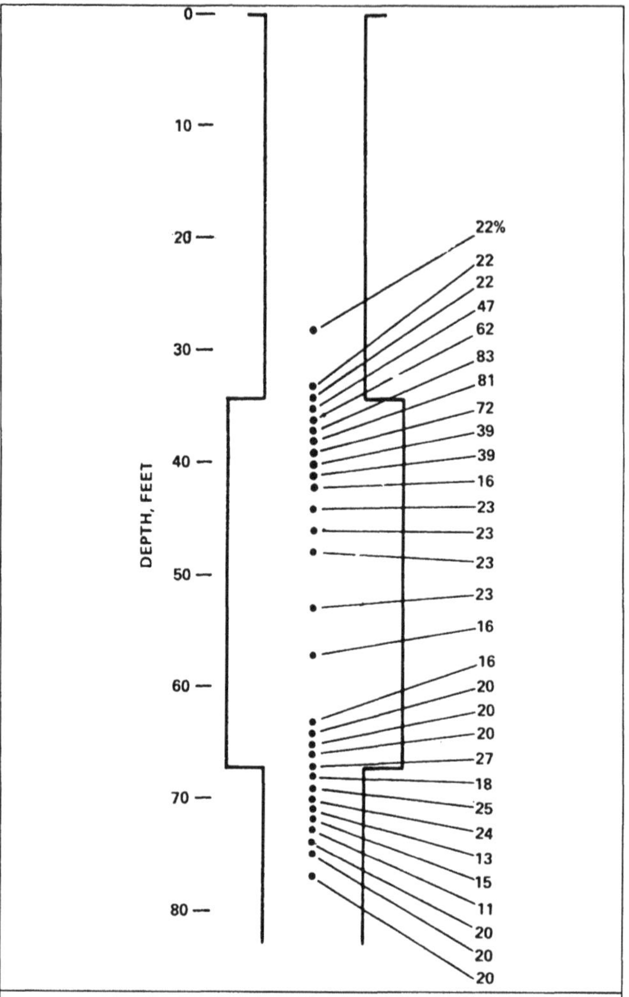

Fig. 5.13—Percentage by which tracer velocity rate exceeded metered rate (from Ref. 3).

The one-half peak-height transit time will be a measure of the maximum velocity in laminar flow and will be a measure of an intermediate velocity between the maximum and average velocities in turbulent flow. Because of uncertainty about which velocity is being measured, the one-half peak-height transit time is best used in laminar flow conditions.

The flow profile is usually calculated from a velocity-shot log by applying Eq. 5.22 to determine the fraction of total flow at each measurement location. Remember that this calculation procedure is based on a constant wellbore cross-sectional area and a constant velocity profile correction factor. It is also good practice to calculate the absolute flow rate (Eq. 5.19), at least in the maximum flow region above the perforations. A comparison of the flow rate above the perforations from the velocity-shot log with that predicted from the surface rate can indicate whether the wellbore cross-sectional area is different from expected or whether a leak in the tubing is present. If there is any question about the flow rate calculated from the velocity-shot log, a caliper log should be run to ensure that the proper cross-sectional area is used in the calculation.

Once the flow rate or fraction of total flow has been calculated from all velocity-shot measurements, these rates must be assigned depths to complete the interpretation. The most common practice is to assign the flow rate to the bottom detector location. In regions of fluid exit, however, the flow profile can be refined by a more careful depth assignment procedure. Anthony[13] and Anthony and Hill[14] showed that, in regions of fluid exit, the interpreted profile from the velocity-shot log will most closely approximate the actual flow profile if the interpreted velocity or fraction of total flow is assigned to the midpoint between the detectors. When logging just above a region of fluid exit, such as a perforated interval, the calculated velocity should be assigned to the bottom detector to max-

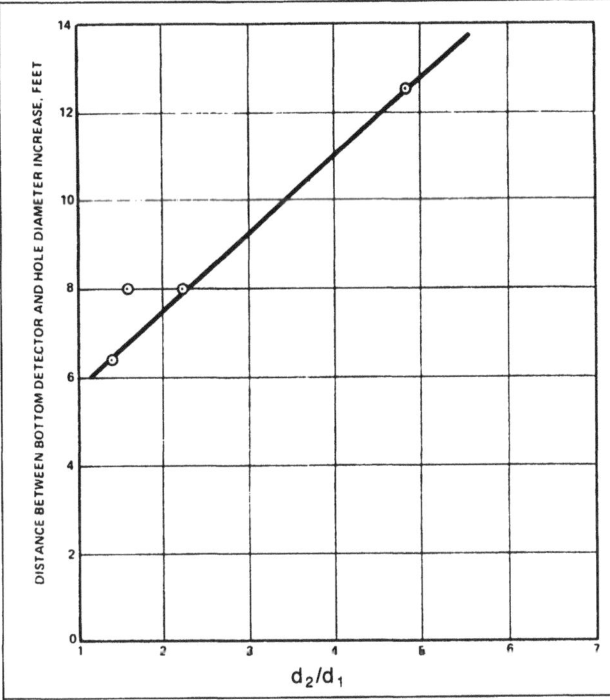

Fig. 5.14—Distance from a diameter change required for undisturbed flow (from Ref. 3).

imize depth resolution. Similarly, the velocity calculated from a shot just below a perforated interval should be assigned to the top detector location.

5.4.3 Effect of Variable Wellbore Cross-Sectional Area on Velocity-Shot Logs. The velocity-shot analysis method is predicated on the assumptions that the cross-sectional area of the wellbore is constant and that the flow rate between detectors is constant (no fluid exits between detectors). The requirement of constant cross-sectional area limits the applicability of the velocity-shot log in many openhole wells and other irregular completions. If wellbore cross-sectional area variations are known, such as from a caliper log, and are not too extreme, a velocity-shot log can still be applied. Bearden *et al.*[3] found that downstream from a diameter increase, the velocity calculated from a velocity shot will be too high because of a jetting effect. Velocity-shot analysis was not affected, however, by decreases in wellbore diameter. The results are illustrated in Fig. 5.13, which shows the percentage by which velocities calculated from velocity-shot measurements exceeded the average velocity. All tracer-calculated velocities exceeded the average velocity because the tracer velocities were calculated from a leading-edge transit time and, thus, should correspond to the maximum velocity. As seen in Fig. 5.13, for several feet downstream from the diameter increase, the interpreted flow rates were much too high. Bearden *et al.*[3] developed an empirical guideline for the distance the bottom detector should be from a diameter enlargement (Fig. 5.14). This correlation is specific for the tool geometry studied but provides a general guideline.

Anthony and Hill[14] extended this work to other diameters and found that the velocity behavior as measured by a tracer behaves as shown in Fig. 5.15. Downstream from a diameter increase, the velocity gradually decreases over a distance s, while upstream from a diameter decrease the velocity increases more abruptly to the velocity in the smaller diameter. Combining the Bearden *et al.* and the Anthony and Hill results leads to the correlation shown by the dashed line in Fig. 5.16 for the distance below a diameter increase required to ensure that the velocity measured with a velocity-shot log corresponds to the average velocity in the larger-diameter section. For accurate logging in the region of a diameter increase, the upper detector should be at least a distance s downstream from the diameter increase. Above a diameter decrease, the bottom detector can be placed quite close (within a foot) of the diameter change

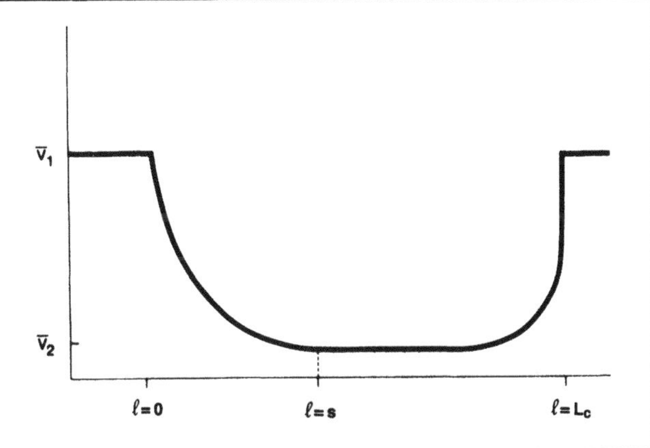

Fig. 5.15—Tracer velocity in regions of diameter change (from Ref. 14).

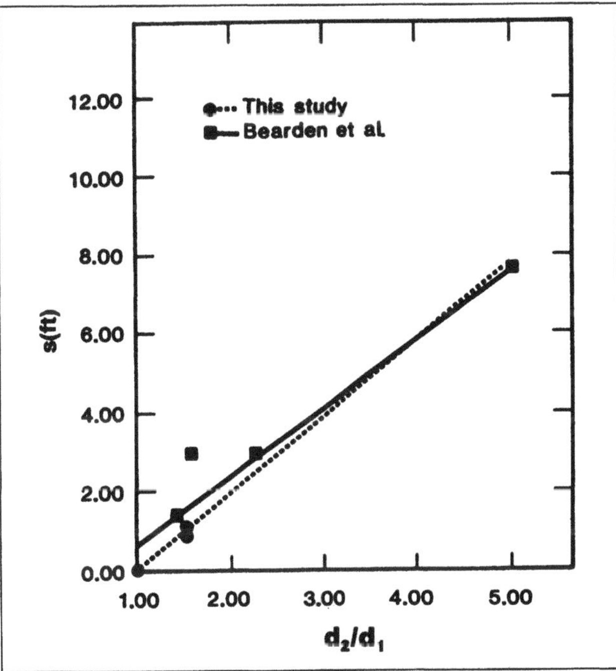

Fig. 5.16—Required distance below a diameter increase for accurate logging (from Ref. 14).

without sacrificing log quality. Obviously, a caliper log is essential to the proper running and interpretation of velocity-shot logs in wells with various cross-sectional areas.

5.4.4 Effect of Fluid Exit Between Detectors on Velocity-Shot Analysis. As shown by Hill and Solares,[5] fluid exit between detectors causes errors in the depth assignment of fluid losses with the velocity-shot method, similar to the "smearing" of fluid loss that occurs with the tracer-loss method. To understand this effect, we first calculate the actual transit time of a particle of fluid moving at the average velocity when fluid exit occurs between detectors. Fig. 5.17 shows two hypothetical cases of fluid exit between detector locations z_1 and z_2. In Fig. 5.17a, one-half of the total flow exits at a point midway between the two detectors. Thus, the average velocity will be q_0/A_w for one-half the distance between detectors, and $q_0/2A_w$ for the remaining distance. The actual transit time of a particle moving at the average velocity will be

$$\Delta t = \frac{A_w(\frac{1}{2}L)}{q_0} + \frac{A_w(\frac{1}{2}L)}{\frac{1}{2}q_0} \qquad (5.23)$$

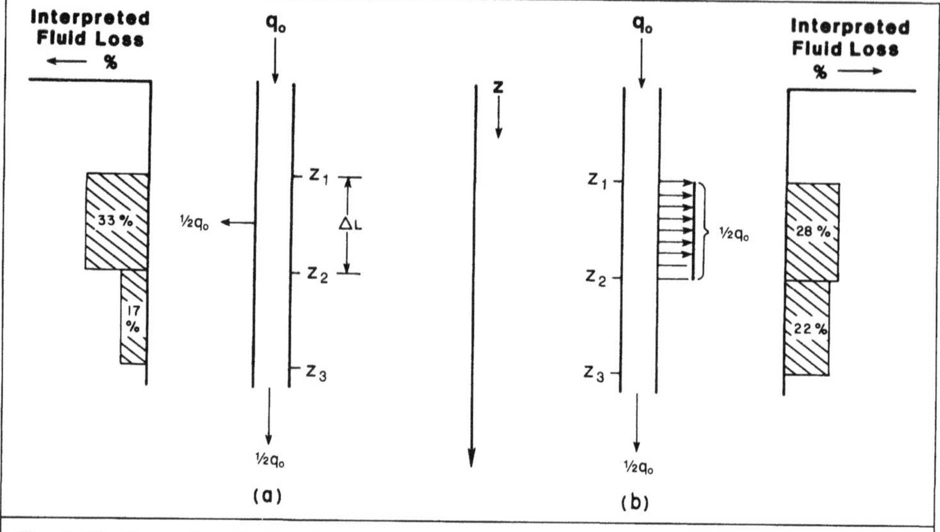

Fig. 5.17—Errors in velocity-shot depth resolution because of fluid exit between detectors (from Ref. 5).

or

$$\Delta t=\frac{3}{2}\frac{A_wL}{q_0}, \quad (5.24)$$

where A_w=cross-sectional area of flow, L=detector spacing, and q_0=volumetric flow rate above the top detector.

In Fig. 5.17b, one-half the total flow exits uniformly over the interval between detectors. In this case, the average velocity is a linear function of depth:

$$v=\frac{dz}{dt}=\frac{q_0-\left(\frac{z-z_1}{z_2-z_1}\right)(½q_0)}{A_w}. \quad (5.25)$$

The transit time for a particle to move from z_1 to z_2 is calculated by integrating Eq. 5.25 as

$$\Delta t=\int_{z_1}^{z_2}\frac{A_w dz}{q_0\left[1-½\left(\frac{z-z_1}{z_2-z_1}\right)\right]} \quad (5.26)$$

or

$$\Delta t=\frac{2A_wL}{q_0}\ln(2). \quad (5.27)$$

For the general case of a fluid loss, q_e, that exits uniformly over a distance L, the transit time is given by

$$\Delta t=\frac{A_wL}{q_e}\ln\left(\frac{q_0}{q_0-q_e}\right). \quad (5.28)$$

Returning to Fig. 5.17, we can now compare the flow rates calculated by velocity-shot analysis to the actual fluid-loss conditions. For the discrete fluid exit (Fig. 5.17a), an actual transit time of $3/2(A_wL/q_0)$ was calculated. If this transit time is then used in a velocity-shot calculation based on average velocity (Eq. 5.17), the flow rate obtained from the velocity-shot analysis would be

$$q=\frac{A_wL}{\frac{3}{2}\frac{A_wL}{q_0}}=\frac{2}{3}q_0.$$

In the standard interpretation, this volumetric flow rate would be assigned to the location of the bottom detector, z_2. Thus, the velocity-shot analysis assigns a fluid loss of one-third of the total flow instead of one-half of the total flow to the interval z_1 to z_2. Another velocity-shot measurement with the top detector at z_2 and the bottom detector at z_3 would lead to the remaining $\frac{1}{6}q_0$ fluid loss being assigned to this interval.

A similar analysis for the case of uniform exit of ½ q_0 between z_1 and z_2 would yield 0.72 q_0 as the volumetric flow rate assigned at the bottom detection location. A 28% fluid loss is interpreted to be occurring in the interval z_1 to z_2 as opposed to the actual 50%; the remaining 22% would be assigned to lower depths based on subsequent velocity-shot measurements.

These illustrations show that the depth resolution for velocity-shot analysis is about twice the detector spacing when fluid loss occurs between detectors. Thus, the detector spacing should be as small as possible while maintaining reasonably accurate Δt measurements. The best means of improving accuracy for a fixed detector spacing is to use the interval method described by Lichtenberger[2] and Cobb *et al.*[15]

5.4.5 The Interval Method. One method of improving the resolution of the velocity-shot method is to take readings in overlapping intervals by moving the tool only a short distance (a distance less than the spacing between detectors) between stations. The analysis of such data is called the interval method. By comparing the transit times from overlapping measurements, transit times for shorter intervals can be determined, thus increasing the depth resolution of the log.

To illustrate the method, consider the example presented by Lichtenberger. In this example, the well is perforated over the 5,026- to 5,044-ft [1531- to 1537-m] interval.

The detector spacing on the logging tool is 6 ft [1.8 m]. Table 5.2 gives the velocity-shot data obtained on the well with the tool moved either 2 or 4 ft [0.6 or 1.2 m] between each shot. Fig. 5.18 shows the travel times measured over each 6-ft [1.8-m] interval and their locations in the wellbore.

First, we will analyze the data using the standard velocity-shot method. Thus, we are assuming that the velocity is constant over each interval in which a measurement was taken. If we also assume that the cross-sectional area of the wellbore is constant, then

$$\frac{q_i}{q_0}=\frac{\Delta t_{100}}{\Delta t_i},$$

where q_0 and Δt_{100}=volumetric flow rate and travel time above the perforations and q_i and Δt_i=volumetric flow rate and travel time at any other location.

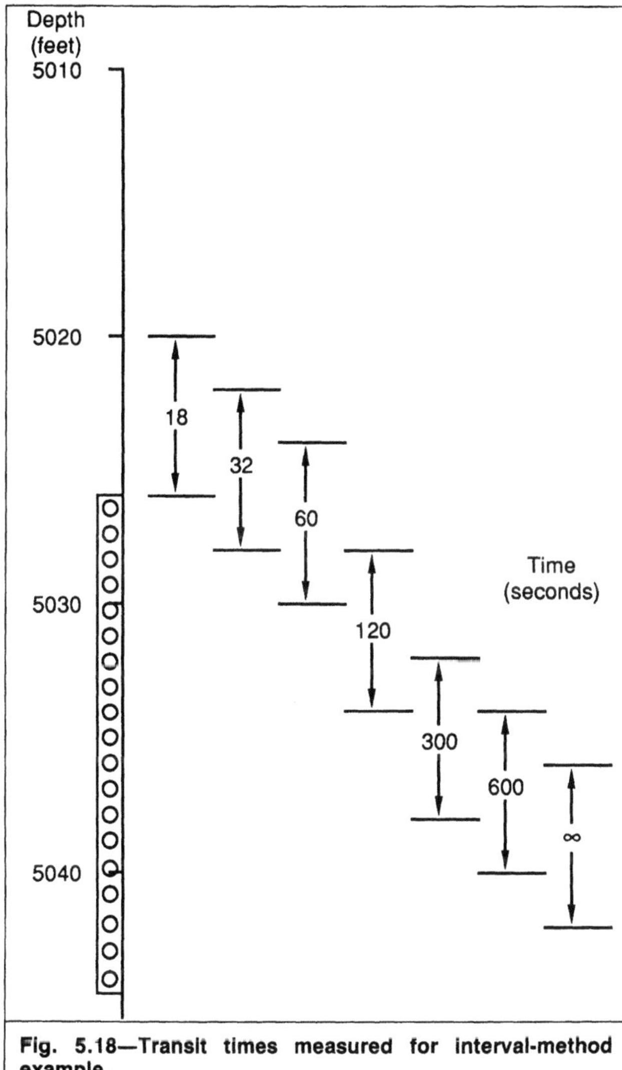

Fig. 5.18—Transit times measured for interval-method example.

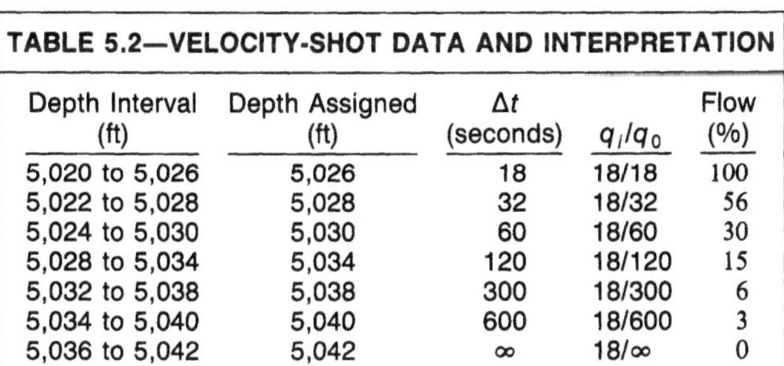

TABLE 5.2—VELOCITY-SHOT DATA AND INTERPRETATION

Depth Interval (ft)	Depth Assigned (ft)	Δt (seconds)	q_i/q_0	Flow (%)
5,020 to 5,026	5,026	18	18/18	100
5,022 to 5,028	5,028	32	18/32	56
5,024 to 5,030	5,030	60	18/60	30
5,028 to 5,034	5,034	120	18/120	15
5,032 to 5,038	5,038	300	18/300	6
5,034 to 5,040	5,040	600	18/600	3
5,036 to 5,042	5,042	∞	18/∞	0

Using this equation and assigning the flow rates to the location of the bottom detector yields the results shown in the last two columns of Table 5.2. Assigning the flow rates to the locations of the bottom detector gives the injection profile shown by the solid line in Fig. 5.19.

Now, we will use the information given to us by the overlapping of the logs to refine the interpretation. From the data at 5,020 to 5,026 ft [1530 to 1532 m] (above the perforations), where the velocity should be constant, we see that the travel time is 3 sec/ft [9 s/m]. Thus, 12 seconds are required for the tracer to travel from 5,022 to 5,026 ft [1531 to 1532 m]. Comparing this with the run made at 5,022 to 5,028 ft [1531 to 1533 m], the travel time from 5,026 to 5,028 ft [1532 to 1533 m] is then $\Delta t = 32 - 12 = 20$ seconds.

We continue to move down the wellbore, calculating travel times for each 2-ft [0.6-m] interval. The 2-ft [0.6-m] travel times are shown in Fig. 5.20. When we have insufficient data to pinpoint a 2-ft [0.6-m] travel time, such as between 5,030 and 5,032 ft [1533 and 1534 m], we must assume that the velocity is constant over a 4-ft [1.2-m] interval and divide the 4-ft [1.2-m] travel time by two.

Using the 2-ft [0.6-m] travel times, we can now calculate flow rates by

$$\frac{q_i}{q_0} = \frac{(3\ \text{sec/ft})(2\ \text{ft})}{\Delta t_i} = \frac{6}{\Delta t_i},$$

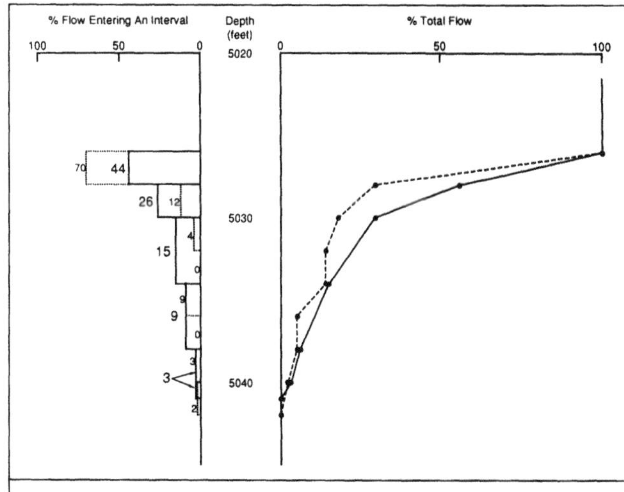

Fig. 5.19—Injection profiles from velocity-shot and interval-method interpretations.

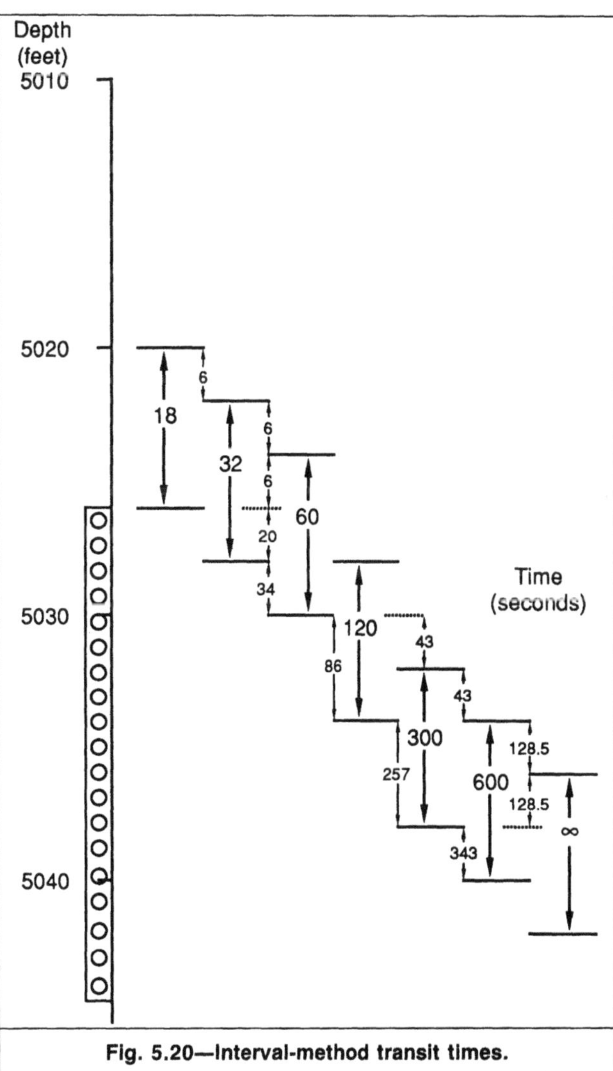

Fig. 5.20—Interval-method transit times.

TABLE 5.3—INTERVAL METHOD RESULTS

Depth (ft)	Δt_i (seconds)	q_i/q_0
5,026	6	1.0
5,028	20	0.30
5,030	34	0.18
5,032	43	0.14
5,034	43	0.14
5,036	128	0.05
5,038	128	0.05
5,040	343	0.02
5,042	∞	0

because Δt_{100} is 3 sec/ft [9 s/m]. The results are shown in Table 5.3.

The profile determined by the interval method is shown by the dashed lines in Fig. 5.19, and it differs substantially from that determined by the usual velocity-shot method.

By using the interval method, the depth resolution of the log has been improved from approximately 12 ft [3.6 m] (twice the detector spacing) to about 4 ft [1.2 m] (twice the interval for which transit times are determined). A good logging practice for velocity-shot logging is to run an initial velocity-shot log with course spacing between measurement locations, then run overlapping measurements in regions of significant fluid loss so that the interval method can be applied in these areas.

Example—Velocity-Shot Log Interpretation. This example presents a typical velocity-shot log run to measure the injection profile of a water injection well. In this well, 460 B/D [73 m^3/d] of water are injected on vacuum. As shown in the completion diagram (Fig. 5.21), 2⅜-in. [6-cm] tubing is set at 3,695 ft [1126 m] in the 4½-in. [11.4-cm], 9.5-lbm [4.3-kg] casing set to a total depth of 4,179 ft [1274 m]. Ten small sets of perforations located between 3,789 and 4,015 ft [1155 and 1224 m] provide communication with the formation.

The configuration of the logging tool used in this well is shown in Fig. 5.22. The important features of the tool for interpretation of the log are the tool OD of 1⅜ in. [3.5 cm] and the detector spacing of 5 ft [1.5 m]. A tool sketch should always be included on the log so that these values can be checked by anyone interpreting the log.

Sixteen velocity-shot measurements were made in this well at the depth locations shown in Table 5.4. Because many small perforated intervals were spread over a relatively long distance, the velocity shots were spaced fairly far apart. Shot 16 was logged in a manner different from the typical velocity shot. At this location, a tracer shot was ejected and logged through with the recorder on depth drive. After a few minutes' wait, the tracer slug was again logged through. The lack of tracer slug movement, other than slight dispersion, indicates no flow at this location in the well. This technique, called a drag on depth, is commonly used to identify the uppermost depth where no flow occurs in the wellbore.

The first step in interpreting the log is to determine the transit time, Δt, from the detector responses at each station. Fig. 5.23 shows the detector responses for Shot 3. Using a leading-edge transit time, Δt for this shot is 4.5 chart divisions, which, with the scale being 3 seconds/division, is 13.5 seconds. Note that a sharp change in gamma ray activity occurs with the first arrival of tracer, making the leading-edge transit time relatively easy to measure. This response is characteristic of turbulent flow.

Farther down the well, the wellbore flow rate has diminished sufficiently for the flow to become laminar. Fig. 5.24 shows the gamma ray responses for Shot 14. The extremely dispersed nature of the tracer slug at this location indicates that the flow is laminar, and the leading-edge transit time is more difficult to locate. Use of the intersection of a tangent to the gamma ray response and the baseline value as the definition of the first arrival points is helpful in measuring the leading-edge transit time. A peak-to-peak transit time could not be used on this shot because the peak was not measured at the second detector.

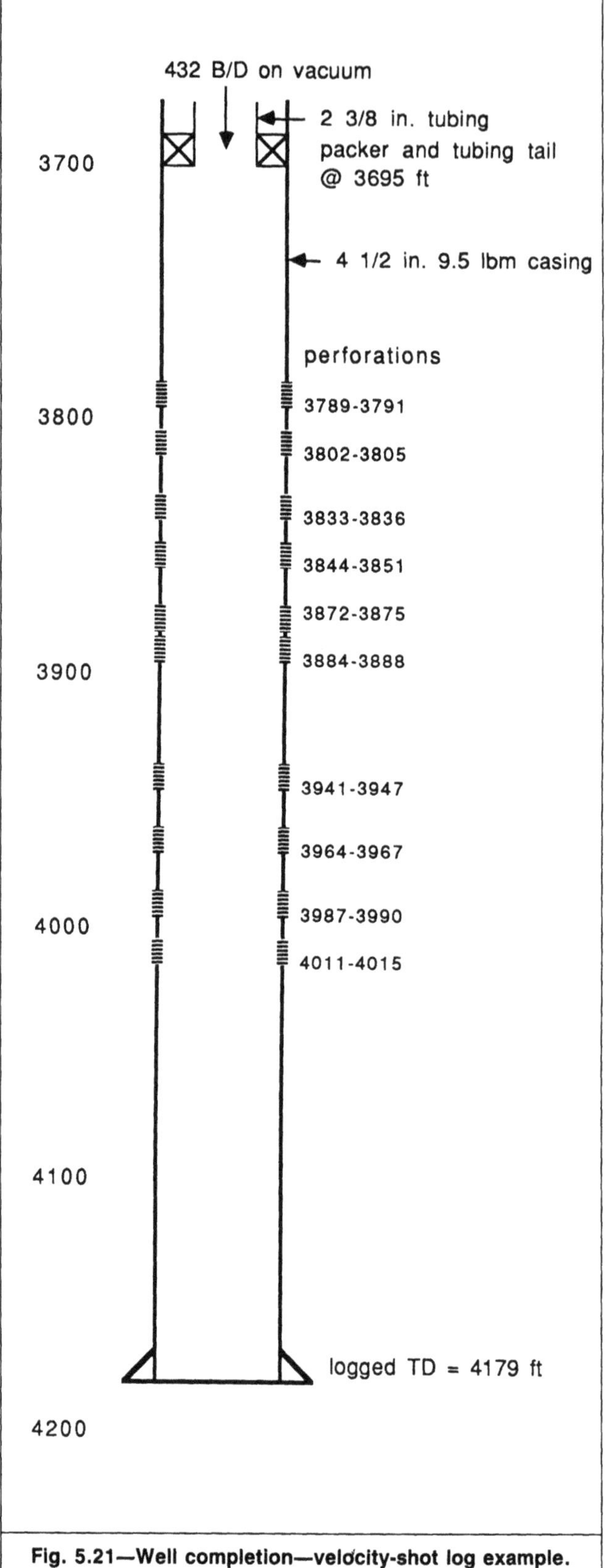

Fig. 5.21—Well completion—velocity-shot log example.

The transit times measured at each station are presented in Table 5.4. Because the wellbore cross-sectional area is presumably constant in this well, the fraction of total flow at each location where the flow is turbulent can be calculated with Eq. 5.22. When the flow is laminar, the fraction of total flow must be corrected for the different ratios of $\bar{v}/v_{max}$ in turbulent and laminar flow, according to Eq. 5.33 (see Sec. 5.6). The resultant fractional flows are presented in Table 5.4.

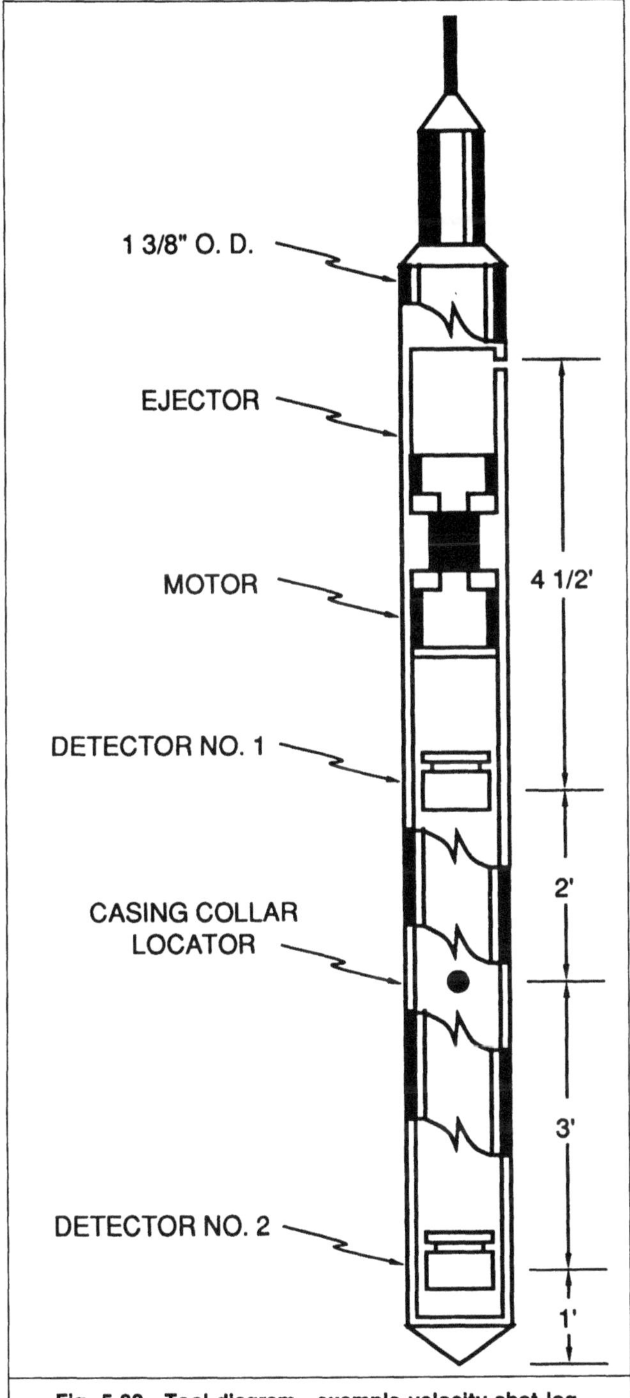

Fig. 5.22—Tool diagram—example velocity-shot log.

TABLE 5.4—VELOCITY-SHOT ANALYSIS RESULTS

Shot	Depth—Top Detector (ft)	Δt (seconds)	$\Delta t_0/\Delta t_i$	q_i/q_0	Assigned Depth (ft)
1	3,715	13.5	1.00	1.00	3,720
2	3,745	14.25	0.95	0.95	3,750
3	3,777	13.5	1.00	1.00	3,782
4	3,795	13.5	1.00	1.00	3,800
5	3,807	15.0	0.90	0.90	3,807
6	3,827	13.5	1.00	1.00	3,832
7	3,830	15.9	0.85	0.85	3,835.5
8	3,835	16.5	0.82	0.82	3,837.5
9	3,840	18.3	0.74	0.74	3,845
10	3,853	22.2	0.61	0.61	3,853
11	3,865	21.0	0.64	0.64	3,870
12	3,877	21.0	0.64	0.64	3,882
13	3,933	36.0	0.38	0.38	3,938
14	3,957	234.0	0.06	0.04	3,962
15	3,979	294.0	0.05	0.04	3,984
16	3,999	Drag on Depth—No Flow		0.00	3,999

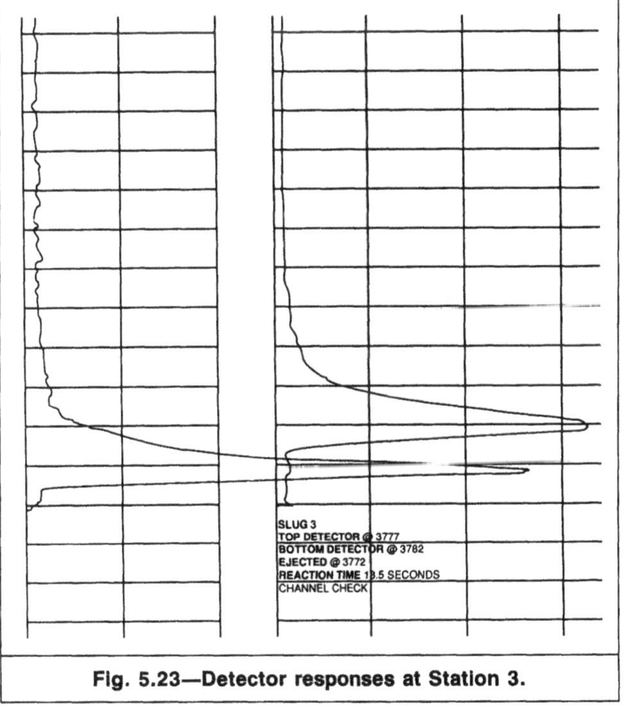

Fig. 5.23—Detector responses at Station 3.

The flow profile interpreted from this velocity-shot log is shown in Fig. 5.25. An anomalous result is immediately evident. At three locations, the velocity apparently decreases and then increases farther down the well, which is impossible without additional fluid entering the wellbore at these locations. There are three possible explanations for this phenomenon. First, the wellbore cross-sectional area could vary so that velocity is actually decreasing and increasing with no change in volumetric flow rate. If this is suspected (e.g., if scaling is common in the area) a caliper log should be run to use in conjunction with the velocity-shot log.

Another possibility is that the apparent changes in velocity simply result from errors in the transit-time measurements. In this log, the most severe discrepancy appears in Shot 5, which has a transit time of 15 seconds compared with transit times of 13.5 seconds measured above and below this shot. With a scale of 3 seconds/division, the 1.5-second difference in these measurements is only one-half a chart division, a relatively small absolute error. An anomalous reading, such as the one for Shot 5, should be carefully repeated to distinguish between measurement error and actual velocity change.

The third possible cause of the apparent velocity increases is actual changes in the well's injection rate. During the time required to obtain all the velocity-shot data, fluctuations in injection rate may have occurred, leading to the anomalous result.

It is important that the anomalous behavior be presented on the interpreted log and not "smoothed" to make a more reasonable interpretation. The possibility of variations in wellbore cross-sectional area is important information to the production engineer. If the variations simply result from transit-time errors, this information helps to determine the accuracy of the interpreted injection profile.

The depths to which the interpreted flow rates are assigned are chosen to maximize the resolution of the log. When a velocity shot is run just above a set of perforations, assign the flow rate to the bottom detector location; when a velocity shot is measured just

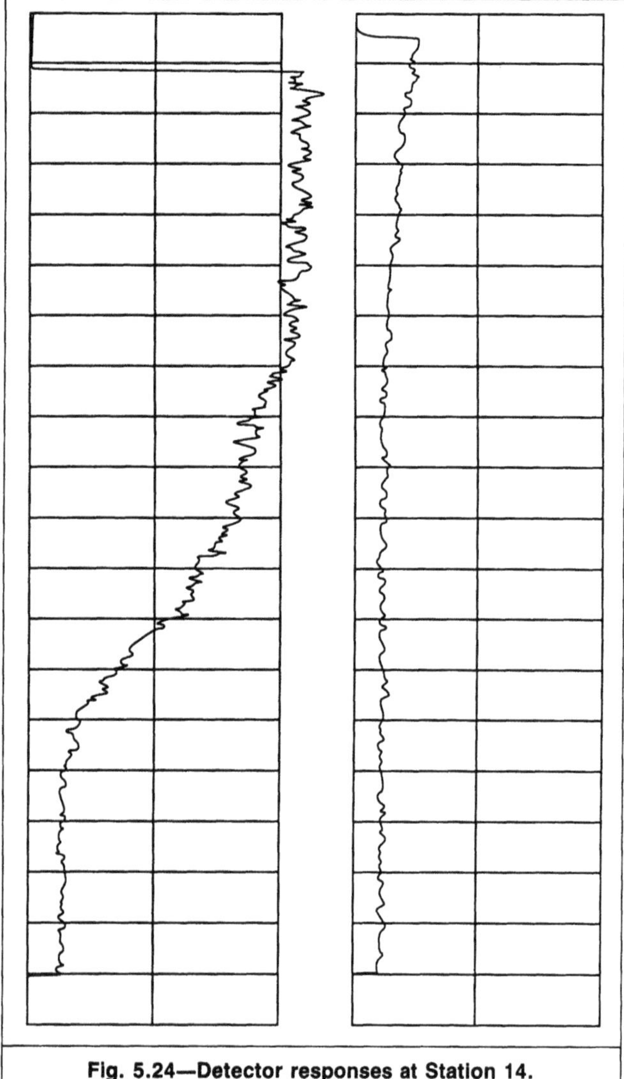

Fig. 5.24—Detector responses at Station 14.

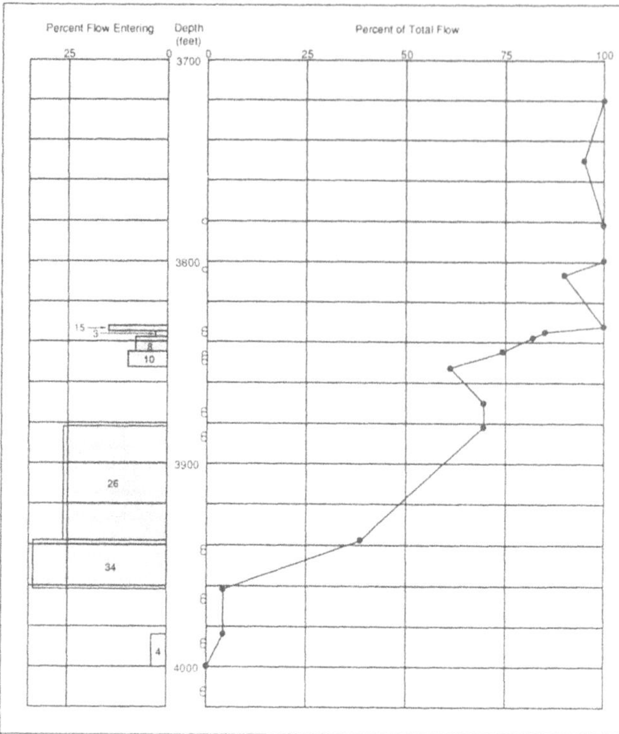

Fig. 5.25—Interpreted flow profile from example velocity-shot log.

below perforations, assign the depth to the top detector location; and when opposite the perforations, choose the midpoint between detectors.

The absolute volumetric flow rate should always be calculated from a velocity-shot measurement made above the perforations according to Eq. 5.18 or 5.19. When the transit time of 13.5 seconds at Station 1 and $B=0.85$ are used, the flow rate is

$$q=\frac{\pi(4.09^2-1.375^2 \text{ in.}^2)(5 \text{ ft})(1/144 \text{ ft}^2/\text{in.}^2)}{4(13.5 \text{ sec})(1/86{,}400 \text{ D/sec})(5.615 \text{ ft}^3/\text{bbl})}(0.85)$$

$$=392 \text{ B/D } [62.3 \text{ m}^3/\text{d}].$$

This compares fairly well with the surface rate of 460 B/D [73 m^3/d], indicating no significant leaks in the tubing.

The log presented is a reasonably good velocity-shot log. The injection profile could have been more precisely defined by running a few more shots in the vicinity of the major fluid exits. Measurements made just below the perforations at 3,884 to 3,888 ft [1184 to 1185 m] and just below the perforations at 3,940 to 3,944 ft [1201 to 1202 m] would more accurately define the profile.

5.5 Tracer Placement

Both the tracer-loss and the velocity-shot logs are affected by the distribution of the tracer in the wellbore fluid—a tracer-loss log depends on the uniform distribution of the tracer in the carrying fluid, while a velocity-shot log requires a distinct pulse of tracer for accurate results. Thus, the manner in which the tracer is placed in the flow stream is an important consideration in radioactive-tracer logging. With most current tools, the tracer is ejected from a port perpendicularly into the flow stream, with the only control on the tracer placement being some control over the duration of the ejection. Furthermore, if the logging tool is not centralized, the ejection port may be against the casing wall, likely leading to a highly nonuniform tracer slug. Studies in laminar flow[16-19] and in turbulent flow[20] show that tracer placement can have a profound effect on tracer-log quality.

From observations of tracer placement in laminar flow, Akers[18] found that, depending on the momentum of the injected tracer, the early distribution of tracer in the wellbore can vary from a trickle down the side of the logging tool to a cloud of tracer that impacts and rebounds off the casing wall (Fig. 5.26). For a velocity-shot log, the optimal behavior would be that shown by the intermediate case in Fig. 5.26, where the bulk of the tracer was placed in the center of the tool/casing annulus. Placing most of the tracer in the highest-velocity flowstreams maintains a more distinct tracer slug.

Akers and Hill[19] reported the effects of four parameters on tracer placement—(1) ejection rate, (2) shot time, (3) nozzle size, and (4) wellbore-fluid velocity. Hill *et al.*[20] extended this work to include tracer ejection into a turbulent flow stream. Figs. 5.27 through 5.30 show the effects of these parameters on the distance across the flow stream the tracer would penetrate; a dimensionless penetration of one indicates that the tracer reached the opposite side of the pipe, while a value near zero shows that the tracer did not penetrate much of the wellbore flowstream. A dimensionless penetration of about 0.5 would be optimal for velocity-shot logging. These results show that with a tool that allows for control of ejection rate and duration, tracer placement can be optimized to yield improved velocity-shot logging results.

5.6 Radioactive-Tracer Logging in Laminar Flow

Laminar flow conditions are found in polymer injection wells, low-rate water injection wells, and in the lower depths of higher-rate injection wells. Radioactive-tracer logging is the primary technique available for quantitatively measuring injection profiles in these conditions because spinner flowmeters respond poorly in the low flow rate or high-viscosity fluids characteristic of laminar flow. Radioactive-tracer logs run in laminar flow, however, are often

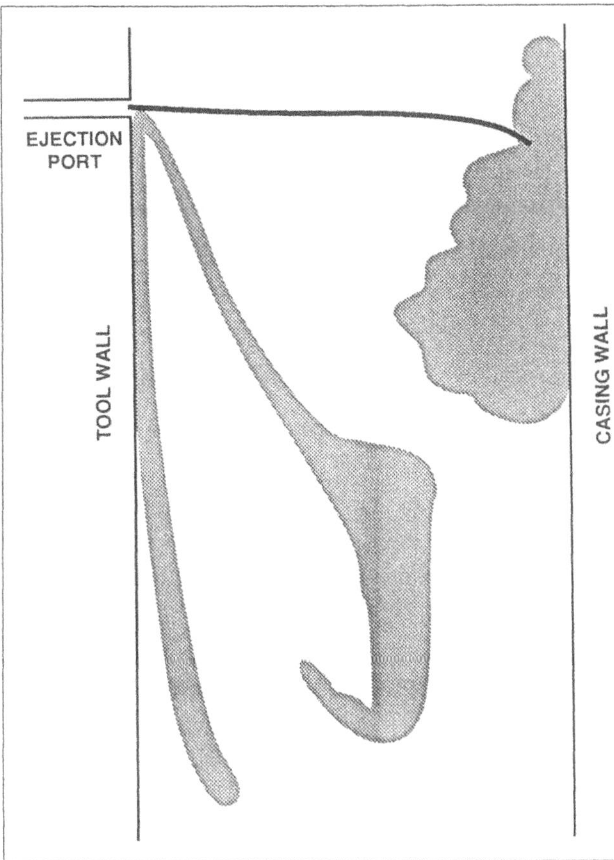

Fig. 5.26—Three general types of tracer placement (from Ref. 18, courtesy T.J. Akers).

difficult to interpret because of the highly dispersed tracer slugs usually observed. Tracer dispersion is so severe that the tracer-loss log cannot normally be applied in laminar flow; therefore, the rest of this section deals with the behavior of velocity-shot logs in laminar flow.

5.6.1 Radioactive-Tracer Logging in Laminar Flow of Water. A typical velocity-shot response in laminar flow of water is shown in Fig. 5.31. The tracer slug is highly dispersed with an indistinct peak and leading edge and with considerable noise in the signal. Akers and Hill[19] found that the tracer dispersion observed in laminar flow results from the parabolic velocity profile that is characteristic of laminar flow, the placement of significant amounts of tracer in low-velocity regions of the wellbore cross section, and the penetration of gamma rays a significant distance beyond the actual tracer material. Taylor[21] first analyzed the dispersion of a tracer that initially is uniformly distributed in laminar flow. His theory showed that after a time l_i/v_{max}, where l_i=initial length of the tracer slug and v_{max}=maximum velocity, the tracer will be distributed as shown in Fig. 5.32, with ramps at the front and back of the slug and a plateau with a concentration that decreases as it moves down the pipe. This distribution is given by the equations (for $t>l_i/v_{max}$)

$$c=c_i(\ell/v_{max}t) \text{ for } 0<\ell<l_i, \quad (5.29)$$

$$c=c_i(l_i/v_{max}t) \text{ for } l_i<\ell<v_{max}t, \quad (5.30)$$

and

$$c=c_i[(l_i+v_{max}t-\ell)/v_{max}t] \text{ for } v_{max}t<\ell<v_{max}t+l_i. \quad (5.31)$$

In these equations, c_i=initial tracer concentration, c=tracer concentration at any later time, ℓ=distance from the initial tracer position, l_i=initial length of the tracer slug, v_{max}=maximum fluid velocity, and t=time. These equations show that the length of the slug is propagating at a velocity v_{max}, the concentration of the plateau region is decreasing with time, and the front ramp has a constant length and moves at the velocity v_{max}. It can also easily be shown that the centroid of the tracer slug moves at the average velocity, $v_{max}/2$, because the position of the centroid at any time X is

$$X=\tfrac{1}{2}(v_{max}t+l_i). \quad (5.32)$$

Taylor's[21] equations describe the shape of a tracer slug as it moves down the pipe. In a velocity-shot log, however, the tracer is measured by a gamma ray detector at a fixed point, with time being the dynamic variable. Fig. 5.33 illustrates the gamma ray response that results from the passage of a dispersing tracer slug in laminar flow. From this gamma ray response, Akers and Hill

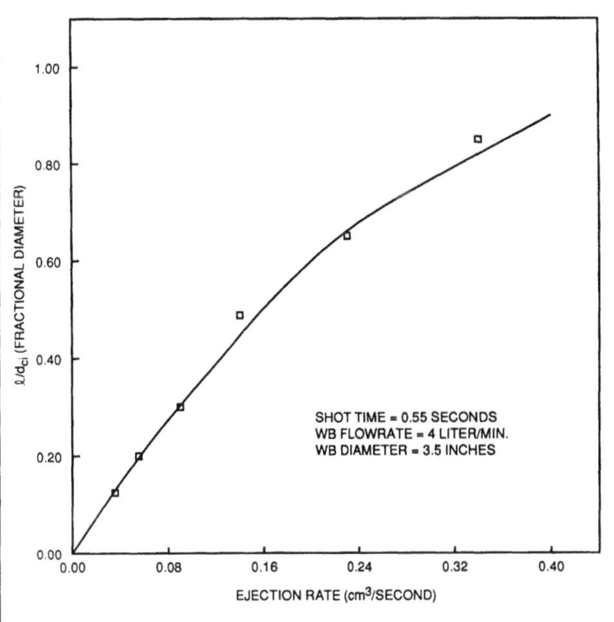

Fig. 5.27—Effect of ejection rate on tracer placement (from Ref. 19, courtesy Canadian Well Logging Society).

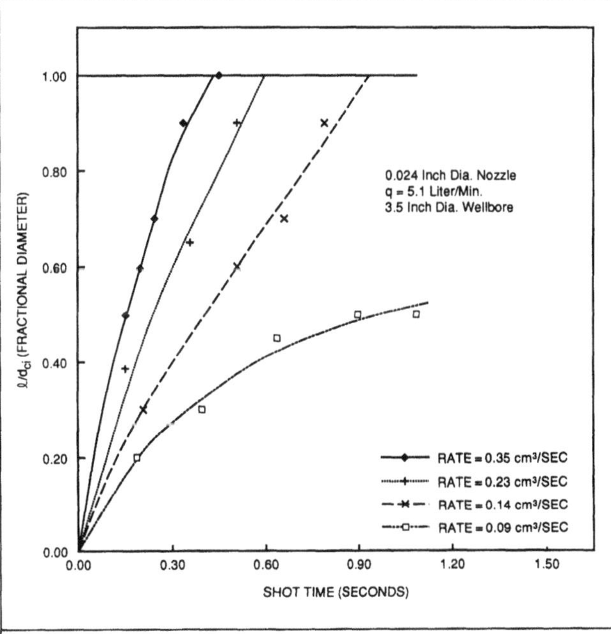

Fig. 5.28—Effect of shot time on tracer placement (from Ref. 19, courtesy Canadian Well Logging Society).

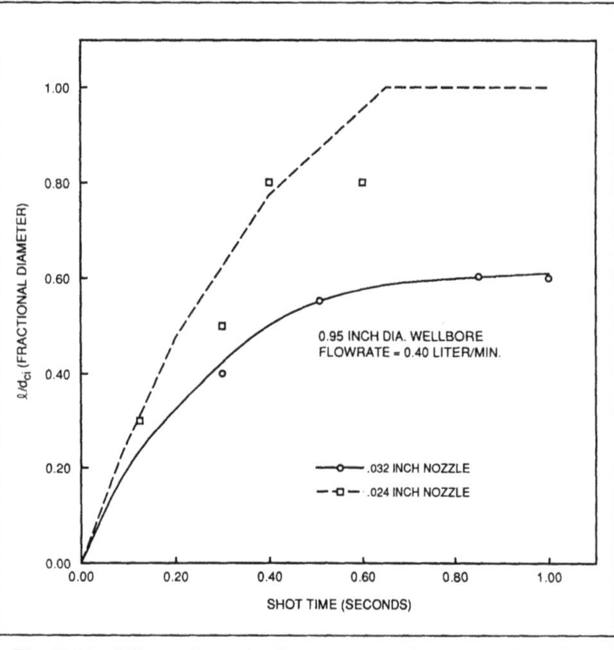

Fig. 5.29—Effect of nozzle size on tracer placement (from Ref. 19, courtesy Canadian Well Logging Society).

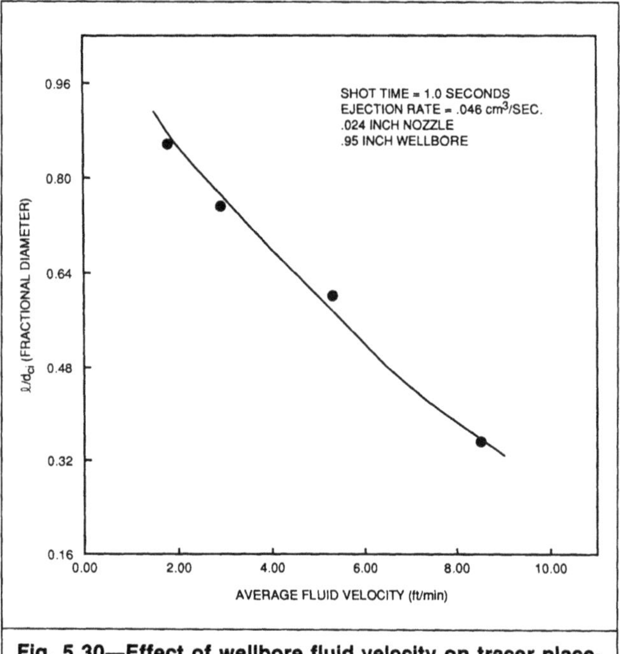

Fig. 5.30—Effect of wellbore fluid velocity on tracer placement (from Ref. 19, courtesy Canadian Well Logging Society).

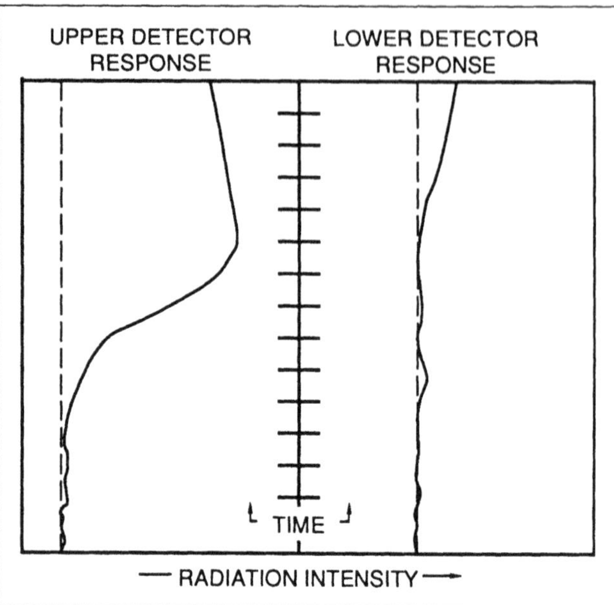

Fig. 5.31—Velocity-shot response in laminar flow (from Ref. 19, courtesy Canadian Well Logging Society).

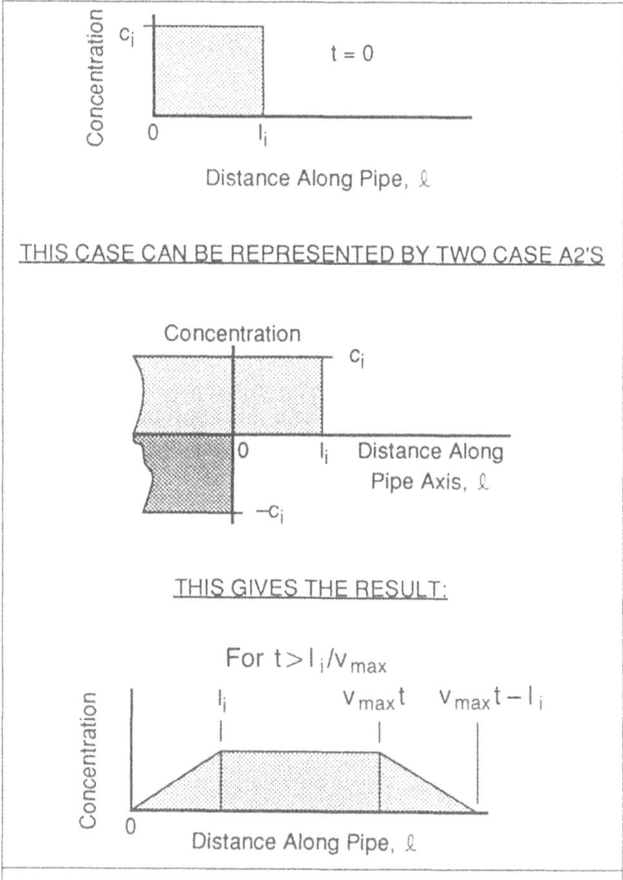

Fig. 5.32—Dispersion of an initially uniform slug of tracer in laminar flow (from Ref. 21).

found that both leading-edge and peak-to-peak transit times yield the maximum velocity when velocity-shot data in laminar flow are analyzed.

Another factor that contributes to the dispersed response observed in tracer logging in laminar flow, particularly influencing the indistinct leading edge typically seen, is the penetration of gamma rays ahead of the tracer itself. Because the tracer is moving slowly in laminar flow, the gamma ray detector will measure a gradually increasing gamma ray level as the leading edge of tracer approaches the detector location. Akers and Hill showed that the gamma ray detector response to the tracer slug begins when the tracer is about 1 ft [0.3 m] from the detector and that this effect contributes significantly to the vague tracer leading edge often observed in laminar flow.

A final factor that affects velocity-shot log quality in laminar flow is the distribution of the tracer in the wellbore. As previously discussed in Sec. 5.5, tracer usually will not be uniformly distributed into the flow stream. If a large portion of the tracer is placed in the low-velocity regions near the tool or casing wall, the tracer slug will become more dispersed than if it were initially uniformly distributed. Simulations of the gamma ray response 4 ft [1.2 m] downstream of the ejector illustrate the effect of the distribution of the tracer in the flowstream (Figs. 5.34 and 5.35). When the majority

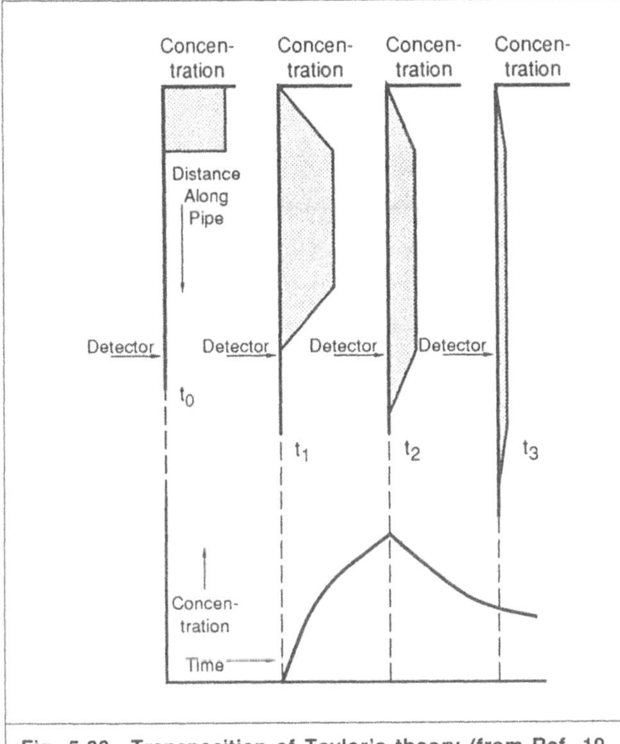

Fig. 5.33—Transposition of Taylor's theory (from Ref. 19, courtesy Canadian Well Logging Society).

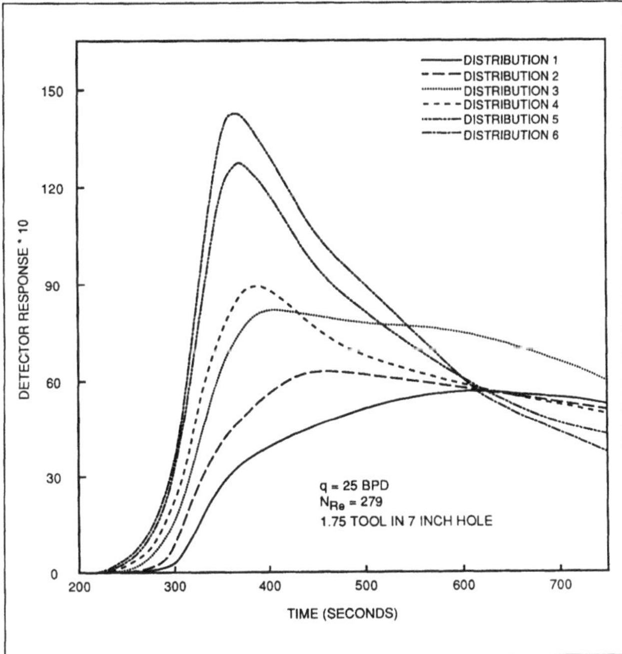

Fig. 5.34—The effect of tracer concentration distribution on detector response (from Ref. 19, courtesy Canadian Well Logging Society).

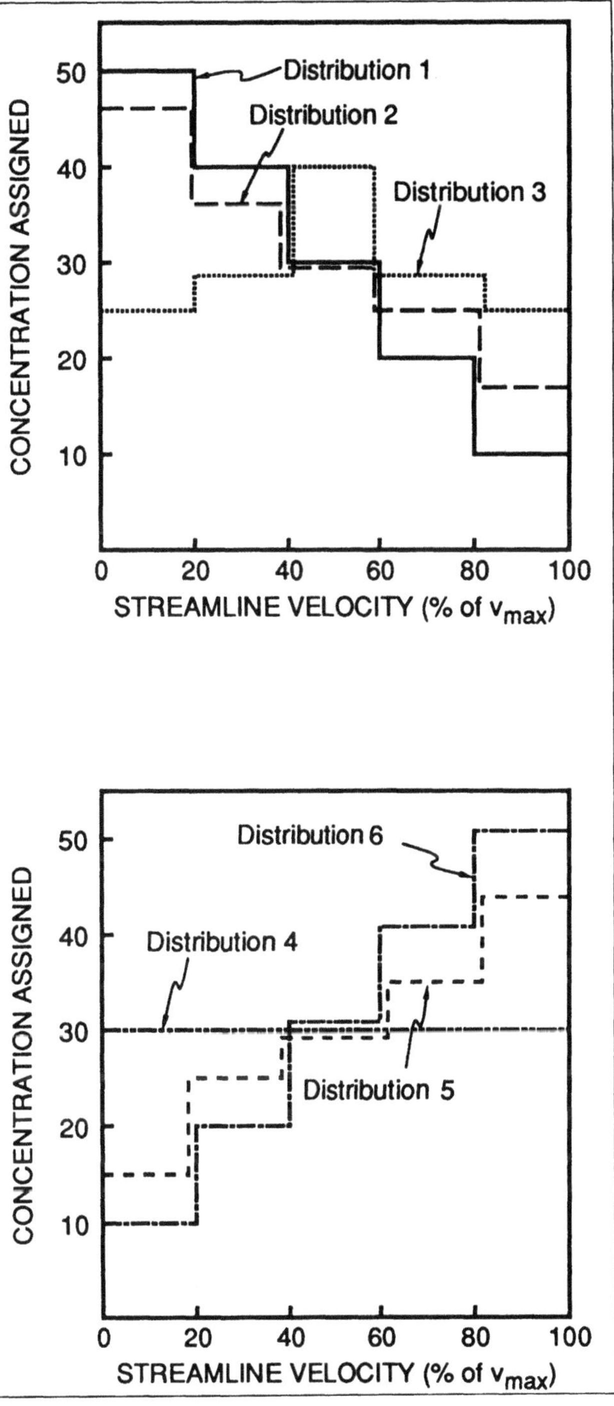

Fig. 5.35—Concentration distributions for Fig. 5.34 (from Ref. 19, courtesy Canadian Well Logging Society).

of the tracer is placed in the high-velocity region of flow (Distribution 6), a fairly sharp gamma ray response results, while placement of a large portion of the tracer in low-velocity regions leads to a very dispersed detector response (Distribution 1). With current logging tools, some experimentation with shot size to find the shot size that yields the sharpest detector responses would improve velocity-shot log quality in laminar flow.

A common interpretation error made when flow changes from turbulent to laminar flow in the lower portion of a well is the use of the wrong velocity profile correction factor, $\bar{v}/v_{max}$. At some point in every well the velocity becomes low enough that laminar flow occurs, while the flow is turbulent above this point. If all velocity-shot data are interpreted with use of the simple ratio of transit time to transit time above the perforations to yield fractional flow (Eq. 5.22), the interpreter has tacitly assumed that the ratio $\bar{v}/v_{max}$ is constant throughout the well. This is not true because of the transition to laminar flow that occurs at some point in the well, and the interpreted profile must be corrected. Furthermore, the correction required depends on which transit time is used in the turbulent-flow portion of the well. If a peak-to-peak transit time is used, the average velocity is determined in the turbulent-flow portion of the well, while the maximum velocity is calculated in laminar flow. With use of a leading-edge transit time, the maximum velocity is the result in both the turbulent and laminar flow sections of the well. Because the ratio $\bar{v}/v_{max}$ is about 0.85 in turbulent flow

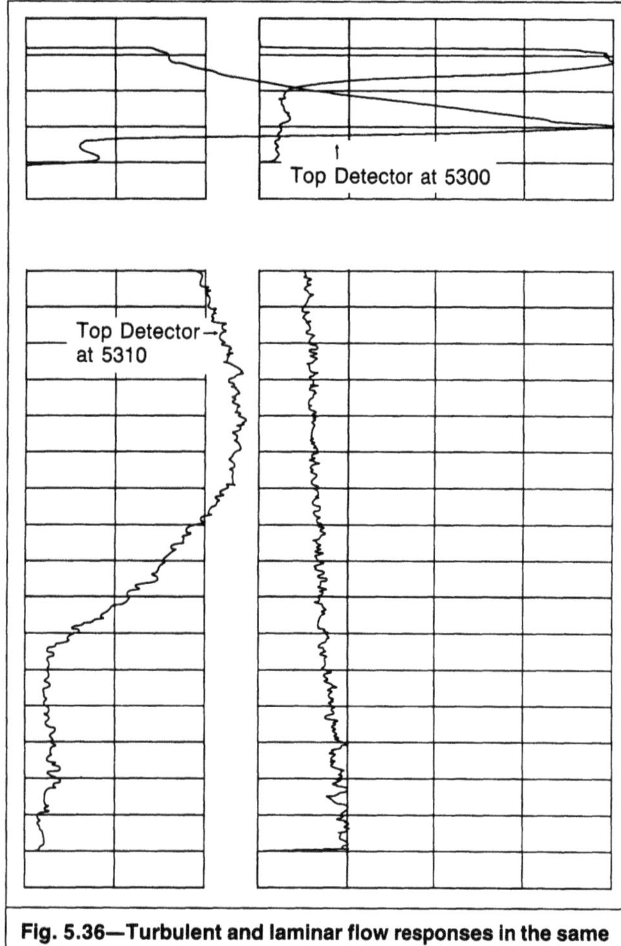

Fig. 5.36—Turbulent and laminar flow responses in the same well.

and about 0.65 in laminar flow for the typical pipe and tool sizes, corrections are needed in both cases. To account for the change to laminar flow, the following equations should be used to calculate the flow fraction in the laminar flow portion of the well.

If peak-to-peak Δt is used,

$$\frac{q_i}{q_0}=\frac{\Delta t_{100}}{\Delta t_i}(0.65). \qquad (5.33)$$

If leading-edge Δt is used,

$$\frac{q_i}{q_0}=\frac{\Delta t_{100}}{\Delta t_i}\left(\frac{0.65}{0.85}\right). \qquad (5.34)$$

The location at which the flow has changed from turbulent to laminar can usually be readily determined from the shape of the gamma ray responses as the responses become noticeably more dispersed in laminar flow. This is illustrated in Fig. 5.36. With the top detector at 5,300 ft [1615 m] in this well, the sharp, distinct slug typical of turbulent flow is recorded. After the tool is moved 10 ft [3 m] down the well, the velocity has decreased sufficiently for the flow to be laminar, and the top detector response is very dispersed.

5.6.2 Radioactive-Tracer Logging in Viscous Solutions. With the increasing applications of polymer and chemical flooding, demands for injection profiling in viscous solutions are increasing. In viscous-fluid injection wells, the flow is typically laminar because of low flow rate and high apparent viscosity. Spinner flowmeters do not typically respond well in laminar flow, so radioactive-tracer logging is often applied in these wells.

Tracer placement can be a more severe problem in viscous fluids than in the laminar flow of water. In logging a microemulsion injection well, Bragg *et al.*[16] found that the injection profile obtained with a standard velocity-shot logging tool differed markedly from the injection profile to brine in the same well. When logging during microemulsion injection, the tracer did not move a significant distance after ejection, leading to the interpreted injection profile shown in Fig. 5.37. Fig. 5.38 shows the injection profile to brine for this well. The authors suspected that the tracer remained in the stagnant layer along the casing wall at the point of injection. Subsequent studies[17] in a flow loop supported the conclusion of poor dispersion because it was found that the tracer material was ejected through the polymer stream and that most of the tracer material remained in the low-velocity layer near the wall (Fig. 5.39). In contrast, the tracer ejected into water was rapidly dispersed, as seen in Fig. 5.40.

To log successfully with radioactive tracer in viscous fluids, the tracer must be ejected so that the bulk of the tracer will remain in the highest-velocity region of the flow stream. Bragg *et al.*[16] accomplished this by increasing the number of tracer ejector ports so that the tracer-ejection velocity was decreased. Special tools have also been developed to overcome the tracer placement problems that occur with perpendicular ejection into viscous fluids.[22,23] Fig.

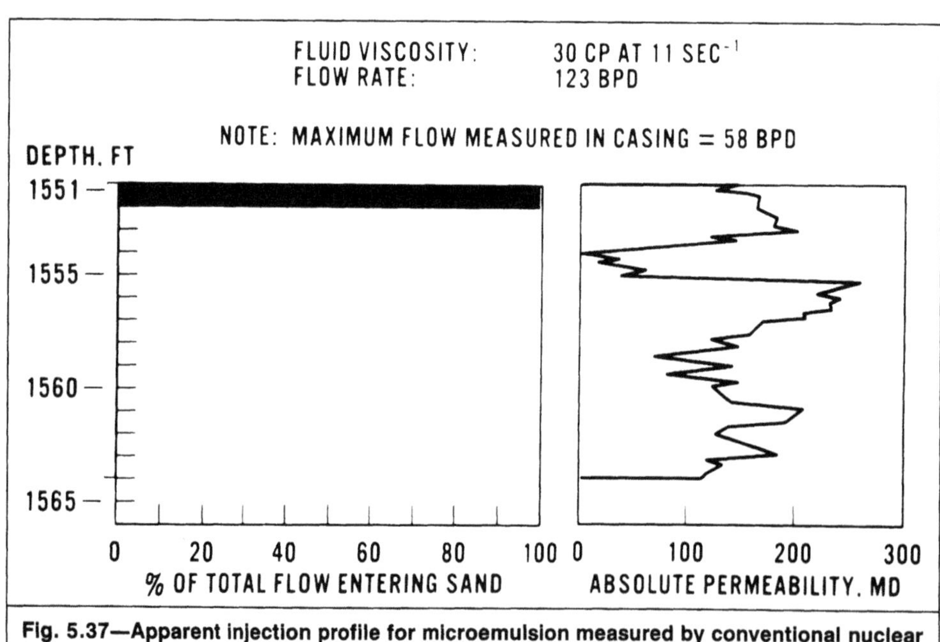

Fig. 5.37—Apparent injection profile for microemulsion measured by conventional nuclear flolog (from Ref. 16).

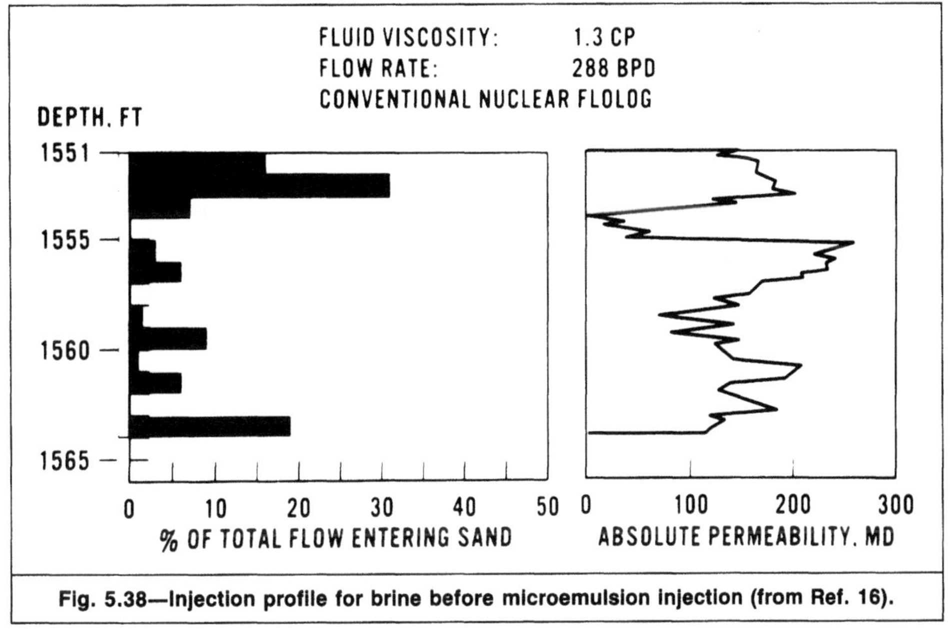

Fig. 5.38—Injection profile for brine before microemulsion injection (from Ref. 16).

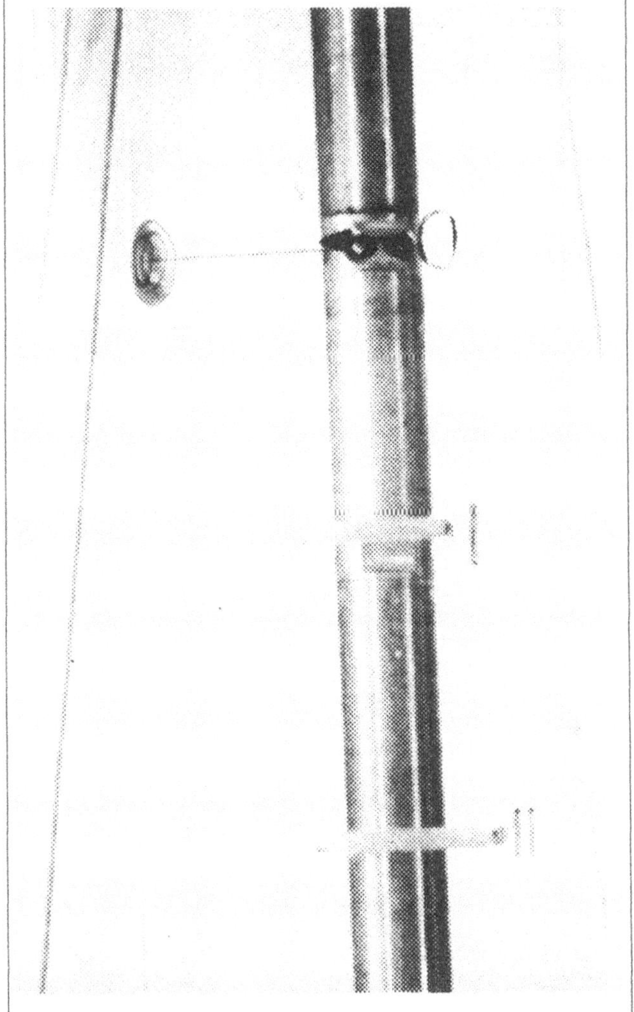

Fig. 5.39—Injected tracer dye sticking on flow-loop wall and forming filament (from Ref. 17).

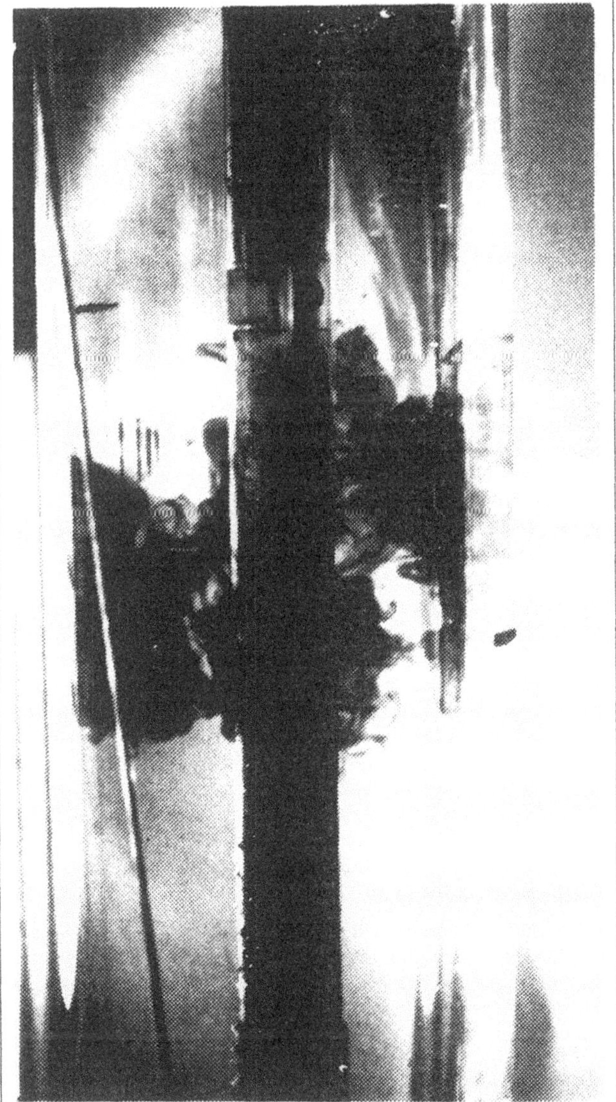

Fig. 5.40—Injected tracer dye dispersing rapidly into water stream (from Ref. 17).

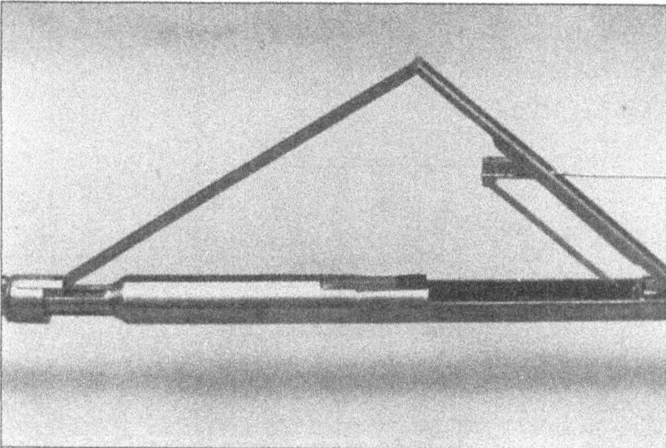

Fig. 5.41—Swing-arm tracer instrument (From Ref. 22. Photo courtesy Atlas Wireline Services, Western Atlas Intl. Inc.)

5.41 illustrates one such device, the swing-arm tracer tool, which ejects tracer axially to place more of the tracer in the high-velocity portion of the flow stream.

5.7 Two-Pulse Tracer Logging

A new method of tracer logging, the two-pulse method, has been proposed by Hill and Solares,[5] Anthony and Hill,[14] and Hill.[24] In this technique, two slugs of tracer are ejected a known distance apart, above all fluid exits, either by use of two ejectors or by the sequential ejection of two slugs with one ejector and logging through the slugs with a gamma ray detector to determine the initial spacing between the pulses. The log is then run in the same manner as a tracer-loss log. The distance between the peaks of the tracer pulses is measured as a function of depth as they move down the wellbore by repeatedly logging through the pulses with a gamma ray detector. It is shown[5,14] that the volumetric flow rate at any point in the well, q_i, is related to the spacing between the slugs, L_i, by

$$\frac{q_i}{q_0}=\frac{A_{wi}L_i}{A_{w0}L_0}, \qquad (5.35)$$

where q_0 and L_0=volumetric flow rate and initial slug spacing above any fluid exits, respectively, and A_{wi} and A_{w0}=cross-sectional areas of the wellbore at the two measurement locations. The two-pulse analysis is independent of cross-sectional-area variations between the measurement locations; only the cross-sectional areas at the measurement locations are needed.

A hypothetical log and interpreted flow profile using the two-pulse method are shown in Fig. 5.42. This method is seen mainly as a way to measure a flow profile quickly with more accuracy than is possible with a tracer-loss log. Additionally, the two-pulse log is less sensitive than tracer-loss or velocity-shot logs to changes in cross-sectional area. A drawback to the method that has not been fully explored is the dispersion of the tracer slugs. Because two distinct slugs of tracer are required, smearing of the slugs could make the technique unwieldy if tracer placement is not controlled well.

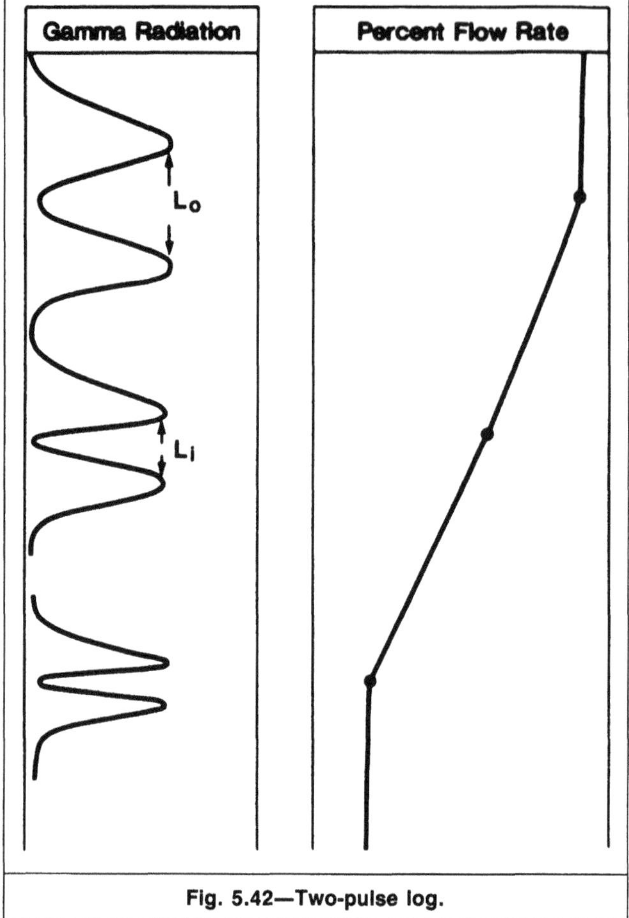

Fig. 5.42—Two-pulse log.

5.8 Guidelines for Running and Interpreting Radioactive-Tracer Logs

When run and interpreted properly, radioacti... logs are reliable for determining injection profiles. However, many pitfalls must be avoided to obtain a good radioactive-tracer log. This section presents general guidelines for obtaining high-quality radioactive-tracer logs.

5.8.1 Tracer-Loss Logs.

1. Before a tracer-loss log is run, a baseline gamma ray log should be run. This baseline log should be run with high sensitivity to help distinguish secondary peaks on the tracer-loss log.

2. If possible, when a tracer-loss log is run, the tracer slug should be logged through two or three times before it reaches the first fluid-exit zone. By averaging the areas of the slugs logged above fluid exits, a better estimate of the initial slug area can be obtained. This is important because the rest of the log analysis is based on the initial slug area.

3. When logging up through a tracer slug, the tool should be passed completely through the tracer slug and continued up the wellbore a short distance to locate any secondary peaks of tracer that indicate channeling.

4. Movement of a secondary peak on a tracer-loss log is conclusive evidence of a secondary flow path. This path could be a channel or vertical flow in the formation near the wellbore when the reservoir vertical permeability is high. The secondary peak, however, cannot be analyzed quantitatively to determine the flow rate in the channel.

5. The area method should be used to analyze a tracer-loss log. The areas of the tracer slugs should be determined by planimetry or equally accurate means. Approximating the tracer slugs with triangles to determine areas can introduce significant errors in the interpretation.

6. A tracer-loss log is not an accurate means of obtaining an injection profile. It should be used primarily as a guide for the running of subsequent velocity-shot logs.

7. The times when the tracer peaks are logged through should be recorded so that a timed-slug analysis can be performed.

5.8.2 Velocity-Shot Logs.

1. When a velocity-shot log is run, the detectors should be spaced as closely as possible while an accurate transit-time measurement is maintained in the highest-velocity portion of the well. With typical tracer tools, a spacing that yields a transit time of about 10 seconds above the perforations is recommended.

2. A few shots should be made initially to determine the shot size (and ejection rate, if variable) that will yield a sharp peak and first arrival at both gamma ray detectors.

3. Above the perforations, several velocity shots should be run. If the transit times are not all equal, a problem (e.g., a tool, variable

injection rate, or a variable well-cross-sectional-area problem) is indicated.

4. A velocity-shot log must be used with caution in a well with a varying cross-sectional area. A good caliper log is needed in this case to determine where to run the velocity shots and to interpret the log.

5. In regions of high fluid loss, velocity shots should be spaced closely together. Overlapping shots so that the interval method can be applied will improve depth resolution in such regions.

6. In laminar flow, particularly with viscous fluids, tracer placement is critical to log quality. If poor results are obtained with conventional tools, specialty tools such as the swing-arm tracer tool may improve results.

7. A different value of $\bar{v}/v_{max}$ in the laminar flow portion of a well as compared with the turbulent flow section should be used in the log interpretation.

8. Above all fluid exits, the absolute flow rate, not just the percentage of flow, should be calculated from the velocity-shot log. A large discrepancy between the calculated rate and the surface rate indicates that (1) a measurement error was made, (2) the injection rate has changed, (3) the cross-sectional area of the wellbore is different from what it was thought to be, or (4) a tubing leak is occurring.

9. Either a leading-edge or a peak-to-peak transit time can be used to interpret a velocity-shot log with similar accuracy if both can be read from the log with similar accuracy.

10. Without overlapping shots, the depth resolution of a velocity-shot log is approximately two detector spacings. When shots are overlapped so that the interval method can be applied, the depth resolution is improved to approximately twice the interval length. The log interpreter must keep in mind that fluid-loss zones may be displaced in depth by this amount and not read too much into the interpreted depth of fluid-loss zones.

11. Depth resolution can be improved by careful depth assignment. The velocity obtained from a shot with the logging tool just above a set of perforations should be assigned to the bottom detector location. A velocity obtained from a shot with the logging tool just below a set of perforations should be assigned to the top detector location. When the logging tool is opposite a perforated zone, the velocity should be assigned to a depth midway between the detector locations.

5.8.3 General Recommendations.

1. Whenever possible, tracer-logging tools should be centralized so that tracer is not ejected directly against the casing wall. This is particularly important for velocity-shot logs.

2. A radioactive-tracer log should include a well diagram, a tool sketch, the surface well conditions, and a tabulation of the chart speeds used (e.g., how many seconds per division on time drive). The log should provide *all* information needed for a complete, independent log analysis.

3. Anomalous results, such as the flow rate apparently decreasing then increasing farther downhole, should be presented, not smoothed, on the interpreted log. Such results are either showing a real physical effect or giving a useful indication of log quality.

4. Channeling is sometimes indicated by a large difference between the profile determined with the tracer-loss log and that determined by the velocity-shot log. When the tracer-loss log shows a significantly higher flow rate at certain depth locations than the velocity-shot log, flow outside the casing can explain the discrepancy. This interpretation should be applied cautiously, however, because of the inaccuracy of the tracer-loss log.

5. The two-pulse log may be preferable to a tracer-loss log, especially in wellbores with varying cross-sectional areas.

Nomenclature

A_w = wellbore cross-sectional area, ft^2 [m^2], L^2
A_{wi} = wellbore cross-sectional area at Position i, ft^2 [m^2], L^2
A_{w0} = wellbore cross-sectional area at Position 0, ft^2 [m^2], L^2
$A_{\gamma i}$ = area under the gamma-ray-intensity-vs.-depth curve, chart divisions-ft [chart divisions·m]
$A_{\gamma 100}$ = area under the gamma-ray-intensity-vs.-depth curve above all fluid exits, chart divisions-ft [chart divisions·m]
B = velocity profile correction factor ($\bar{v}/v_{max}$)
c = tracer concentration, lbm/ft^3 [kg/m^3], m/L^3
c_b = background tracer concentration, lbm/ft^3 [kg/m^3], m/L^3
c_i = initial tracer concentration, lbm/ft^3 [kg/m^3], m/L^3
d_{ci} = casing ID, ft [m], L
d_{Te} = tool OD, ft [m], L
d_1 = smaller wellbore diameter in region of diameter change, ft [m], L
d_2 = larger wellbore diameter in region of diameter change, ft [m], L
f_i = fraction of total flow at Location i
G = a constant
l_i = initial length of tracer slug, ft [m], L
l_m = measured slug length, ft [m], L
l_s = actual slug length, ft [m], L
ℓ = distance from initial tracer position, ft [m], L
L = detector spacing, ft [m], L
L_i = spacing between tracer slugs at Position i, ft [m], L
L_o = spacing between tracer slugs at Position o, ft [m], L
L_0 = length of a tracer slug above a fluid exit, ft [m], L
m_e = mass of tracer exiting the wellbore, lbm [kg], m
m_i = mass of tracer below a fluid exit, lbm [kg], m
m_0 = mass of tracer above a fluid exit, lbm [kg], m
q = volumetric flow rate, ft^3/sec [m^3/s], L^3/t
q_e = volumetric flow rate exiting the wellbore, B/D [m^3/d], L^3/t
q_i = volumetric flow rate at Position i, B/D [m^3/d], L^3/t
q_0 = volumetric flow rate above a fluid exit, B/D [m^3/d], L^3/t
q_{100} = volumetric flow rate above all fluid exits, ft^3/sec [m^3/s], L^3/t
s = distance, ft [m], L
t = time, seconds, t
Δt = tracer transit time, seconds, t
Δt_i = transit time measured at Location i, seconds, t
Δt_{le} = leading edge transit time, seconds, t
Δt_{pp} = peak-to-peak transit time, seconds, t
Δt_s = time required for passage of a tracer slug, seconds, t
t_γ = gamma ray detector response time, seconds, t
Δt_{100} = transit time measured above all fluid exits, seconds, t
$\bar{v}$ = average velocity, ft/sec [m/s], L/t
v_f = fluid velocity, ft/sec [m/s], L/t
v_{max} = maximum velocity, ft/sec [m/s], L/t
v_s = slug velocity, ft/sec [m/s], L/t
v_T = tool velocity, ft/sec [m/s], L/t
X = position of the tracer slug centroid, ft [m], L
z = position in the wellbore, ft [m], L
z_1 = Position 1, ft [m], L
z_2 = Position 2, ft [m], L
z_3 = Position 3, ft [m], L
γ_i = gamma ray intensity, chart divisions
γ_{it} = true gamma ray intensity, chart divisions

References

1. *Interpretive Methods for Production Logs*, second edition, Atlas Wireline Services, Western Atlas Intl. Inc., Houston (1982) 35–39.
2. Lichtenberger, G.J.: "A Primer on Radioactive Tracer Injection Profiling," *Proc.*, Southwestern Petroleum Short Course (1981) 251–63.
3. Bearden, W.G. *et al.*: "Interpretation of Injectivity Profiles in Irregular Boreholes," *JPT* (Sept. 1970) 1089–97.

4. Wiley, R. and Cocanower, R.D.: "A Quantitative Technique for Determining Injectivity Profiles Using Radioactive Tracers," paper SPE 5513 presented at the 1975 SPE Annual Technical Conference and Exhibition, Dallas, Sept. 28–Oct. 1.
5. Hill, A.D. and Solares, J.R.: "Improved Analysis Methods for Radioactive Tracer Injection Profile Logging," *JPT* (March 1985) 511–20.
6. Self, C. and Dillingham, M.: "A New Fluid Flow Analysis Technique for Determining Borehole Conditions," paper SPE 1752 presented at the 1976 SPE Symposium on the Mechanical Engineering Aspects of Drilling and Production, Fort Worth, March 5–7.
7. *Production Log Interpretation,* Schlumberger Ltd., Houston (1973) 21–22.
8. Ford, W.O. Jr.: "How New Injectivity Profiling Method Works," *World Oil* (Aug. 1, 1962) 43–47.
9. Kelldorf, W.F.N.: "Radioactive Tracer Surveying—A Comprehensive Report," *JPT* (June 1970) 661–69.
10. Taylor, G.I.: "The Dispersion of Matter in Turbulent Flow Through a Pipe," *Proc.,* Royal Society (1954) **A223,** No. 1155, 446–68.
11. Aris, R.: "On the Dispersion of a Solute in a Fluid Flowing Through a Tube," *Proc.*, Royal Society (1956) **A235,** No. 1200, 67–77.
12. Knudsen, J.G. and Katz, D.L.: *Fluid Dynamics and Heat Transfer,* McGraw-Hill Book Co., New York City (1958) 192.
13. Anthony, J.L.: "Two-Pulse Tracer Logging: An Experimental and Theoretical Study," MS thesis, U. of Texas, Austin (1984).
14. Anthony, J.L. and Hill, A.D.: "An Extended Analysis Method for Two-Pulse Tracer Logging," *SPEPE* (March 1986) 117–24; *Trans.*, AIME, **281.**
15. Cobb, L. *et al.*: "System for Determining Fluid Flow Rate in Boreholes," U.S. Patent No. 3,255,347 (1966).
16. Bragg, J.R., Roesner, R.E., and Strassner, J.E.: "Measuring Well Injection Profiles of Polymer-Containing Fluids," paper SPE 10690 presented at the 1982 SPE/DOE Symposium on Enhanced Oil Recovery, Tulsa, April 4–7.
17. Roesner, R.E. *et al.*: "Visual Flow Loop Investigation of Nuclear Flolog Performance in Non-Newtonian Fluids," *Proc.,* 1982 SPWLA Annual Logging Symposium, Corpus Christi, July 6–9.
18. Akers, T.J.: "Radioactive Tracer Logging in Laminar Flow," MS thesis, U. of Texas, Austin (1985).
19. Akers, T.J. and Hill, A.D.: "Radioactive Tracer Logging in Laminar Flow," *Proc.*, 1985 Cdn. Well Logging Soc. Formation Evaluation Symposium, Calgary, Sept. 29–Oct. 2.
20. Hill, A.D., Boehm, K.E., and Akers, T.J.: "Tracer Placement Techniques for Improved Radioactive Tracer Logging," *JPT* (Nov. 1988) 1484–92.
21. Taylor, G.I.: "Dispersion of Soluble Matter in Solvent Flowing Slowly Through a Tube," *Proc.*, Royal Society (1953) **A219,** No. 1137, 186–203.
22. Roesner, R.E., Sloan, M.L., and Turney, R.A.: "New Logging Instruments for Polymer and Water Injection Wells," *Proc.,* 1983 SPWLA Annual Logging Symposium, Calgary, June 27–30.
23. Knight, B.L. and Davarzani, M.J.: "Injection Well Logging Using Viscous EOR Fluids," *SPEFE* (June 1986) 300–08.
24. Hill, A.D.: "Two-Pulse Tracer Ejection Method for Determining Injection Profiles in Wells," U.S. Patent No. 4,622,463 (1986).

SI Metric Conversion Factors

ft	× 3.048*	E−01	=	m
in.	× 2.54*	E+00	=	cm
lbm	× 4.535 924	E−01	=	kg

*Conversion factor is exact.

Chapter 6
Spinner-Flowmeter Logging

6.1 Introduction

A spinner flowmeter is an impeller that is placed in the well to measure fluid velocity in the same manner that a turbine meter measures flow rate in a pipeline. Like a turbine meter, the force of the moving fluid causes the spinner to rotate. The rotational velocity of the spinner is assumed linearly proportional to fluid velocity, and electronic means are incorporated into the tool to monitor rotational velocity and sometimes direction. A significant difference between a spinner flowmeter and a turbine meter is that the spinner impeller does not span the entire cross section of flow whereas the turbine meter impeller does, with a small clearance between the impeller and the pipe wall.

Throughout the 1940's and 1950's, several attempts were made to adapt surface fluid-metering techniques to the downhole measurement of flow profiles. Among the tools developed were rotameters,[1] hot-wire anemometers,[2-4] a conductivity probe for monitoring the interface between fresh and salt water,[5] and the first spinner flowmeters.[6-8] The spinner flowmeters gradually emerged as the most versatile and reliable of these devices for measuring downhole fluid velocities. The first spinner flowmeters all had a packer or deflector device (called a fluid trap) to direct all or most of the wellbore flow past the impeller. These gradually disappeared, probably because of the added operational difficulties they caused and because of reductions in the bearing friction in spinner flowmeters. Interestingly, the deflector-type flowmeters have had a resurgence in recent years because of advantages they have in multiphase flow.[9]

A properly run spinner-flowmeter log should yield a reliable flow profile in single-phase flow in a constant-diameter wellbore. The spinner-flowmeter, however, is susceptible to mechanical problems, and the quality of the log depends strongly on the logging procedure and the care taken in running the log. If the wellbore cross-sectional area is variable, such as in an openhole completion, a caliper log is needed to interpret a spinner-flowmeter log. In multiphase flow, a spinner flowmeter is often a very poor log, though in some instances it will perform well. The behavior of spinner flowmeters in multiphase flow is discussed thoroughly in Chap. 9.

This chapter treats spinner flowmeters in single-phase flow. It begins with descriptions of the types of spinner flowmeters in common use. Procedures for running spinner flowmeters are then presented, followed by a section on the theory of spinner-flowmeter responses. This is followed by a discussion of spinner-flowmeter log interpretation methods and a description of some special types of spinner flowmeters. The chapter concludes with guidelines for running and interpreting spinner-flowmeter logs.

6.2 Tools and Operations

Two designs of spinner flowmeters—the helical and the vane-type—are prevalent. The more common is the helical spinner, such as Gearhart Industries' spinner flowmeter (Fig. 6.1). Helical spinners are typically longer than they are wide, with a varying blade pitch. The spinner blade is housed in an open cage that extends from the main body of the tool to allow fluid flowing past the tool to impact the spinner. The spinner blade is usually suspended between jewel bearings to allow rotation with little friction.

The Schlumberger Fullbore Spinner (Fig. 6.2), a vane-type flowmeter, has four metal blades that mechanically retract to a tool diameter of $1^{11}/_{16}$ in. [4.3 cm] for running in the tubing. When the tool enters a larger diameter, a spring action extends the blades and the centralizer bands open, making the tool ready for operation. By this mechanism, the Fullbore Spinner is able to sample a larger portion of the wellbore than a helical spinner, whose size is restricted by tubing size.

Deflector flowmeters, such as Atlas Wireline Services' basket flowmeter (Fig. 6.3), have a funnel or some other device to direct the flow past the spinner located inside the tool. Deflector flowmeters have advantages over standard spinner flowmeters in low-flow-rate wells and in certain multiphase flow applications (see Chap. 9).

Some means to measure the rotational velocity of the spinner must be provided in any spinner flowmeter. One common method is to embed a small magnet in the impeller shaft; electronic pulses are generated as the magnet passes switches that surround the impeller shaft. Three to eight pulses are generated by each rotation of the spinner and are counted by surface electronics to give spinner speed. Some spinners are also able to indicate direction of rotation, discriminating direction either by pulse height (amplitude) or by polarity. In a normal, single-phase-flow situation, direction sensing, while not required, is helpful in log interpretation. In multiphase flow, where fluids can be moving up and down simultaneously, and in certain special instances, such as logging to detect crossflow in a shut-in well, the ability to sense the direction of rotation of the spinner and hence the direction of fluid flow can be very important to the log interpretation.

6.3 Running a Spinner Flowmeter

Care must be applied in the actual running of a spinner flowmeter to ensure useful data. The tool must be checked for proper operation before logging, the well conditions must be suitable for using a spinner flowmeter, and the log must be run correctly.

The logging company is, of course, responsible for providing a tool that is in good working order. Nonetheless, someone—either the logging engineer or the company representative—should see that the spinner flowmeter is tested at the wellsite before running a log.

Tool Operation. Check that the spinner turns freely because spinner friction adds nonlinearity to the flowmeter response. A typical helical spinner is suspended between jewel bearings, one of which is adjustable. The bearings should be adjusted so that the spinner blade rotates with the least possible friction without the spinner blade wobbling. The spinner should rotate freely without wobbling when gently blown. The bearings should always be inspected before logging to ensure that they are clean and unbroken. These checks of tool operation should be done in the shop, but should also be repeated at the wellsite immediately before the log is run.

Surface Electronics. This can be easily checked by spinning the impeller on the surface and observing the response.

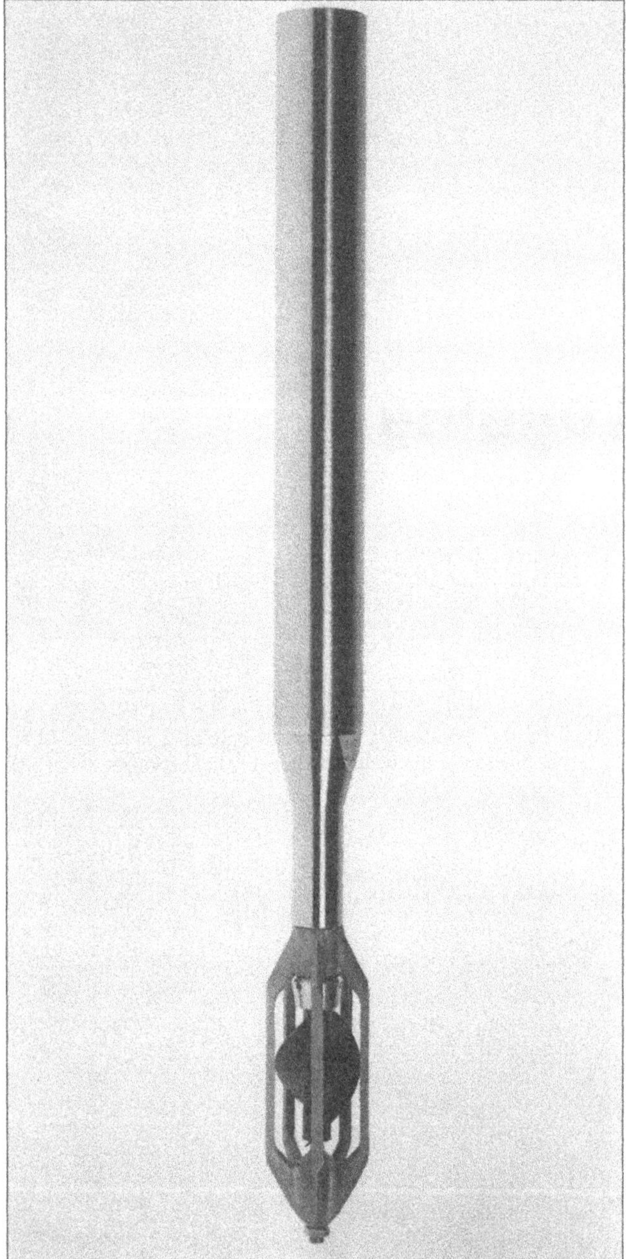

Fig. 6.1—Gearhart Industries helical spinner flowmeter (courtesy Gearhart Industries, Ft. Worth, TX).

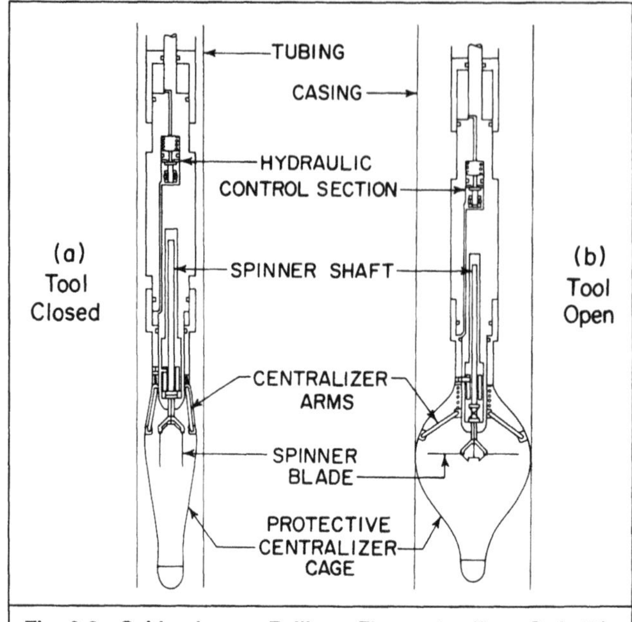

Fig. 6.2—Schlumberger Fullbore Flowmeter (from Ref. 10).

The engineer should decide before running a spinner-flowmeter log whether the well conditions are such that a useful log can be expected. He or she should consider (1) constant flow rate, (2) sufficient flow rate, (3) physical condition of the well, and (4) sand production.

Constant Flow Rate. Maintain as constant a flow rate as possible while a spinner flowmeter is run. This condition seems simple to maintain, but this is not always the case (e.g., when injection is from a manifold common with several wells, when production is to a manifold, or when the well has been shut in for a period before logging). Check the stability of the well conditions during logging by making a few repeat measurements in the upper part of the well, above all perforations. If possible, place a surface meter on the well during logging to provide an independent measure of the well flow rate.

Sufficient Flow Rate. Flow rates must be high enough that the flow is turbulent in most of the well. At low flow rates, spinner response is not linear with flow rate and cannot be easily interpreted. When a spinner flowmeter is run dynamically (the logging tool is moving), the fluid velocity must be large enough compared to the tool velocity to provide reasonable resolution. The fluid velocity should contribute sufficiently to be recognizable above the signal noise level from tool motion. In a low-rate well, though a spinner response can be obtained by moving the tool, the sensitivity to fluid velocity may be low.

Physical Condition of Well. Spinner-flowmeter-log interpretation is based on a constant wellbore cross-sectional area. Do not use spinner flowmeters in some openhole wells or other irregular completions. When they are run in openhole wells, a companion caliper log is essential to interpret the spinner-flowmeter log.

Sand Production. When spinner flowmeters are used in single-phase production wells, such as gas wells or new injection wells, the production must be free of entrained particulates that will jam the spinner bearings. In any well, clean fluids are needed for proper operation of a spinner flowmeter. With spinners that use a magnetic pickup to measure spinner rotations, rust or other magnetic particles can collect around the spinner shaft, reducing the response and changing it through the course of a logging run.

When a spinner flowmeter is run, always include centralizers in the tool string. Because of the velocity profile, a spinner centered in the flow stream will give the highest and most reproducible response. In running a spinner log, measurements are usually made dynamically—i.e., the tool is moved through the wellbore. It is important that cable speed be as constant as possible during each run. As is discussed in Sec. 6.5, making runs at several cable speeds, moving the tool both up and down, is desirable. Also, make measurements with the tool stationary at several locations in the well. These stationary readings are a means of determining the tool's actual threshold velocity, the minimum velocity needed for the spinner to turn. Stationary readings are also a measure of the log accuracy. If the spinner response varies significantly with time during a stationary reading, the dynamic runs will have a similar level of inaccuracy. Unstable stationary readings can be caused by unstable well conditions, poor tool performance, or multiphase-flow effects. Whatever the cause, they are a clear indication of log inaccuracy. Stationary runs are also a check on the depth control of the dynamic runs. In some cases, an offset between the depth scale of up and down runs will be caused by the high cable speeds and finite response time of the spinner or by use of a running average as the spinner response. Stationary readings can indicate this offset so that the depth scale of the dynamic runs can be corrected. Fig. 6.4 shows a log in which the up and down runs were each offset by 6 ft [1.8 m]. Correcting the depth scale changed the interpreted location of the lowest injection by this amount.

Finally, make a few repeat runs, particularly in the area of the well above all perforations, to check the stability of the total flow

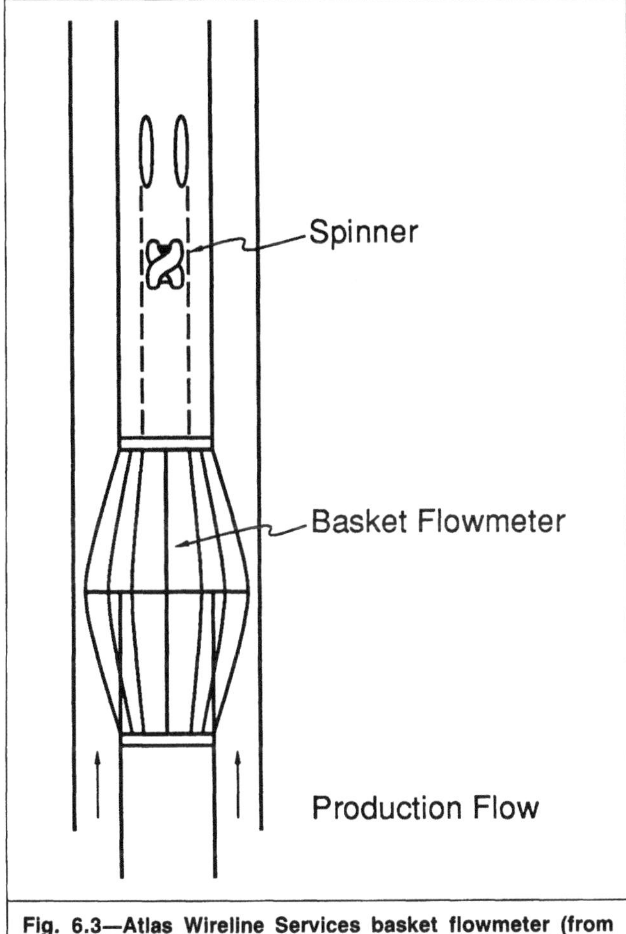

Fig. 6.3—Atlas Wireline Services basket flowmeter (from Ref. 11).

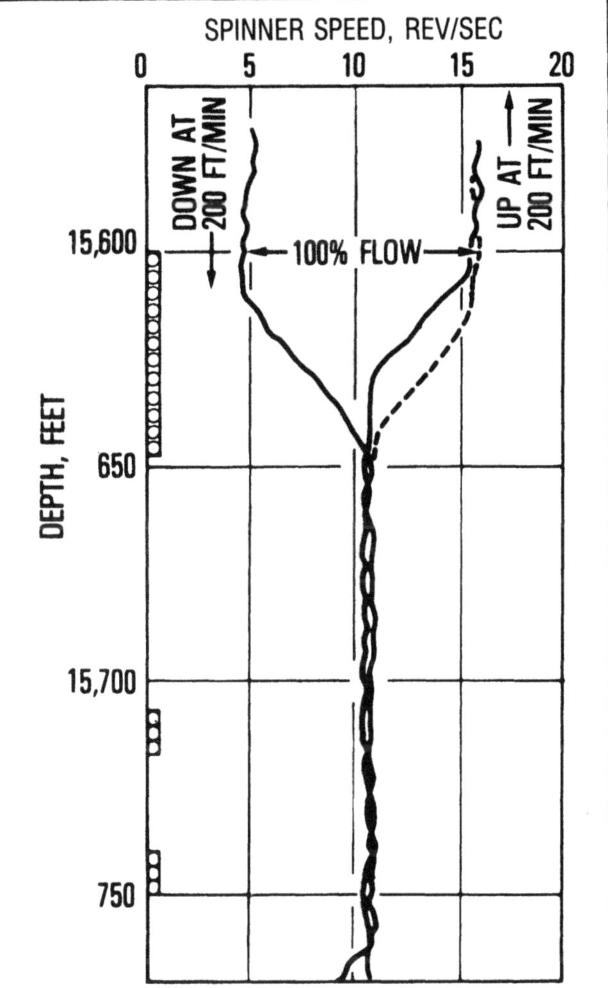

Fig. 6.4—Depth offset of spinner, up and down passes (from Ref. 12).

rate from the well. A few repeat runs should be made throughout the well; if runs repeated at the same cable velocity vary significantly, either the tool is not operating properly or the well conditions are not stable. In either case, the log is not likely to provide an accurate flow profile.

6.4 Theory of Spinner Response

Spinner log interpretation depends on a linear relationship between spinner rotational velocity and fluid velocity. In this section we examine the basis for this assumption, following the work of Leach *et al.*[10]

Consider a vane-type flowmeter, such as the Fullbore Spinner. The blade is angled to the direction of fluid flow so that if we sliced through a blade some distance r from the center and looked at it from the end of the blade, it would appear as in Fig. 6.5.

The fluid motion exerts an upward force on the spinner blade. Because the spinner cannot move upward, the force of the fluid causes a rotational motion of the spinner. Assuming a frictionless spinner and a fluid with zero viscosity, the spinner blades will move just fast enough so that a particle of fluid will pass by unimpeded—i.e., the streamlines will not be deflected. From the geometry of the system, this assumption relates the spinner blade velocity to fluid velocity as

$$v_{bl}=v_f \tan \alpha_{bl}. \quad (6.1)$$

The blade velocity is also related to the frequency of rotation for this ideal case, f_i, by

$$v_{bl}=2\pi r f_i, \quad (6.2)$$

where r=radial distance from the center of the spinner.

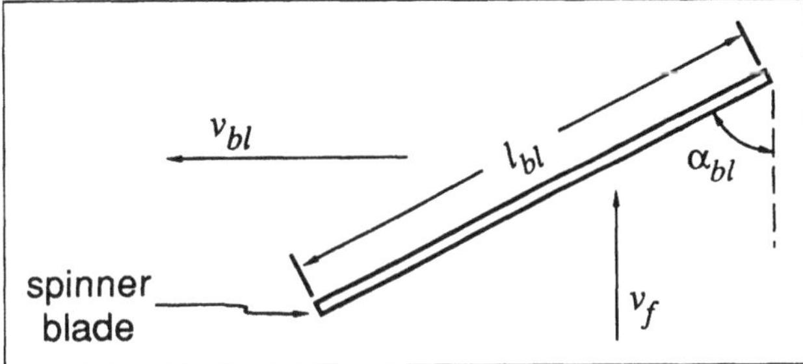

Fig. 6.5—Relation of spinner blade movement to fluid movement.

Equating Eqs. 6.1 and 6.2,

$$f_i=\frac{v_f \tan \alpha_{bl}}{2\pi r}. \quad (6.3)$$

Spinner flowmeters are constructed so that the angle of incidence, α_{bl}, varies with radius to keep $\tan \alpha_{bl}/r$ constant. Thus,

$$f_i=Cv_f, \quad (6.4)$$

and in the ideal case of no friction or viscosity, the spinner frequency varies linearly with fluid velocity. The ideal response slope, C, is the inverse of the pitch of the spinner blade, where pitch is

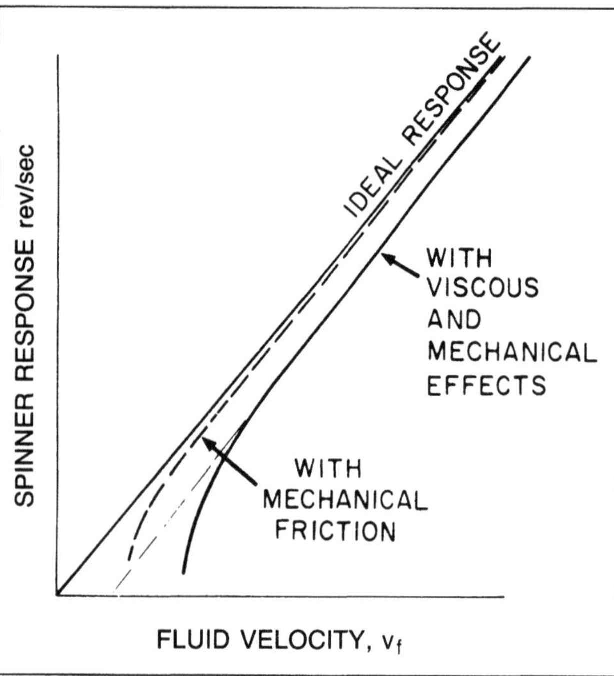

Fig. 6.6—Spinner response vs. fluid velocity schematic (from Ref. 10).

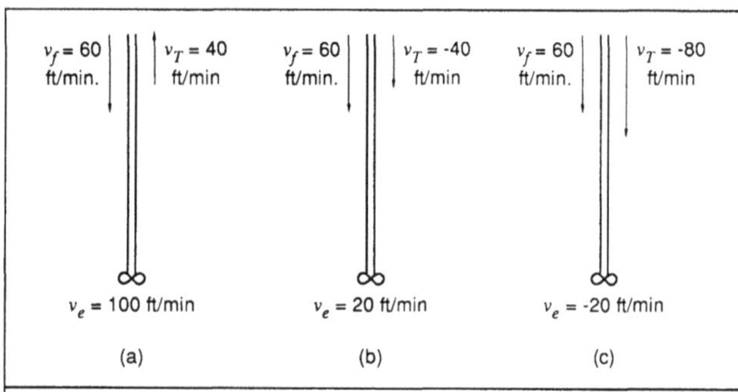

Fig. 6.7—Additive nature of tool and fluid velocities.

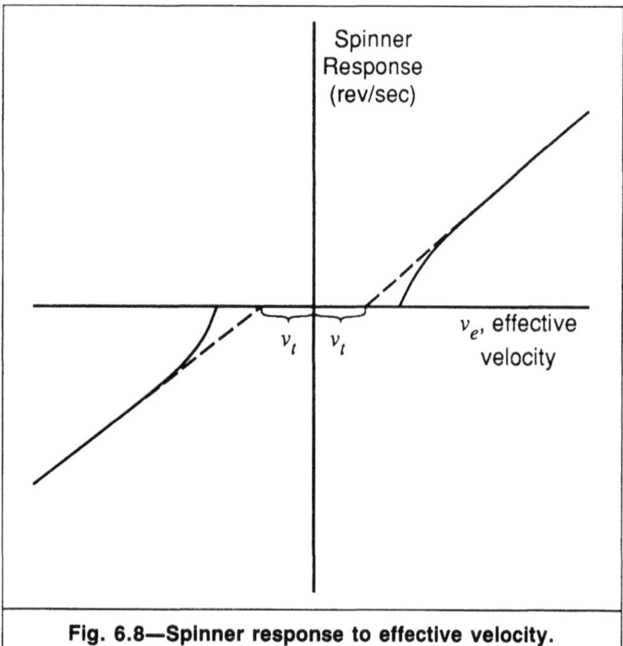

Fig. 6.8—Spinner response to effective velocity.

defined as the distance a fluid particle will travel for one rotation of the spinner. For example, for a spinner with a pitch of 4 in. [10 cm] and fluid velocity in feet per minute, the ideal response slope is 0.05 (rev/sec)/(ft/min) [0.015 (rev/sec)/(m/min)].

Of course, in the real case, spinners are not frictionless and fluids have finite viscosity. Considering these effects separately, Leach *et al.*[10] showed that spinner friction causes the frequency response to be of the form

$$f=\frac{v_f \tan \alpha_{bl}}{2\pi r}-\frac{F_T}{4\pi^2 \rho_f v_f r^3 h_{bl}}, \quad \text{(6.5)}$$

where F_T=retarding torque, ρ_f=fluid density, and h_{bl}=blade thickness. Thus, spinner friction adds another term to the ideal response equation; if the retarding torque, F_T, is sufficiently large, this second term will be significant, and the spinner frequency response will not be linear with the fluid velocity.

The effect of fluid viscosity also reduces the frequency response of the spinner. The viscosity effects, also called Reynolds number effects, yield a frequency response of the form[10]

$$f=\frac{v_f \tan \alpha_{bl}}{2\pi r}-\frac{K_D}{4\pi}, \quad \text{(6.6)}$$

where K_D=drag coefficient. For a blade Reynolds number less than $\sim 5\times 10^5$, the drag coefficient is

$$K_D=\frac{G}{\sqrt{N_{Rebl}}}=G\sqrt{\mu_f/(\rho_f v_f l_{bl})}, \quad \text{(6.7)}$$

where μ_f=fluid velocity, G=constant, and l_{bl}=blade length (Fig. 6.5). The spinner response in a viscous fluid can be written for these conditions as

$$f=\frac{v_f \tan \alpha_{bl}}{2\pi r}-\frac{G}{4\pi}\sqrt{\frac{\mu_f}{\rho_f v_f l_{bl}}}, \quad \text{(6.8)}$$

where G=constant. The effect of fluid viscosity, like the effect of mechanical friction, reduces the frequency response and introduces a nonlinear dependence on fluid velocity. These effects are most predominant at low fluid velocities. Fig. 6.6 compares the ideal spinner frequency response with that observed with mechanical friction present and with viscous and mechanical effects. At higher flow rates, these effects reduce the spinner response. However, this response is nearly linear with fluid velocity, with about the same slope as in the ideal case.

6.5 Spinner-Flowmeter-Log Interpretation

6.5.1 The Effective Velocity. All spinner-flowmeter-log interpretation is based on the spinner response being a linear function of fluid velocity. In dynamic runs, it is further assumed that the fluid and tool velocities are additive so that the spinner responds to an effective velocity, v_e, given by

$$v_e=v_f+v_T, \quad \text{(6.9)}$$

where v_f=fluid velocity and v_T=tool velocity. A sign convention is needed with this equation. We will assume that v_f is positive. v_T is also positive when v_f and v_T are in opposite directions and is negative when the tool and the fluid are moving in the same direction. Fig. 6.7 illustrates this convention. In this example, the fluid velocity is 60 ft/min [18 m/min] downward. In Fig. 6.7a, the tool moves 40 ft/min [12 m/min] upward; thus the effective velocity seen by the spinner flowmeter is

$$v_e=+60 \text{ ft/min}+(+40 \text{ ft/min})$$

$$=+100 \text{ ft/min } [+30 \text{ m/min}].$$

When the tool moves in the same direction as flow, as in Fig. 6.7b, the effective velocity is

$$v_e = +60 \text{ ft/min} + (-40 \text{ ft/min})$$
$$= +20 \text{ ft/min } [+6 \text{ m/min}].$$

Finally, if the tool moves downward at a velocity greater than the flow velocity (Fig. 6.7c), Eq. 6.9 yields

$$v_e = +60 \text{ ft/min} + (-80 \text{ ft/min})$$
$$= -20 \text{ ft/min } [-6 \text{ m/min}].$$

The change in sign of v_e indicates that the effective velocity at the spinner is now in the upward direction; the spinner would rotate in the direction opposite that in the first two instances.

Spinner response to effective velocity will be linear for high enough values of v_e. At lower velocities, the actual spinner response will drop until, at some velocity, the spinner ceases to turn, as shown by the solid line in Fig. 6.8. In spinner-log interpretation, the linear portion of the response curve is extrapolated to the zero spinner-response line; the intercept of the extrapolated line on the effective velocity axis is called the threshold velocity, v_t. Threshold velocity is sometimes referred to as bypass velocity and is a hypothetical minimum velocity required to start spinner rotation, if the response is completely linear. Incorporating the threshold velocity, the spinner response can be written

$$f = m_p(v_e - v_t) \quad (v_e > v_t) \quad \text{(6.10)}$$

and

$$f = m_n(v_e + v_t) \quad (v_e < -v_t), \quad \text{(6.11)}$$

where m_p and m_n = slopes of the response curve for positive and negative spinner responses, respectively, and v_t = a positive number. The two slopes, m_p and m_n, are not generally equal because the tool body shields the spinner when the net flow direction is downward, while the spinner contacts the flowstream before the tool body when the net flow direction is upward. m_p and m_n are not, however, absolutely related to rotation direction. In some tools, the spinner signal is electronically compensated to yield the same response for spinner rotation in either direction. It is a good practice to make a few spinner runs to determine m_p and m_n unambiguously when the well is shut in.

The response slopes, m_p and m_n, will be less than the ideal response slope as calculated from the pitch of the spinner because of friction and fluid viscosity (see Sec. 6.4). When a straight-line fit for the spinner response is used, the nonlinear dependence on fluid velocity that arises from frictional and viscous drag is neglected. The threshold velocity also increases as viscous and frictional drag increases because the spinner response is decreased (see Eqs. 6.5 and 6.8).

Typical values for the response slope are from 0.04 to 0.05 (rev/sec)/(ft/min) [0.012 to 0.015 (rev/sec)/(m/min)] for many modern spinners, while the threshold velocity usually ranges from 3 to 6 ft/min [0.9 to 1.8 m/min] in liquid and from 12 to 15 ft/min [3.7 to 4.6 m/min] in gas. The higher threshold velocity in gas is probably because of the low density and high velocity of the medium. As shown in Eq. 6.5, a combination of low density and high velocity leads to high frictional drag, which reduces the spinner response from the ideal response. This reduction results in a high threshold velocity.

The linear response to effective velocity given by Eqs. 6.10 and 6.11 is the basis for all quantitative spinner-flowmeter-log interpretation. Next, we examine three particular interpretation methods based on this response—the multipass, the two-pass, and the single-pass methods.

6.5.2 Multipass Method. The multipass or in-situ calibration method[13] is the most accurate technique of spinner-flowmeter evaluation because the spinner response characteristics are determined under in-situ conditions. As the name implies, multiple passes in the well at different tool speeds and directions are needed to apply the method. Stable well conditions must exist during all the logging passes for the multipass method to be applied.

TABLE 6.1—SPINNER FLOWMETER RESPONSE FOR A DOWNWARD FLUID VELOCITY OF 60 ft/min

Cable Speed (ft/min)	Effective Velocity (ft/min)	Spinner Response (rev/sec)
+40	+100	3.6
+20	+80	2.8
0	+60	2.0
−40	+20	0.4
−80	−20	−0.4
−120	−60	−2.0

The multipass interpretation is based on the linear spinner response,

$$f = m_p(v_f + v_T - v_t). \quad \text{(6.12)}$$

Solving for fluid velocity, v_f, when tool velocity is zero yields

$$v_f = \frac{f_0}{m_p} + v_t, \quad \text{(6.13)}$$

where f_0 = intercept of response curve with $v_T=0$ axis. Thus, if the response line (v_t and m_p) is determined from multiple passes of the spinner, the fluid velocity can be calculated from the intercept of the response curve at $v_T=0$. Alternatively, if f is set to zero in the spinner response equation (Eq. 6.12), the fluid velocity is

$$v_f = -v_{T_0} + v_t. \quad \text{(6.14)}$$

The intercept of the response curve with the cable speed axis can thus also be used to calculate the fluid velocity.

The final step in the interpretation is to convert the fluid velocity to volumetric flow rate. The velocity measured by the spinner flowmeter, if it is centralized, will be biased toward the maximum velocity in the center of the pipe. This velocity must be converted to average velocity by multiplying by the ratio v/v_{max}, which for turbulent flow is about 0.83. Thus, at each station,

$$q = BA_w v_f, \quad \text{(6.15)}$$

where q = volumetric flow rate, A_w = cross-sectional area of the wellbore, B = velocity profile correction factor, and v_f = fluid velocity from the multipass interpretation.

As an illustration, consider the spinner response to cable speed for a given fluid velocity (60 ft/min [18 m/min]) given in Table 6.1 and shown in Fig. 6.9. As seen in the figure, the spinner response is defined by two lines, one for positive and one for negative spinner responses. Calculating the slopes of these lines yields a value of 0.04 (rev/sec)/(ft/min) [0.12 (rev/sec)/(m/min)] for both m_p and m_n. Note that although the response slopes, m_p and m_n, are equal in this example, they are often different with an actual spinner tool. The intercepts of the two response lines on the cable speed axis yield the threshold velocity, as

$$(v_{Tp} - v_{Tn}) = 2v_t, \quad \text{(6.16)}$$

where v_{Tp} and v_{Tn} = intercepts of the positive and negative spinner-response curves with the $f=0$ axis, respectively. This equation is obtained by substituting $f=0$ into the positive and negative spinner response equations and subtracting one from the other. In the example, the threshold velocity is found to be 10 ft/min [3 m/min].

Because the threshold velocity is found by taking the difference between two experimentally determined lines, it is very sensitive to any errors or fluctuations in the spinner response. If the well flow rate is not stable or if two-phase flow effects cause a noisy

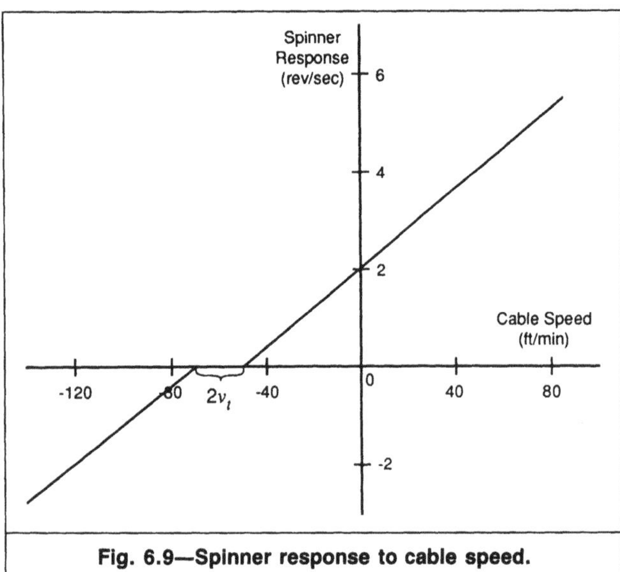

Fig. 6.9—Spinner response to cable speed.

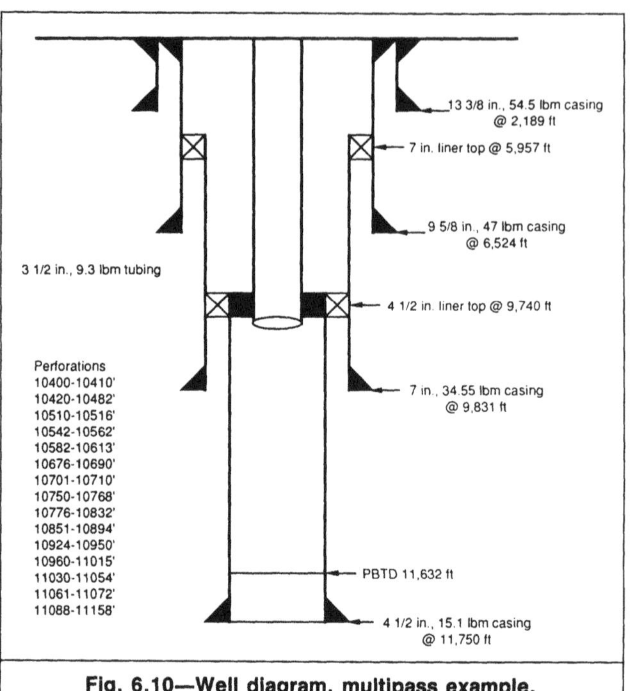

Fig. 6.10—Well diagram, multipass example.

spinner response, the threshold velocity cannot be accurately determined in the manner just described. In these situations, the threshold velocity may be obtained by logging in the rathole or with the well shut in. In a gas production well, however, this technique will most likely yield the threshold velocity in liquid, which is typically significantly different from that in gas. However the threshold velocity is obtained, it should be compared with the threshold velocity predicted by the tool supplier; if it is significantly higher than expected, the spinner is fouled with debris or the bearings are not adjusted properly.

To apply the multipass method, logging runs must be made at enough different cable speeds to define both the positive and negative spinner response lines (at least two positive and two negative spinner responses are needed). The cable speeds used should be different enough to give as large a spread in the data as possible—a difference of 30 ft/min [9 m/min] between each pass is a reasonable target within the constraint of practical logging speed. In logging a well, the multipass method is applied at a number of depth locations or stations in the well. At each station, the response line slope, m_p, and threshold velocity, v_t, are determined (if possible), and the fluid velocity is then calculated from the intercept of the response line at $v_T=0$ from Eq. 6.13 or from the intercept of the response line at $f=0$ with Eq. 6.14.

The steps in the multipass method are summarized below.

1. Choose the stations (depth locations) at which the flow rates will be calculated (as a minimum, choose one station between each of the perforated intervals).
2. Read the spinner responses at different cable speeds at each station.
3. For each station, plot f vs. v_T (cable speed).
4. Calculate the slope, m_p or m_n, for each response line.
5. For each station where positive and negative spinner responses occur, determine the threshold velocity, v_t, from Eq. 6.16.
6. Calculate the v_f at each station by applying Eq. 6.13 or 6.14.
7. Convert the fluid velocities to volumetric flow rates with Eq. 6.15.

Example 1—Multipass Method, Gas Production Well. The well shown in Fig. 6.10 was producing 12 MMscf/D [0.34×10^6 std m^3/d] of gas and 500 B/D [79.5 m^3/d] of condensate from a long interval with 15 separate perforated zones. The well diagram, which should be included with any production log, shows the locations of the perforations and contains information about casing and tubing sizes that is needed to compare surface rates with downhole rates computed from the log. Though there is significant liquid production from this well, we will analyze the spinner-flowmeter log as if it were single-phase gas flow. With a GOR of 24,000 scf/bbl [4323 std m^3/m^3], this may be a reasonable assumption; if it is not, we may see anomalous results in the log interpretation because of two-phase-flow effects. It is also almost certain that the rathole will contain liquid, so results obtained by logging in the rathole, such as measurement of the threshold velocity, may not be applicable for other parts of the well.

The well was logged with a Fullbore flowmeter, with four up and four down passes made, as shown on the log (Fig. 6.11). Cable speeds of 32, 61, 93, and 125 ft/min [9.8, 18.6, 28.3, and 38.1 m/min] were used on the down runs, and the up passes were made at 33, 62, 96, and 129 ft/min [10.1, 18.9, 29.3, and 39.3 m/min]. To make the log more readable, the up and down runs are plotted on separate tracks on the log, with each track on a scale of 0 to 40 rev/sec. Note that stationary measurements were made between each set of perforations; this provides a good check on the interpretation from the multipass method. Because the Fullbore flowmeter does not sense direction of rotation, all spinner responses are positive.

The first step in the interpretation is to choose the stations (depth locations) for which flow rate will be calculated and to read the spinner responses at each cable speed at these stations. In this example, stations were chosen between perforated intervals. This is generally a good practice because effects of nonaxial flow on the spinner response are not included. In a well with long perforated intervals, however, stations should also be chosen within perforated zones to give better definition of the flow profile. Stations within perforated intervals should be selected at depths where the spinner response is fairly flat and steady.

The spinner responses read from the log at each station are shown in Table 6.2. The cable speeds for the up runs are negative because flow is upward in this production well. According to the sign convention for computing effective velocity, cable speed will be positive when moving opposite the direction of flow and negative when moving in the direction of flow. Note that for Stations 12 through 16, the spinner responses for the up runs are negative. This is because the spinner has reversed direction of rotation compared with the up-run responses higher in the well. Reversal of the spinner is indicated by the response decreasing to near zero, then increasing again as the tool is moved up the well. Looking at the up run at 62 ft/min [18.9 m/min], for example, the response diminishes to near zero at 11,040 ft [3665 m], remains near zero for about 100 ft [30 m], then increases above 10,940 ft [3335 m]. Such a response

TABLE 6.2—SPINNER FLOWMETER RESPONSES

		Cable Speed (ft/min)							
	Depth	−129	−96	−62	−33	32	61	93	125
Station	(ft)	Spinner Responses (rev/sec)							
1	10,380	22.5	24.5	26.5	28	31.5	33.5	35	37
2	10,415	23	25	26.5	28.5	32	33.5	35	37
3	10,500	22	24	26	27.5	31	32.5	34.5	36
4	10,540	22	24	26	27.5	31	32.5	34	35.5
5	10,570	21.5	23.5	25.5	27	30.5	32	34	35.5
6	10,650	21.5	23	25	27	30.5	32	34	35.5
7	10,700	21	23	25	26.5	30	31.5	33.5	35
8	10,730	20.5	23	24.5	26.5	30	31.5	33	35
9	10,772	13	15	17	18.5	21.5	23	25	27
10	10,840	4.5	6	8	9.5	12.5	14	16	18
11	10,910	2.5	4.5	6.5	8	10.5	12	13.5	15.5
12	10,965	−6	−3.5	−1.5	−1	4	5.5	7.5	9
13	11,020	−5.5	−3	−1	−1	3.5	5	7	8
14	11,060	−8.5	−6.5	−4	−2	1.5	3	5	6.5
15	11,080	−8	−6.5	−4	−2	1	3	5	7
16	11,170	−8	−8	−4	−2	2	3.5	5.5	7

must result from a reversal in spinner rotation direction or from a fluid exit from this production well, which is unlikely. Another indication of the spinner reversal is the change in the relative magnitudes of the spinner responses at different cable speeds. Above the reversal point, the up run at 33 ft/min [10.1 m/min] has the largest spinner response and the run at 129 ft/min [39.3 m/min] has the lowest response, while below the reversal point, the highest cable speed produces the highest response and the lowest cable speed yields the lowest response. On this log, the spinner response does not reach zero when the reversal takes place because the spinner response on this run is actually the average of the raw spinner output over an 8-ft [2.4-m] distance. If the spinner turns at all over an 8-ft [2.4-m] distance, a nonzero response will result. Any rectifying spinner exhibits this behavior.

The next step in the interpretation is to plot the spinner responses as a function of tool velocity (Fig. 6.12) to generate the in-situ calibration plot. In Fig. 6.12, only six stations have been plotted because all other stations yielded data coinciding closely with one of these stations. A plot should be made for each station, of course. On the in-situ calibration plot, the data for each station should make a straight line and the lines for each station should be close to parallel. These requirements are satisfied by this log. If the response slopes do not agree closely, or if the points at each station do not give a straight line, the flowmeter is not responding linearly to ef-

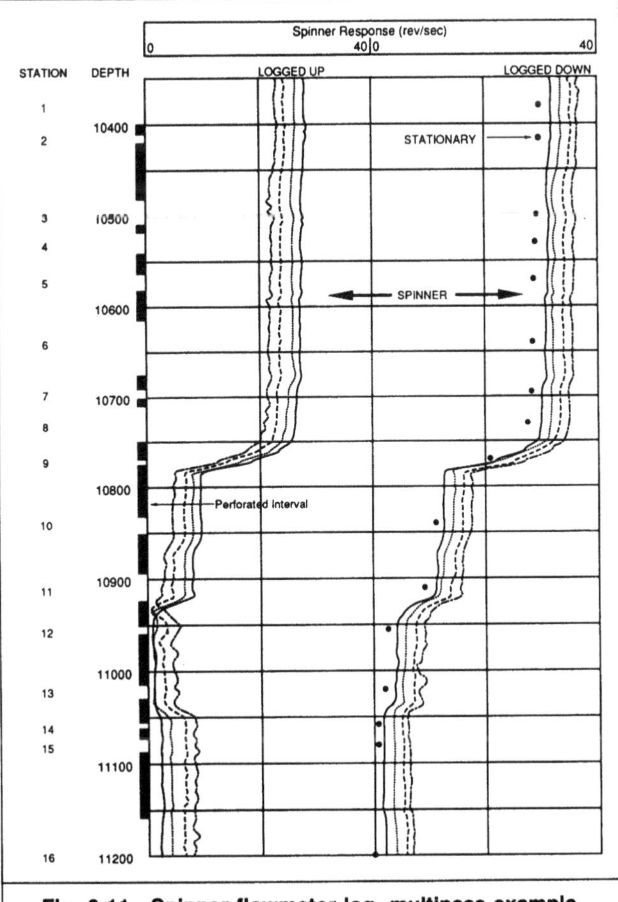

Fig. 6.11—Spinner-flowmeter log, multipass example.

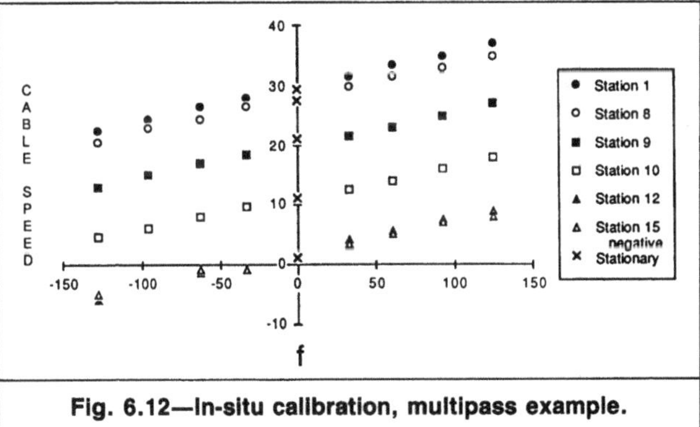

Fig. 6.12—In-situ calibration, multipass example.

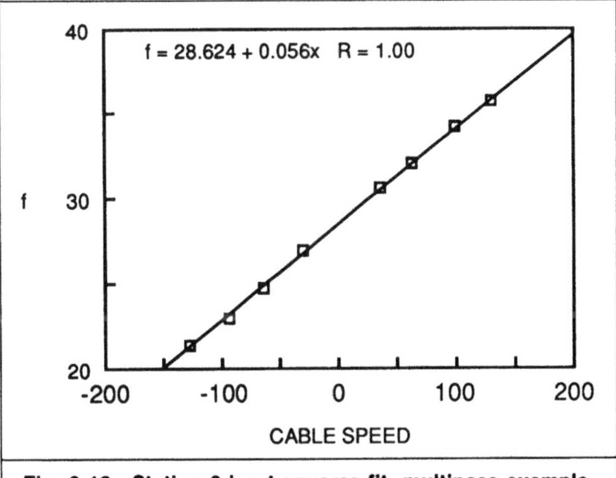

Fig. 6.13—Station 6 least-squares fit, multipass example.

TABLE 6.3—SPINNER FLOWMETER INTERPRETATION

Station	m_p (rev/sec)/(ft/min)	f_0 (rev/sec)	v_t (ft/min)	$\bar{v}$ (ft/min)	q (Mcf/D)	Total Flow (%)	Flow Entering (%)
1	0.057	29.9	530	440	50.6	100	−4
2	0.055	30.1	553	459	52.7	104	3
3	0.055	29.2	536	445	51.2	101	−3
4	0.053	29.1	554	460	52.9	105	5
5	0.055	28.7	527	438	50.3	99	2
6	0.056	28.6	516	428	49.3	97	0
7	0.055	28.2	518	430	49.4	98	2
8	0.056	28.1	507	421	48.4	96	23
9	0.053	20.1	385	319	36.7	73	31
10	0.052	11.1	219	182	20.9	41	5
11	0.049	9.2	193	160	18.4	36	35
12	0.055	0.2	9	8	0.9	2	−7
13	0.05	2	45	38	4.3	9	9
14	0.055	−0.3	0	0	0.0	0	2
15	0.064	−1	−10	−8	−1.0	−2	−4
16	0.055	0.2	9	8	0.9	2	2

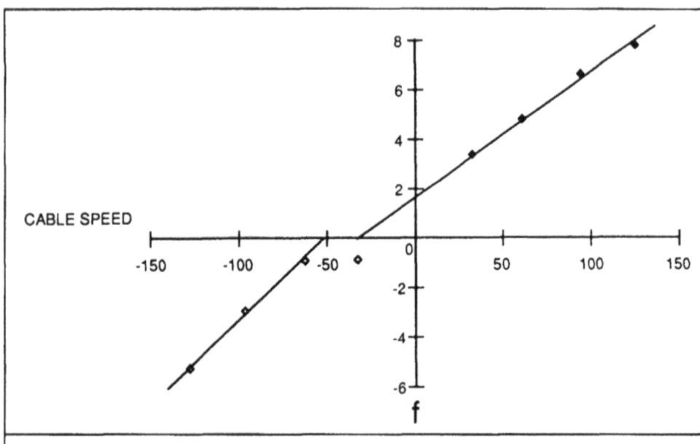

Fig. 6.14—Station 13 data, multipass example.

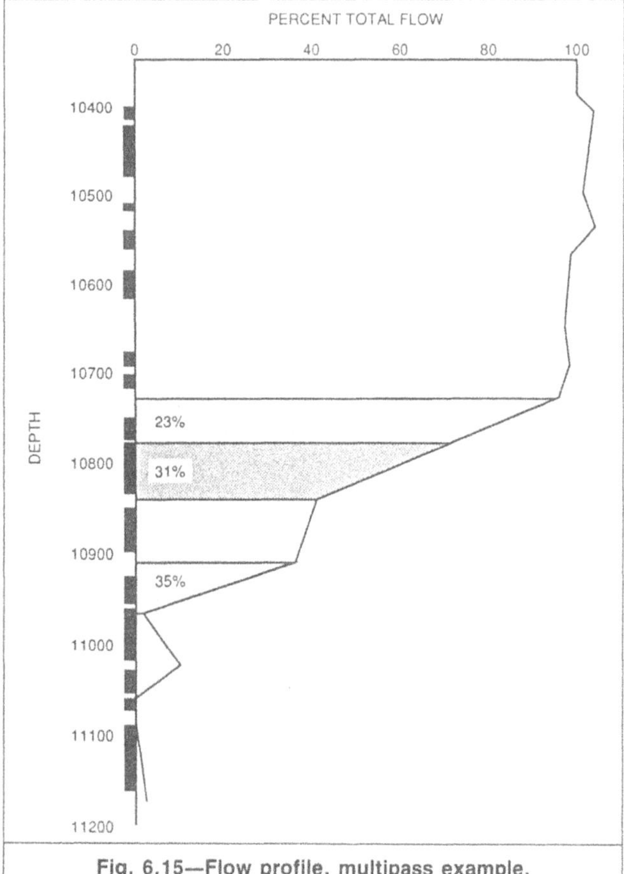

Fig. 6.15—Flow profile, multipass example.

fective velocity, and the log cannot be considered an accurate representation of well conditions.

Straight lines are then fit to the spinner responses at each station. If sufficient data points exist, a least-squares fit should be used.[14,15] Fig. 6.13 shows the least-squares fit to the data from Station 6, with the equation for the best straight-line fit shown. From the least-squares fit, the slope of the response line at this station is 0.056 (rev/sec)/(ft/min) [0.184 (rev/sec)/(m/min)]. The response slopes for all other stations are determined in a similar fashion and are given in Table 6.3. The responses agree reasonably well and are in the range expected for this flowmeter. These slopes are only for the positive spinner responses. Because of the nonlinearity of the spinner at low velocities, the positive and negative responses will not lie on the same line.

The threshold velocity can be determined from any station where positive and negative spinner responses occurred. In the example shown, negative spinner responses were recorded at Stations 12 through 16. The positive and negative response lines for Station 13 are shown in Fig. 6.14. Note that the spinner response at −33 ft/min [−10.1 m/min] was not included in the negative response line because the spinner was apparently stalled at this cable speed. It also appears that the negative response line has a greater slope than the positive response line. This is contrary to the expected behavior. For the negative responses, the net-flow direction is downward past the tool body, and shielding of the spinner by the tool typically would diminish the tool response compared with a net upward flow. This anomalous behavior may result from two-phase flow effects in this low part of the well.

The threshold velocity is now calculated according to Eq. 6.16—e.g., for Station 13, the positive response is

$$f=1.99+0.05v_T$$

and the negative response is

$$f=3.31+0.068v_T.$$

Therefore,

$$(v_{Tu}-v_{Td}) \text{ at } f=0=[-1.99/0.05-(-3.31/0.068)],$$

$$2v_t=8.8 \text{ ft/min [2.7 m/min]},$$

and

$$v_t=4.4 \text{ ft/min [1.3 m/min]}.$$

This threshold velocity is typical of a liquid, not a gas flow stream. Similar calculations at Stations 14 and 15 yield values of 4.1 and 7.8 ft/min [1.2 and 2.4 m/min], respectively, for threshold velocities. Averaging the results from Stations 13 through 15, the threshold velocity is estimated to be 5.4 ft/min [1.6 m/min]. At

Station 12, the calculated threshold velocity is unreasonably small, and that calculated for Station 16 is actually negative, which is impossible.

These results for threshold velocity indicate that liquid is probably present in the lower part of the well and that the rathole fluid may be a dense liquid that is causing anomalous spinner responses. Because we interpret the log as if all flow were single-phase gas, some error is introduced by the two-phase flow conditions in the lower part of the well. In particular, the threshold velocity obtained in this region may not be appropriate for the upper part of the well, where the flow is more likely to be single-phase gas. Fortunately, velocities are high enough in this well that an error in the threshold velocity will not seriously affect the flow profile interpretation.

The fluid velocity at each station can now be calculated from the response line intercept at $v_T=0$—e.g., for Station 8, f at $v_T=0=$ 28.1 rev/sec.

Then, using Eq. 6.13 at Station 8,

$$v_f=\frac{28.1\ \text{(rev/sec)}}{0.056\ \text{(rev/sec)/(ft/min)}}+5.4\ \text{ft/min}$$

$$=507\ \text{ft/min [155 m/min]}.$$

This calculation is repeated at all stations; the results are presented in Table 6.3. The flow profile is determined by calculating the volumetric flow rate at each station. Using Eq. 6.15 at Station 8,

$$q=BA_wv_f$$

and

$$A_w=0.0798\ \text{ft}^2\ [0.0074\ \text{m}^2]$$

for 4½-in., 15.1-lbm [11.4-cm, 6.8-kg] casing. Therefore,

$$q=0.83(0.0798\ \text{ft}^2)(507\ \text{ft/min})$$

$$=33.58\ \text{ft}^3/\text{min}=48.4\ \text{Mcf/D}\ [137\times10^3\ \text{m}^3/\text{d}].$$

The results for all stations are given in Table 6.3.

To check the integrity of the tubing and packer, it is good practice to compare the flow rate calculated from the log above all perforations with the surface flow rate. To do so for this gas well, the flow rate at Station 1 must be converted to standard conditions. For a rough approximation, the compressibility factor of the gas can be ignored, so that

$$q_{sc}=q_{bh}\left(\frac{p_{bh}}{p_{sc}}\right)\left(\frac{T_{sc}}{T_{bh}}\right),$$

where the subscript sc=standard conditions and the subscript bh= bottomhole conditions at Station 1 above all perforations. In this well, bottomhole conditions are a pressure of 4,050 psia [28 MPa] and a temperature of 211°F [99°C]. The downhole total flow rate at standard conditions is thus

$$q_{sc}=50.6\times10^3\ \text{ft}^3/\text{D}\left(\frac{4{,}050\ \text{psia}}{14.7\ \text{psia}}\right)\left(\frac{520°\text{R}}{671°\text{R}}\right)$$

$$=10.8\ \text{MMscf/D}\ [0.31\times10^6\ \text{std m}^3/\text{d}].$$

This rate compares reasonably well with the surface rate of 12 MMscf/D [0.34×10^6 std m^3/d]; the neglect of the gas compressibility factor could explain the discrepancy.

The flow profile is usually presented as the percent of total flow rate as a function of depth. With use of the flow rate at the highest station as the total flow rate, the flow profile shown in Table 6.3 and plotted in Fig. 6.15 is obtained. In the upper part of the well, where little production is occurring, the interpreted flow rate varies somewhat, with the flow rate at some stations apparently exceeding the flow rate at the uppermost station. The results in Table 6.3 show that these variations result from slight variations in the response slope and are an indication of the error level in the log interpretation. In spinner-flowmeter interpretion, such variations often will be eliminated by use of a single average value for the response slope or by the average of the spinner response over some distance around each station. These smoothing techniques remove apparent anomalies in the interpreted flow profile but mask the errors involved in spinner-flowmeter interpretation.

From this log, it is clear that the top seven perforated zones are not contributing significantly to the production from the well and that about 90% of the flow is coming from only three zones. The anomalous behavior in the lower part of the well, below 10,960 ft [3341 m], could be caused by the presence of significant amounts of liquid in this region of low gas rate. Recall that anomalous results were observed in this part of the well, particularly at Station 16 below the bottom set of perforations, when threshold velocity was computed. A series of stationary measurements in this part of the well would help to determine the validity of the threshold velocity obtained dynamically. The minimum velocity found to rotate the spinner when stationary would provide an upper bound on the dynamic threshold velocity.

6.5.3 Two-Pass Method. Another spinner flowmeter log interpretation technique developed by Schlumberger[13] for use with the Fullbore Flowmeter is the two-pass technique. This method uses two logging runs, an up pass and a down pass, which are superimposed in a region of zero fluid velocity (static column) to determine the flow profile. If the two passes are made at the same cable speed with a tool that does not sense direction, the two passes should overlie each other in the no-flow region if m_p and m_n are equal. To apply this method, the spinner must rotate in opposite directions throughout the well during the two runs.

To derive the two-pass method, write the equations for the spinner response to the up and down runs as

$$f_u=m_p(v_f+v_{Tu})+b_u \quad\text{(6.17)}$$

and

$$f_d=m_n(v_f+v_{Td})+b_d, \quad\text{(6.18)}$$

where f_u and f_d=spinner frequency responses to up and down runs, respectively, and b_u and b_d are constants that contain the threshold velocity. For an injection well, use of the normal sign convention will yield a positive v_{Tu} and a negative v_{Td}. Also, because the spinner must rotate in opposite directions in the two runs, f_d will be negative and f_u will be positive. Recall, however, that for a Schlumberger Fullbore Flowmeter, the tool response will be positive, no matter which direction the spinner is rotating. Thus, the actual down response that would be recorded would be

$$f_d=m_n|v_f+v_{Td}|+b_d \quad\text{(6.19)}$$

or, considering the signs of v_f and v_{Td},

$$f_d=m_n(|v_{Td}|-v_f)+b_d. \quad\text{(6.20)}$$

Now, in a static column ($v_f=0$), the two responses are

$$f_{us}=m_pv_{Tu}+b_u \quad\text{(6.21)}$$

and

$$f_{ds}=m_n|v_{Td}|+b_d. \quad\text{(6.22)}$$

For the logs in the static column to coincide, the down response curve must be shifted by the factor f_u-f_d, given by

$$f_{us}-f_{ds}=m_pv_{Tu}+b_u-m_n|v_{Td}|-b_d. \quad\text{(6.23)}$$

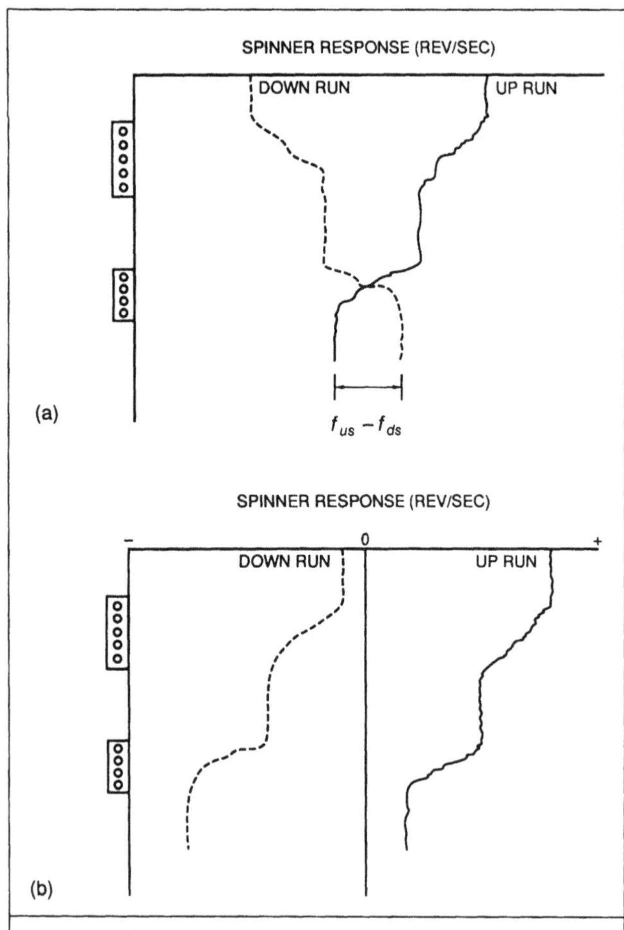

Fig. 6.16—Spinner response (a) without and (b) with direction sensing.

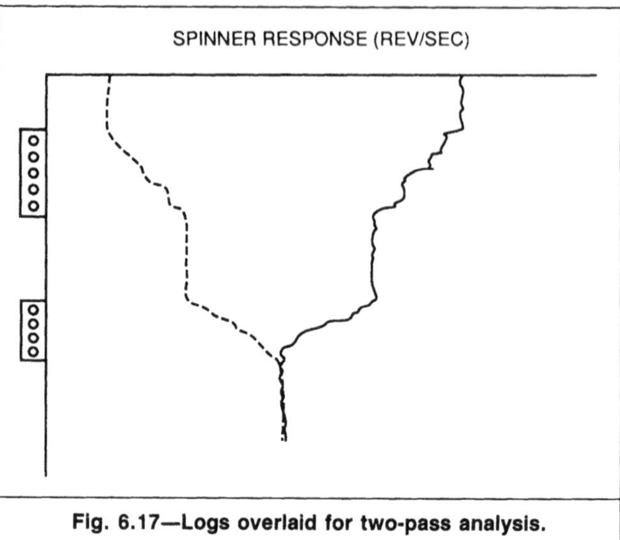

Fig. 6.17—Logs overlaid for two-pass analysis.

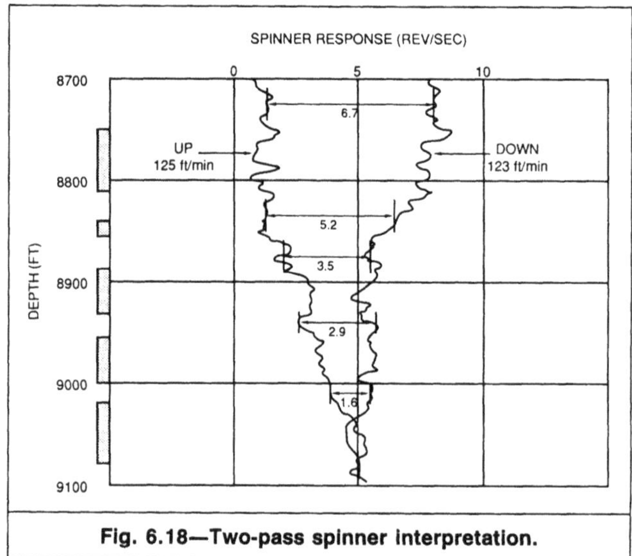

Fig. 6.18—Two-pass spinner interpretation.

The shifted down response, f'_d, is then

$$f'_d = f_d + (f_{us} - f_{ds}), \quad (6.24)$$

$$f'_d = m_n - m_n v_f + m_p v_{Tu} + b_u - m_n, \quad (6.25)$$

or

$$f'_d = m_p v_{Tu} - m_n v_f + b_u. \quad (6.26)$$

Finally, the difference between the up response and the shifted down response, Δf, is

$$\Delta f = f_u - f'_d = m_p + m_p v_f - m_p + m_n v_f \quad (6.27)$$

or, solving for v_f,

$$v_f = \frac{\Delta f}{m_p + m_n}. \quad (6.28)$$

Thus, in the two-pass method, the fluid velocity is calculated by determining the difference in spinner response between up and down runs at several locations in the wellbore. If the response slopes are constant throughout the well (which is assumed for this method) and if the velocity profile correction factor is constant throughout the well, the ratio of the velocity at any point in the well to the velocity above all the perforations yields

$$\frac{q_i}{q_{100}} = \frac{\Delta f_i}{\Delta f_{100}}, \quad (6.29)$$

where q_i and q_{100} = volumetric flow rates at Position i and above all perforations. The flow profile can then be constructed by applying Eq. 6.29 at a number of depth locations.

It is often not possible to log in a region of static fluid. Fill may be just below the lowest perforations, or the wellbore below the lowest production or injection interval may contain mud or other detritus, preventing the use of a spinner flowmeter. In this case, the two-pass method can still be applied by overlaying the up and down passes in the lowest region of the wellbore that can be logged. The fluid velocity at the lowest point that can be logged is then added to the velocity obtained from the two-pass analysis according to

$$v_f = v_\ell + \frac{\Delta f}{m_p + m_n}, \quad (6.30)$$

where v_ℓ = velocity at the lowest depth that can be logged. This velocity obviously must be obtained independently from the two-pass analysis; the multipass method is the best means for determining v_ℓ.

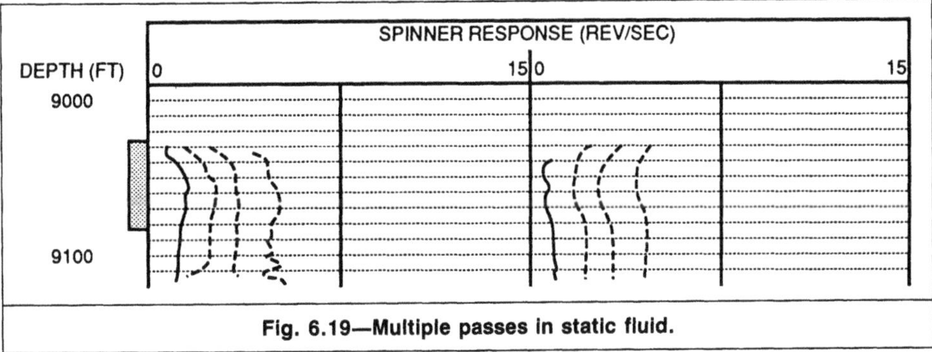

Fig. 6.19—Multiple passes in static fluid.

Recognizing the assumptions and limitations involved in the two-pass method is important. First, the method assumes that the spinner response characteristics, m_p and m_n, are constant throughout the well. Second, the spinner response slopes, m_p and m_n, must still be determined by making multiple passes at some location in the well.

As an example of the two-pass method, consider the hypothetical injection-well logs shown in Fig. 6.16. Fig. 6.16a is a log from a Schlumberger spinner flowmeter that gives positive spinner responses no matter which direction the spinner rotates and Fig. 6.16b is the same log from a tool that senses spinner rotation direction. Because the two-pass method was derived for the Schlumberger log, a log such as that shown in Fig. 6.16b must be converted by taking the absolute value of the down-run (negative) response curve.

To apply the two-pass method, shift the down-run response curve to overlie the up-run response curve in the static-fluid region (see Fig. 6.17). The differences in response curves at several depths are measured, and Eq. 6.28 or 6.29 is applied to define the flow profile. The logs shown here represent a case in which the up and down runs were made at different cable speeds. Most often, the two runs are made at the same cable speed so that the spinner responses measured with a tool that does not sense direction automatically overlie in the region of no flow if the response slopes, m_p and m_n, are equal. If both passes are run at the same cable speed, one step in the interpretation is avoided.

The primary advantages of the two-pass method are that it is quick and that it visually magnifies the spinner response. Another advantage claimed in its development is that it is not affected by changes in wellbore fluid viscosity. This assumes that a viscosity change affects the spinner-response curve by changing only the threshold velocity and not the response curve slope. This is not always valid because fluid viscosity adds a nonlinear dependence on fluid velocity and thus will affect the entire response curve (see Sec. 6.4). Perhaps a more significant factor is that fluid viscosity in the wellbore can be changed drastically only by the influx of a second phase. As will be seen later, two-phase flow may cause severe difficulties in spinner-flowmeter interpretation, no matter what interpretation method is used.

Example—Two-Pass Interpretation, Gas-Condensate Well. Fig. 6.18 shows a two-pass overlay for a well producing 1.85 MMscf/D [0.053×10^6 std m^3/d] of gas and 65 B/D [10.3 m^3/d] of condensate. The well was logged with a Fullbore Flowmeter, so the technique was applied without taking the absolute value of either run because all responses were positive. Choosing stations between the perforated intervals, the differences between the up and down passes shown on the figure and in Table 6.4 were measured. Note that the log is quite noisy and that an average response over a short distance was used. This noisiness is caused by the presence of a significant amount of liquid in the wellbore, even though the production rate of liquid is low.

The flow profile from the two-pass interpretation is constructed by applying Eq. 6.29 at each station; Table 6.4 gives the results. To calculate fluid velocities and volumetric flow rates from the two-pass analysis, obtain the spinner response slopes, m_p and m_n, from multiple runs in some part of the wellbore. Fig. 6.19 shows multiple runs made in the lower part of the wellbore. An in-situ calibration plot for a station at 9,100 ft [2774 m], below the lowest perforations, is shown in Fig. 6.20. The symmetry of the calibration about the origin confirms that there is little or no flow at this depth. Least-squares fits of the up and down response lines yield values of 0.0384 (rev/sec)/(ft/min) [0.0117 (rev/sec)/(m/min)] for m_p and of 0.0382 (rev/sec)/(ft/min) [0.0116 (rev/sec)/(m/min)] for m_n. The fluid velocity at Station 1 can now be calculated with Eq. 6.28.

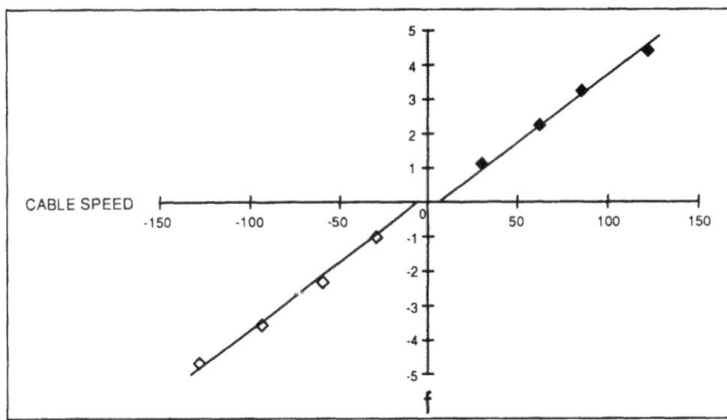

Fig. 6.20—In-situ calibration for station at 9,100 ft [2774 m], two-pass example.

TABLE 6.4—TWO-PASS INTERPRETATION EXAMPLE

Station	Depth (ft)	Δf	Total Flow (%) Two-Pass	Total Flow (%) Multipass
1	8,725	6.7	100	100
2	8,835	5.2	78	85
3	8,875	3.5	52	52
4	8,940	2.9	43	48
5	9,010	1.6	24	33
6	9,100	0.0	0	0

$$v_f=\frac{\Delta f}{m_p+m_n}$$

$$=\frac{6.7\ \text{(rev/sec)}}{(0.0384+0.0382)\text{(rev/sec)/(ft/min)}}$$

$$=87\ \text{ft/min [27 m/min]}.$$

This velocity is near the maximum velocity because the spinner flowmeter was centralized, so the velocity profile correction factor must be applied to obtain the average velocity, $\bar{v}$.

$$\bar{v}=Bv_f$$

$$=0.83(87\ \text{ft/min})$$

$$=72.2\ \text{ft/min [22 m/min]}.$$

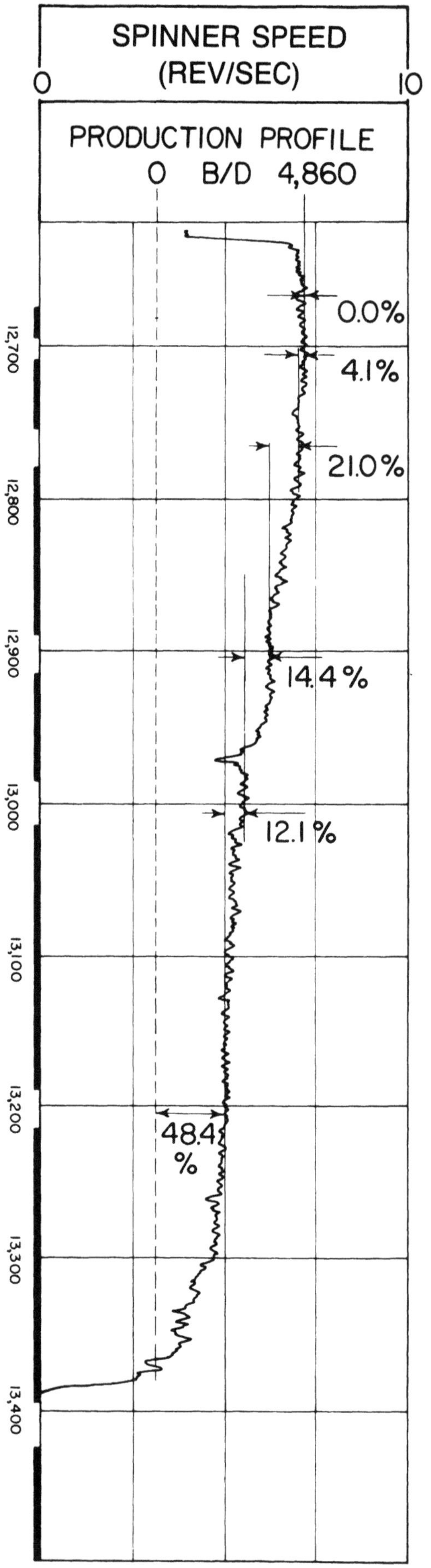

SPINNER SPEED (REV/SEC) 0 10

PRODUCTION PROFILE 0 B/D 4,860

0.0% 4.1% 21.0% 14.4% 12.1% 48.4 %

12,700 12,800 12,900 13,000 13,100 13,200 13,300 13,400

Fig. 6.21—Single-pass spinner interpretation (from Ref. 10).

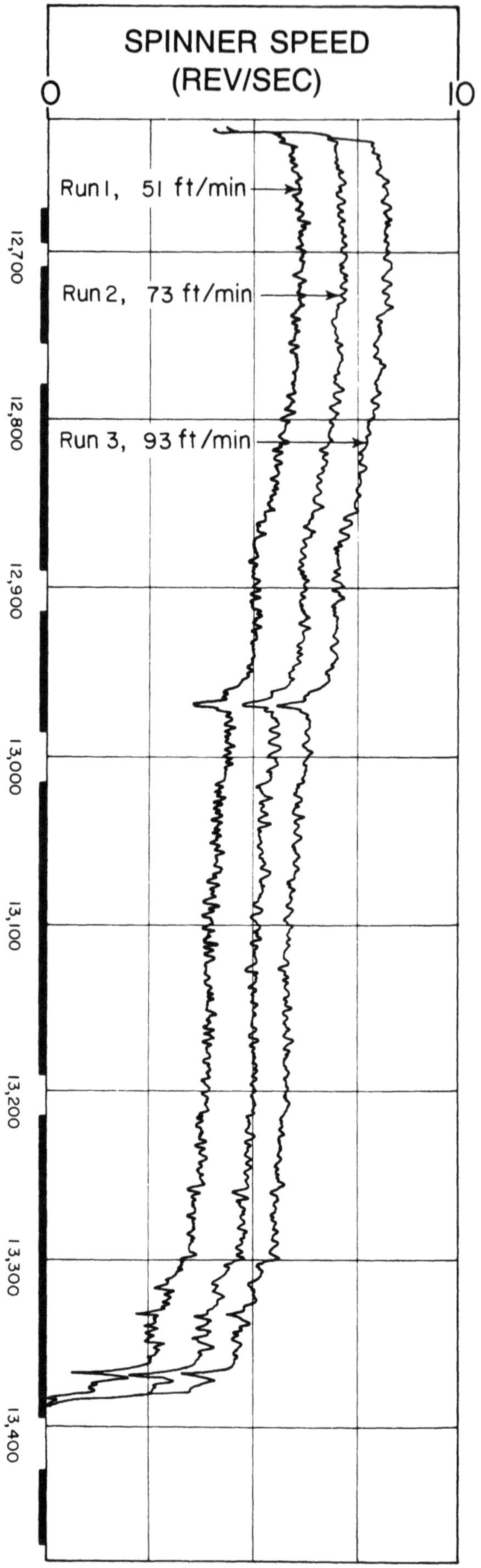

SPINNER SPEED (REV/SEC) 0 10

Run 1, 51 ft/min
Run 2, 73 ft/min
Run 3, 93 ft/min

12,700 12,800 12,900 13,000 13,100 13,200 13,300 13,400

Fig. 6.22—Multiple passes (from Ref. 10).

This well was completed with 7-in. [17.8-cm] casing with a 6.094-in. [15.48-cm] ID, a bottomhole temperature of 175°F [79°C], and a pressure of 1,850 psi [12.8 MPa]. The volumetric rate at standard conditions at Station 1, neglecting the compressibility factor of the gas, is

$$q=\bar{v}A_w$$

$$=(72.2\ \text{ft/min})(0.202\ \text{ft}^2)$$

$$\times\left(\frac{1{,}850\ \text{psia}}{14.7\ \text{psia}}\right)\left(\frac{520°\text{R}}{635°\text{R}}\right)\left(\frac{1{,}440\ \text{minutes}}{\text{day}}\right)$$

$$=2.16\ \text{MMscf/D}\ [0.06\times10^6\ \text{std m}^3/\text{d}].$$

This flow rate agrees reasonably well with the surface production rate of 1.85 MMscf/D [0.05×10^6 std m^3/d].

Complete multiple passes were made on this well so that the multipass interpretation method could be applied. Table 6.4 shows the flow profile obtained with the multipass interpretation, and it agrees reasonably well with the results from the two-pass interpretation. Multiphase-flow effects, as indicated by the noisiness of the log, explain the discrepancies in the two interpretations.

6.5.4 Single-Pass Interpretation. The simplest, but least reliable, method of spinner interpretation uses a single logging run and is based on a linear spinner response to total flow rate. With this method, the highest spinner response (above all perforations) is assigned 100% flow and the lowest spinner response is assumed to be in static fluid and thus is assigned 0% flow. At any point in between, the flow rate is assumed proportional to spinner response, as

$$v_f=v_{100}\left(\frac{f-f_s}{f_{100}-f_s}\right). \qquad (6.31)$$

Thus, the fraction of total flow can be quickly calculated throughout the well. Fig. 6.21 shows an example of this interpretation method. If the spinner response characteristics remain constant throughout the well, this analysis is perfectly valid. Because no in-situ calibration is being performed, however, the log interpreter has no way to evaluate the spinner performance. Also, there is no means to determine whether flow is exiting above the zone logged (such as through a hole in the tubing) with this method because the log is uncalibrated. Finally, choosing the region of static fluid can be difficult and can affect the result significantly.

The log presented by Leach *et al.*[10] (Fig. 6.21) is a good example of the difficulty of the single-pass interpretation. Multiple passes were also run on this well so that the multipass interpretation could be applied and compared with the single-pass interpretation. Fig. 6.22 shows that three down passes were run at 51, 73, and 93 ft/min [15.5, 22.3, and 28.3 m/min]. No up passes were run, so the threshold velocity cannot be obtained in the usual manner; however, threshold velocity can be estimated from down runs in a static-fluid region.

If stations between the perforated intervals, except the lowest station (see Fig. 6.22), are chosen, an in-situ calibration plot for each station (Fig. 6.23) can be used to calculate flow rates. Note that for Station 7, where the spinner response flattens at a depth of about 13,350 ft [4069 m], the calibration line intersects the zero cable speed axis at a negative f value, indicating little or no flow at this depth. Assuming zero flow, the threshold velocity can be estimated as equal to the cable speed axis intercept—in this case about 3 ft/min [0.9 m/min].

The flow profile obtained from the multipass interpretation is compared with that from the single-pass interpretation in Table 6.5. There are significant differences, particularly the interpreted contribution of the bottom layer. This difference is entirely because of the location selected as the zero flow point for the single-pass interpretation; the no-flow point was selected at about 13,370 ft [4075 m] for the single-pass interpretation while the multipass interpretation showed that no flow was occurring at 13,350 ft [4069 m]. The decrease in spinner response below 13,360 ft [4072 m] is likely because of drilling mud, sludge, or other residue in the rathole region or because of slowing of the tool as it neared bottom.

TABLE 6.5—COMPARISON OF SINGLE-PASS AND MULTIPASS INTERPRETATIONS

Production Interval (ft)	Single-Pass Flow (%)	Multipass Flow (%)
12,675 to 12,695	0.0	0
12,710 to 12,756	4.1	6
12,780 to 12,890	21.0	26
12,915 to 12,985	14.4	15
13,015 to 13,190	12.1	19
13,215 to 13,395	48.4	34
13,425 to 13,500	0.0	0

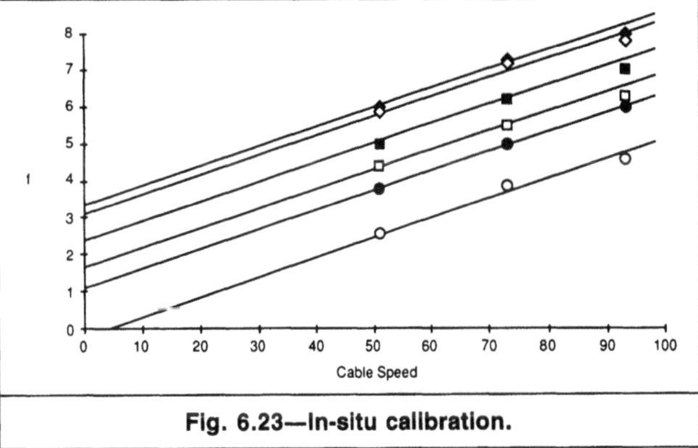

Fig. 6.23—In-situ calibration.

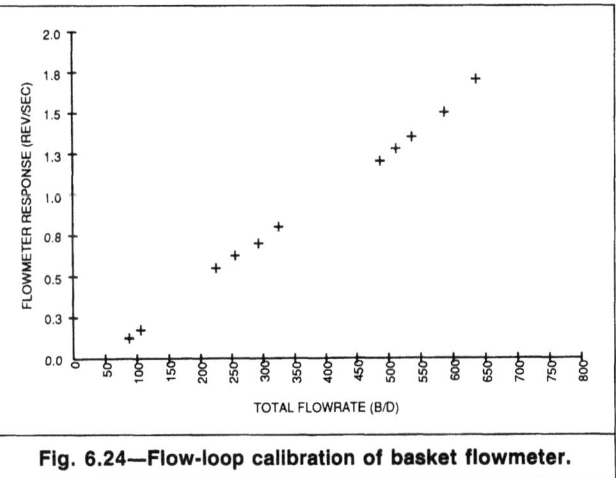

Fig. 6.24—Flow-loop calibration of basket flowmeter.

6.6 Special Tools and Applications

6.6.1 Low-Flow-Rate Wells. If the flow rate in a well is too low, a spinner flowmeter is not sufficiently sensitive for accurate logging. For example, in a well producing 400 B/D [63.5 m^3/d] in a 5-in. [12.7-cm] -ID wellbore, the average velocity above all perforations is 11.4 ft/min [3.5 m/min] and the maximum velocity in the center of the pipe will be about 13 ft/min [4 m/min]. With a typical spinner response of 0.05 (rev/sec)/(ft/min) [0.015 (rev/sec)/ (m/min)], the maximum change in spinner response from the top to the bottom of the well is 0.65 rev/sec. Details of the flow profile are difficult to obtain from such a small change in the spinner response.

A flow-concentrating flowmeter (Fig. 6.3) can extend this range considerably. Fig. 6.24 shows the response characteristics of an Atlas Wireline Services basket flowmeter measured under controlled conditions in a flowloop. A response was obtained at rates below 100 B/D [15.9 m^3/d] and is linear with flow rate above about 100

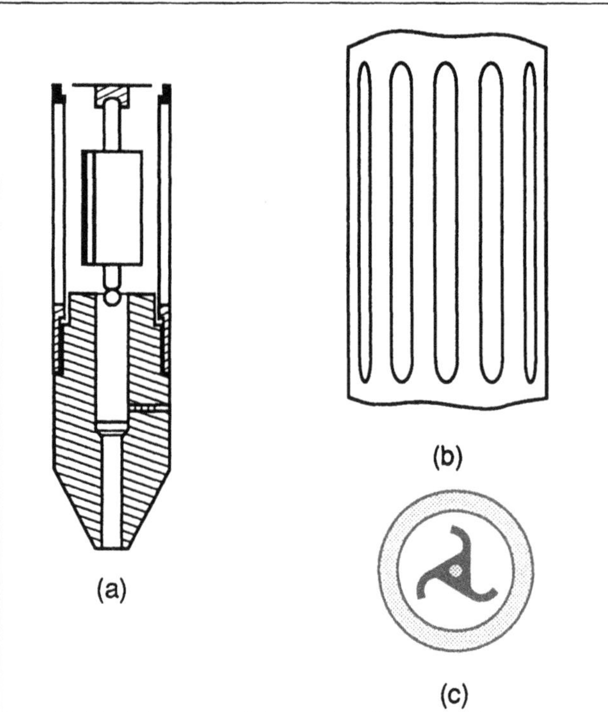

Fig. 6.25—Horizontal spinner (from Ref. 16, courtesy Southwestern Petroleum Short Course).

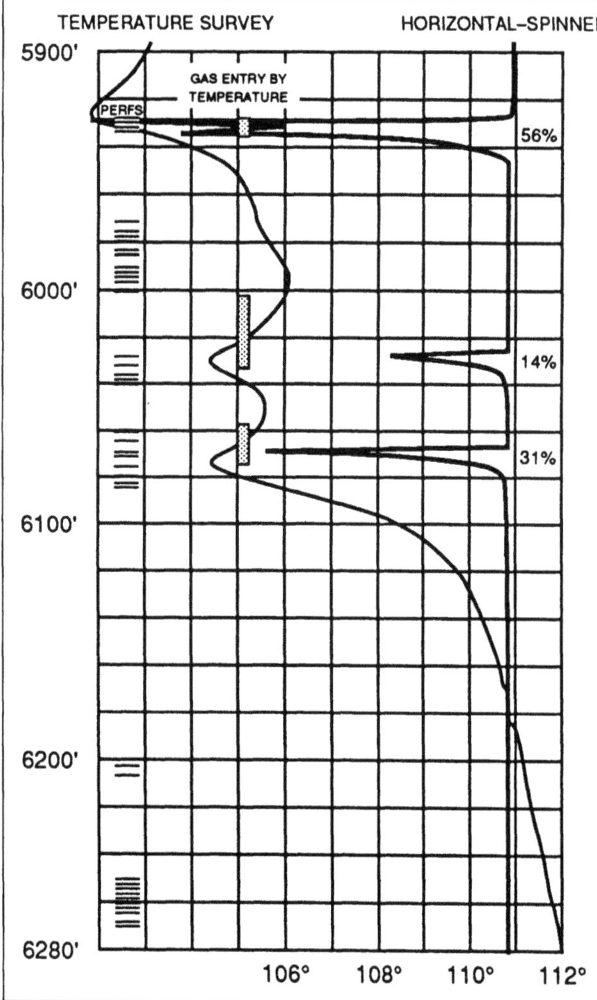

Fig. 6.26—Horizontal spinner response in a gas well (from Ref. 16, courtesy Southwestern Petroleum Short Course).

B/D [15.9 m^3/d]. The low-rate capabilities of a flow-concentrating flowmeter will depend on the internal dimensions of the tool—the smaller the passage for flow inside the tool, the lower the flow rate that can be accurately measured. For measurements at rates below the capabilities of either spinner or flow-concentrating flowmeters, radioactive-tracer logs are the best choice for flow profiling, at least in injection wells.

6.6.2 High-Rate Wells. Wells in which fluid velocities exceed the upper bounds of accuracy for a spinner flowmeter are occasionally encountered. This is most likely to occur in high-flow-rate gas wells. In some instances, the rotational velocity of the spinner can be high enough to damage the bearings that suspend the spinner. In such a case, log response can sometimes be improved by shielding part of the spinner from the flow stream by covering part of the cage housing the spinner. A surface calibration of such a shielded spinner would be necessary to ensure that it maintained a linear response to fluid velocity.

6.6.3 Horizontal Spinners. A novel approach that has been taken to flow profiling is the horizontal spinner.[16] This device (Fig. 6.25) contains a turbine blade turned 90° from the normal spinner orientation so that it will respond to the horizontal flow issuing from a perforation rather than the vertical fluid flow in the wellbore. Fig. 6.26 shows a horizontal spinner log in a gas well, for which the spinner response agrees qualitatively with a temperature log. The application of a horizontal spinner would be restricted primarily to gas wells, because the velocity from a perforation producing liquid is fairly low, and to horizontal wells. In a horizontal wellbore, the perforations are usually along the low side of the casing, so a horizontal spinner that is run decentralized may pass directly over the perforations. Multiphase flow in the wellbore would greatly complicate the response of such a device because substantial horizontal flow components exist in many multiphase flow regimes. Quantitative analysis of horizontal spinner responses would be difficult even in a single-phase flow because the velocity field of a perpendicular jet entering a turbulent flow stream is complex and not well understood.[17]

6.6.4 Spinner Flowmeters in Openhole Wells. A spinner flowmeter measures fluid velocity, not volumetric flow rate, and as such, changes in wellbore cross-sectional area must be accounted for in flow profile interpretation. Thus, interpretation of a spinner flowmeter log in an irregular borehole requires that a caliper log be run in conjunction with the spinner so that the proper cross-sectional areas can be used in calculating volumetric flow rate from spinner-obtained velocities. Downstream of a diameter increase, the velocity measured will not be representative of the volumetric flow rate because of the jetting effect that occurs (see Sec. 5.4.3 and Fig. 5.19).

The single-pass interpretation method cannot be used in an irregular borehole because it tacitly assumes a constant cross-sectional area.

6.7 Guidelines for Running and Interpreting Spinner Flowmeters

A properly run and interpreted spinner-flowmeter log should provide a reliable flow profile for a single-phase well. The following guidelines point out procedures that should be followed to ensure log quality.

6.7.1 Running Spinner Flowmeters.

1. Well conditions must be suitable for a spinner flowmeter. As a minimum, the well fluids should be clean (no entrained solids), and the rate should be as stable as possible.
2. Multiphase-flow effects often render a spinner flowmeter useless. Unless the multiphase-stream flow rate is large, other devices are needed for velocity measurements (see Chap. 9).
3. A spinner flowmeter should be checked thoroughly on the surface before it is run in the well. The impeller should rotate freely and all electronics should operate as expected.
4. A spinner flowmeter should always be run centralized.

5. Multiple passes, at several different cable speeds and in both up and down directions, should be made across all zones of interest.
6. Stationary readings should be taken at several wellbore locations.
7. A caliper log is needed if the wellbore cross-sectional area is not constant.
8. Repeat runs should always be made to assess well stability and tool performance.

6.7.2 Interpreting Spinner-Flowmeter Logs.
1. The multipass method of interpretation should be used whenever significant lengths of wellbore exist without fluid exits.
2. The two-pass interpretation is useful for visually accentuating the spinner-flowmeter response. The multipass method should be applied to confirm two-pass interpretations.
3. Single-pass interpretation alone is not recommended.
4. The volumetric flow rate above all zones taking or producing fluid should be calculated from the log and compared with surface flow conditions.
5. Spinner-response characteristics obtained from the log interpretation, such as the response slope and threshold velocity, should be close to that predicted by the tool supplier.
6. Stationary measurements are useful, particularly in openhole completions, where fluid entries may be scattered over the entire completion.

Nomenclature

A_w = wellbore cross-sectional area, ft^2 [m^2], L^2
b_d = spinner response intercept (down run), rev/sec, rev/t
b_u = spinner response intercept (up run), rev/sec, rev/t
B = velocity profile correction factor
C = constant, rev/ft [rev/m], rev/L
f = spinner response, rev/sec, rev/t
f_d = spinner frequency response for down run, rev/sec, rev/t
f_d' = shifted spinner response (down run), rev/sec, rev/t
f_{ds} = spinner response in static fluid (down run), rev/sec, rev/t
f_i = ideal spinner response, rev/sec, rev/t
f_u = spinner response for up run, rev/sec, rev/t
f_{us} = spinner response in static fluid (up run), rev/sec, rev/t
f_s = spinner response in static fluid, rev/sec, rev/t
f_0 = spinner response at $v_T=0$
f_{100} = spinner response above all perforations, rev/sec, rev/t
Δf = difference between up and down spinner responses, two-pass method, rev/sec, rev/t
Δf_i = difference between up and down spinner responses, two-pass method, at Location i, rev/sec, rev/t
Δf_{100} = difference between up and down spinner responses, two-pass method, above all perforations, rev/sec, rev/t
F_T = retarding torque, lbf-ft [kJ], mL^2/t^2
G = constant
h_{bl} = spinner blade thickness, ft [m], L
K_D = blade drag coefficient, rev/sec, rev/t
l_{bl} = blade length, ft [m], L
m_n = spinner response slope for negative responses, rev/ft [rev/m], rev/L
m_p = spinner response slope for positive responses, rev/ft [rev/m], rev/L
N_{Rebl} = blade Reynolds number
p = bottomhole pressure, psi [kPa], m/Lt^2
p_{sc} = standard pressure (14.7 psi), psi [kPa], m/Lt^2
q = volumetric flow rate, ft^3/sec [m^3/s], L^3/t
q_{bh} = volumetric flow rate at bottomhole conditions, ft^3/sec [m^3/s], L^3/t
q_i = volumetric flow rate at Location i, ft^3/sec [m^3/s], L^3/t
q_{sc} = volumetric flow rate at standard conditions, ft^3/sec [m^3/s], L^3/t
q_{100} = volumetric flow rate above all perforations, ft^3/sec [m^3/s], L^3/t
r = radial distance from center of spinner blade, ft [m], L
R = least-squares regression coefficient, dimensionless
T_{bh} = bottomhole temperature, °R [K], T
T_{sc} = standard temperature (60°F), °R [K], T
v_{bl} = spinner blade velocity, ft/sec [m/s], L/t
v_e = effective velocity, ft/sec [m/s], L/t
v_f = fluid velocity, ft/sec [m/s], L/t
v_ℓ = velocity at lowest position logged, ft/sec [m/s], L/t
v_t = threshold velocity, ft/sec [m/s], L/t
v_T = tool velocity, ft/sec [m/s], L/t
v_{T0} = tool velocity at $f=0$, ft/sec [m/s], L/t
$\bar{v}$ = average velocity, ft/sec [m/s], L/t
v_{100} = velocity above all perforations, ft/sec [m/s], L/t
α_{bl} = spinner blade angle, degrees, radians
μ_f = fluid viscosity, lbm/ft-sec [Pa·s], m/Lt
ρ_f = fluid density, lbm/ft^3 [kg/m^3], m/L^3

References

1. Rumble, R.C.: "A Subsurface Flowmeter," *Trans.*, AIME (1955) **204,** 258-61.
2. Morgan, F., Reed, D.W., and Gray, L.L.: "Meter for Measuring Distribution of Gas Flow in Well Bores," *Trans.*, AIME (1947) **174,** 253-68.
3. Vincent, R.P., Leibrock, R.M., and Ziemer, C.W.: "Well Flowmeter for Logging Producing Ability of Gas Sands," *Trans.*, AIME (1947) **174,** 305-14.
4. Piety, R.G. and Wiley, B.F.: "Flowmeter for Water Injectivity Profiling," *World Oil* (1952) **134,** No. 6, 176-88.
5. Pfister, R.J.: "An Improved Water-input Profile Instrument," *Trans.*, AIME (1947) **174,** 269-85.
6. Dale, C.R.: "Bottom Hole Flow Surveys for Determination of Fluid and Gas Movements in Wells," *Trans.*, AIME (1949) **186,** 205-10.
7. Riordan, M.B.: "Surface Indicating Pressure, Temperature, and Flow Equipment," *Trans.*, AIME (1951) **192,** 257-62.
8. Newman, J.L., Waddell, C., and Sauder, H.L.: "A Flowmeter for Measuring Subsurface Flow Rates," *Trans.*, AIME (1956) **207,** 316-19.
9. Hill, A.D. and Oolman, T.: "Production Logging Tool Behavior in Two-Phase Inclined Flow," *JPT* (Oct. 1982) 2432-40.
10. Leach, B.D. *et al.*: "The Full Bore Flowmeter," paper SPE 5089 presented at the 1974 SPE Annual Meeting, Houston, Oct. 6-9.
11. Carlson, N.R. and Johnston, M.: "Importance of Production Logging Suites in Multi-phase Flow," *Proc.*, SPWLA Annual Logging Symposium, Calgary (1983).
12. McKinley, R.M.: "Production Logging," paper SPE 10035 presented at the 1982 SPE Intl. Pet. Exhibition and Technical Symposium, Beijing, March 18-26.
13. Peebler, B.: "Multipass Interpretation of the Full Bore Spinner," Schlumberger, Houston (1982).
14. *Production Log Interpretation*, Schlumberger, Houston (1973) 13-14.
15. *Interpretative Methods for Production Well Logs*, second edition, Atlas Wireline Services, Western Atlas Intl. Inc., Houston (1982) 28.
16. Kading, H.W.: "Horizontal-Spinner, A New Production Logging Technique," *Proc.*, Southwestern Petroleum Short Course, Lubbock, TX (1975) 145-49.
17. Howarth, K.B.: "Radioactive Tracer Logging in Turbulent Flow," MS thesis, U. of Texas, Austin (1986).

SI Metric Conversion Factors

bbl	× 1.589 873	E−01	=	m^3
ft	× 3.048*	E−01	=	m
ft^3	× 2.831 685	E−02	=	m^3
°F	(°F−32)/1.8		=	°C
in.	× 2.54*	E+00	=	cm
lbm	× 4.535 924	E−01	=	kg
°R	× 5/9		=	K

*Conversion factor is exact.

Chapter 7
Single-Phase Flow Profiling Summary

7.1 Introduction

Among the most common and straightforward applications of production logging is the measurement of flow profiles in single-phase-flow wells, primarily injection wells. In Chaps. 4 through 6, the responses of temperature, radioactive-tracer, and spinner-flowmeter logs to a single-phase flow were discussed in detail, but separately. Yet the interpretation of the flow profile in a single-phase well can often be enhanced by running two or more of these logs in conjunction, if for no other reason than to verify log quality because different logs should be consistent in their responses to well conditions. In this chapter, the behavior of temperature, radioactive-tracer, and spinner-flowmeter logs in a single-phase flow are compared to illustrate the synergy that results from a combination of these logs.

7.2 Comparison of Temperature, Radioactive-Tracer, and Spinner-Flowmeter Logs

With temperature, radioactive-tracer, and spinner-flowmeter instruments available, two, and sometimes all three, of these logs typically will be run to measure the flow profile in a single-phase well. The most common combinations are temperature and radioactive-tracer logs, temperature and spinner-flowmeter logs, and all three logs. Furthermore, when a radioactive-tracer log is used, two different logs—the tracer-loss and velocity-shot—are usually run. Thus, the engineer has four logs to compare in determining the flow profile—temperature, tracer-loss, velocity-shot, and spinner-flowmeter logs. With a combination tool string, such as that shown in Fig. 7.1, all these logs can be obtained with a single trip in the hole.

Among these logs, the temperature and tracer-loss logs provide qualitative information about the flow profile but cannot be relied on for an accurate analysis of flow distribution; the velocity-shot and spinner-flowmeter logs yield quantitative results about the flow profile. The temperature and tracer logs can, however, respond to flow both inside and outside the casing; spinner-flowmeter logs are sensitive only to fluid movement in the wellbore. To assess the flow profile properly, a log responding to flow outside the wellbore is needed in conjunction with a log that accurately measures flow inside the well. Running two different logs that essentially measure the same thing, such as velocity-shot and spinner-flowmeter logs, provides a check on the quality of both logs. Thus, at least two logs should be run to measure a flow profile—either a temperature or a tracer-loss log to sense fluid movements outside the wellbore with either a velocity-shot or spinner-flowmeter log for an accurate measure of flow in the wellbore.

Examining the responses of these logs in a series of hypothetical injection wells illustrates the relationships among them. Four cases are considered—(1) a "normal" well, with fluid entering the formation opposite two perforated intervals; (2) a well with a channel upward from the upper perforations to a zone higher up the well; (3) a well with a channel down from the lower perforations to a lower zone; and (4) a well with a casing leak below the lowest perforations.

Normal Well. With all flow exiting through perforations and no channels present, all logs should give a similar picture of the exit profile. First examine the temperature logs (Fig. 7.2). The flowing temperature log is useful primarily for indicating the lowest point of fluid injection. A shut-in temperature log identifies the gross injection interval in a mature well; in a young well, some discrimination between the two injection intervals is seen. Thus, the temperature logs provide a qualitative indication of the flow profile. The velocity-shot and spinner-flowmeter logs (Fig. 7.3) provide the best quantitative flow profile. Note that they agree well, with small discrepancies resulting from errors in the velocity-shot analysis when the tool is opposite an injection interval. The tracer-loss log differs significantly in some locations from the velocity-shot or spinner-flowmeter logs because of the poor depth resolution of the tracer-loss log. Note that the difference between the tracer-loss and the velocity-shot or spinner-flowmeter logs between the perforated intervals results not from channeling but from the inaccuracy of the tracer-loss log.

Upward Channel. The utility of the temperature log becomes apparent when a well with a channel is logged (Fig. 7.4). The temperature log with the well flowing is useful primarily to locate the bottom of injection because the slight inflection at the top of the injection interval is often indiscernible. The shut-in log, however, shows fluid injection above the upper perforations, which must be caused by either a channel or a casing leak. The logs responding to flow in the wellbore, velocity-shot or spinner-flowmeter (Fig. 7.5), distinguish between these two possibilities. Because flow does not exit the wellbore above the upper perforations, a channel outside the casing must exist. The tracer-loss log, when analyzed in the usual manner to determine the flow profile, does not locate the channel. Only if the tracer slug is logged through carefully will this upper channel be detected with the tracer-loss log (see Sec. 5.3.3).

Downward Channel. A channel downward from the lowest perforations will be clearly indicated on the flowing and shut-in temperature logs when it is possible to log below the perforations. As seen in Fig. 7.6, the temperature logs do not break toward the geothermal temperature until well below the lowest perforations, indicating a downward channel or a casing leak below the perforations. The velocity-shot or spinner-flowmeter logs again resolve these two possibilities (Fig. 7.7). The velocity-shot and spinner-flowmeter logs find no flow in the wellbore below the lowest perforations, so the downward flow indicated by the temperature log must be occurring outside the casing. The tracer-loss log responds to this downward channel because the movement of the tracer slug through the channel can often be detected. Comparison of the tracer-loss with the velocity-shot or spinner-flowmeter logs confirms that the flow below the lowest perforations must be outside the casing.

Casing Leak. Fig. 7.8 shows the temperature logs for a well with a casing leak below the lowest set of perforations. These logs are identical to the temperature logs for the well with a downward channel because the temperature response will be similar whether

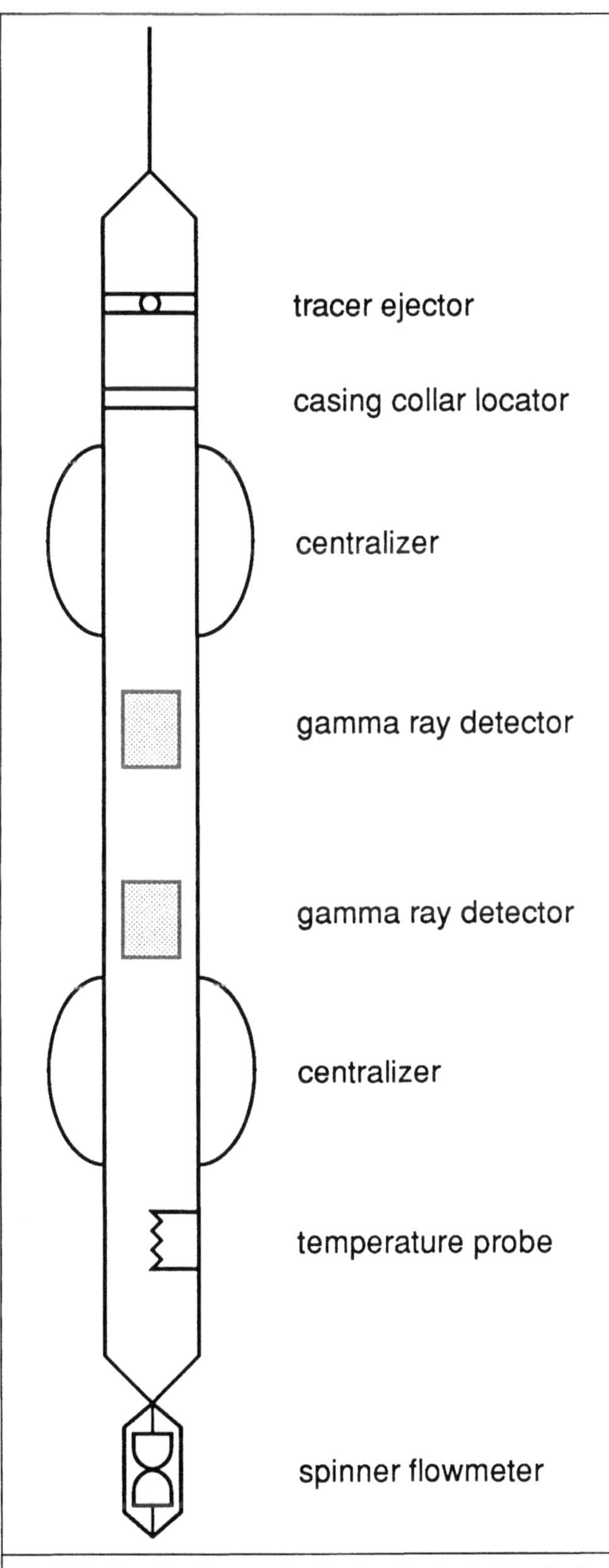

Fig. 7.1—Combination production logging tool string for injection profiling.

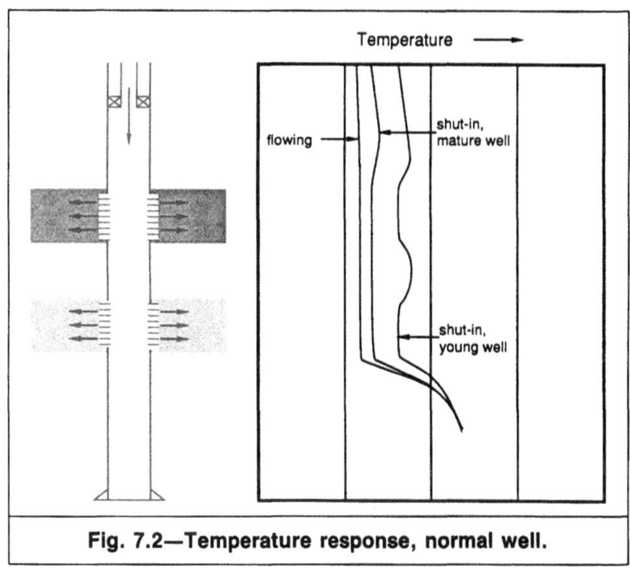

Fig. 7.2—Temperature response, normal well.

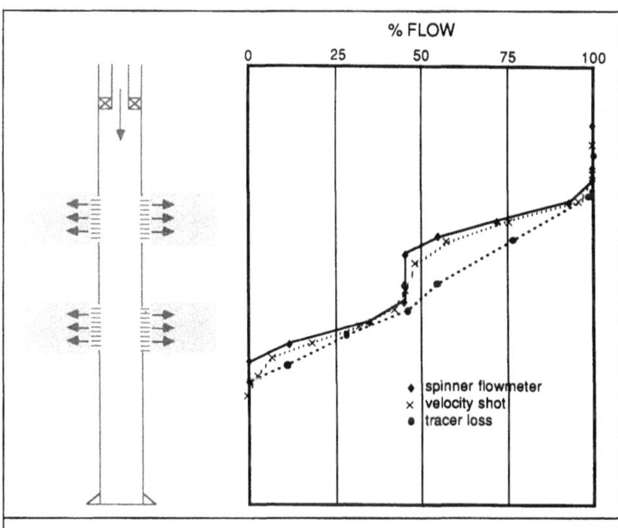

Fig. 7.3—Spinner-flowmeter and radioactive-tracer responses, normal well.

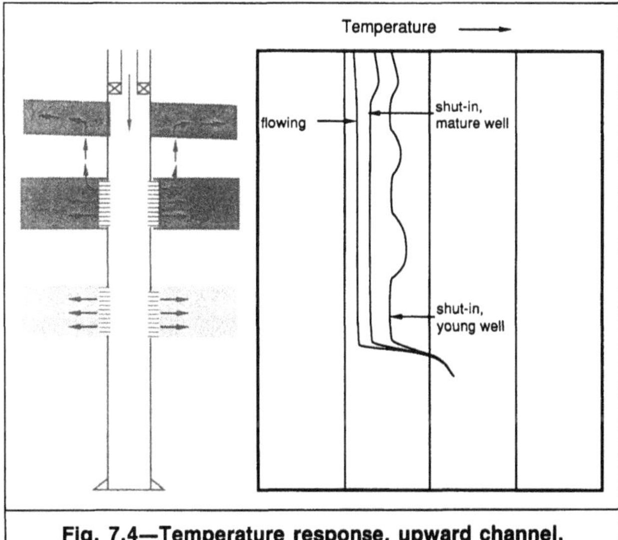

Fig. 7.4—Temperature response, upward channel.

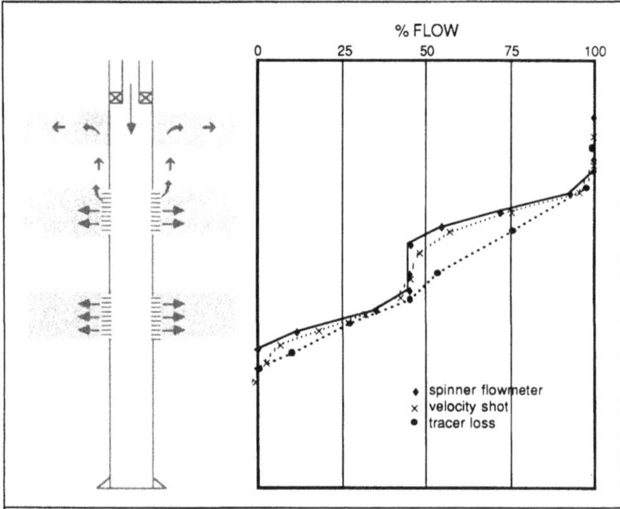

Fig. 7.5—Spinner-flowmeter and radioactive-tracer responses, upward channel.

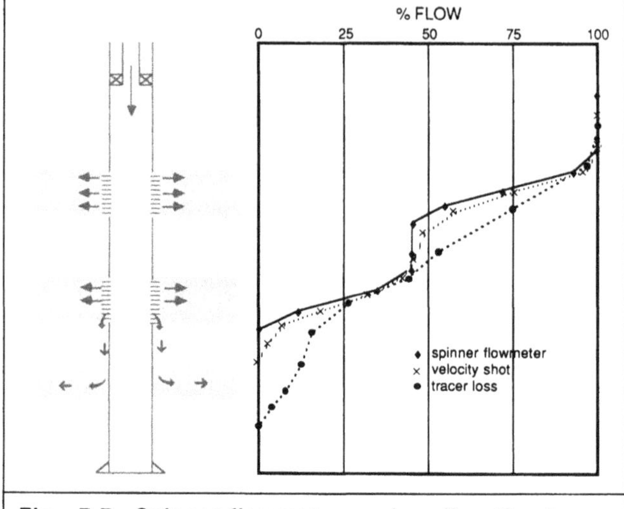

Fig. 7.7—Spinner-flowmeter and radioactive-tracer responses, downward channel.

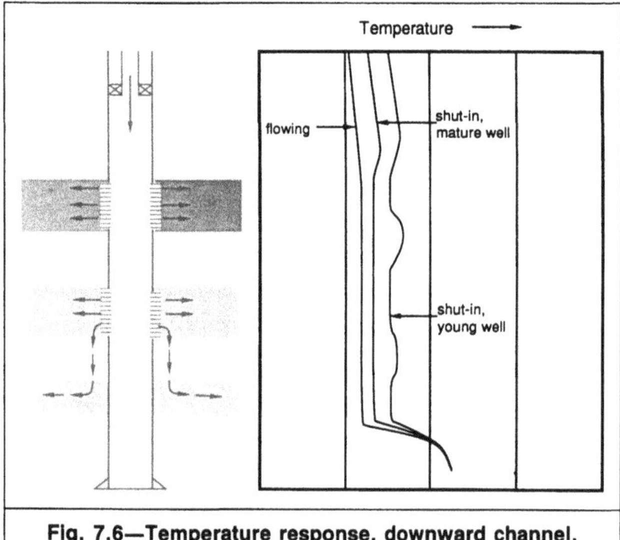

Fig. 7.6—Temperature response, downward channel.

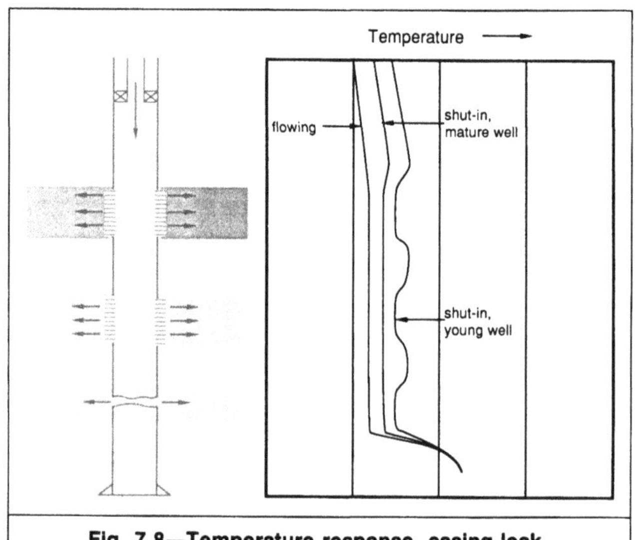

Fig. 7.8—Temperature response, casing leak.

the flow is inside or outside the casing. The velocity-shot and spinner-flowmeter logs (Fig. 7.9) prove that the flow is from a leak in the casing because they detect flow in the wellbore below the lowest perforations to the location of the leak. The tracer-loss log agrees with the velocity-shot or spinner-flowmeter log within its accuracy limits.

These hypothetical examples illustrate that a combination of a log that detects flow outside the casing and one that responds only to flow inside the casing is required to determine conclusively conditions in a well with channeling or casing leaks. Thus, as a minimum, two logs should be run to measure a single-phase flow profile.

7.3 Log Examples

A few examples of multiple logs run in actual injection wells illustrate the relationships among temperature, radioactive-tracer, and spinner-flowmeter logs.

Example 1—Mature Injection Well. The flow profiles interpreted from velocity-shot, tracer-loss, and spinner-flowmeter logs for a well that was injecting 1,600 B/D [255 m^3/d] of water for 20 years are shown in Fig. 7.10. Note that fill covered the lower part of the perforations, making absolute location of the lowest point of injection impossible. The flow profiles interpreted from all three logs agree well. The only apparent discrepancy is the deviation of the tracer-loss results from the other two curves above the perforations. This is caused by the sparsity of the tracer-loss data because the tracer slug was logged through only once (at 5,050 ft [1540 m]) before tracer started exiting the wellbore. The temperature logs for this well (Fig. 7.11) are typical of a mature injection well. Little information is obtained with the dynamic temperature log (well flowing) because fill in the wellbore prevented logging to the bottom of injection. The shut-in log gives a general indication of the top of injection by the gradual shift to higher temperatures above the perforations. It appears that over the life of the well, most of the water has entered the formation below 5,100 ft [1555 m] because there is little separation between the flowing and shut-in logs over this interval. Note that the radioactive-tracer log indicates that about 30% of the flow is now entering the top set of perforations (5,080 to 5,100 ft [1550 to 1555 m]). A comparison of the temperature and tracer results suggests that the profile of this well has changed recently, with the top zone now receiving more injection.

Example 2—Injection Well With Flow Below Perforations. The flow profiles and logs depicted in Fig. 7.12 illustrate the utility of the temperature log, even in an old injection well. In this well, the flow profile interpreted from the tracer-loss log showed flow continuing downward below the lowest indicated perforations, but the tracer slug became too indistinct to locate the bottom of injection. The flowing temperature log, however, indicates by its sharp break toward the geothermal temperature that the lowest injection is at about 4,698 ft [1432 m]. Thus, the downward flow suggested by

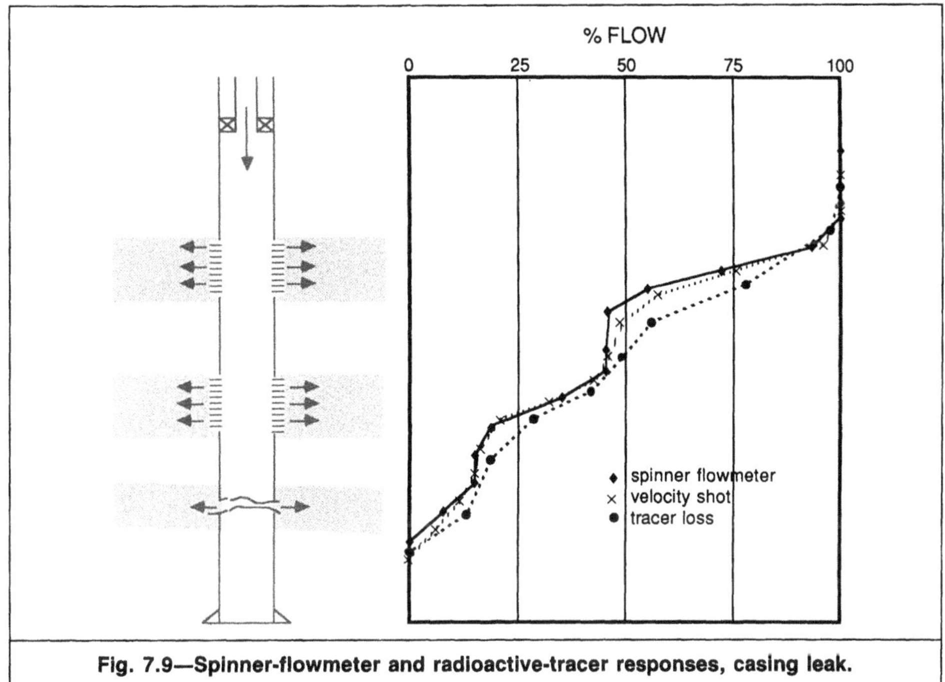

Fig. 7.9—Spinner-flowmeter and radioactive-tracer responses, casing leak.

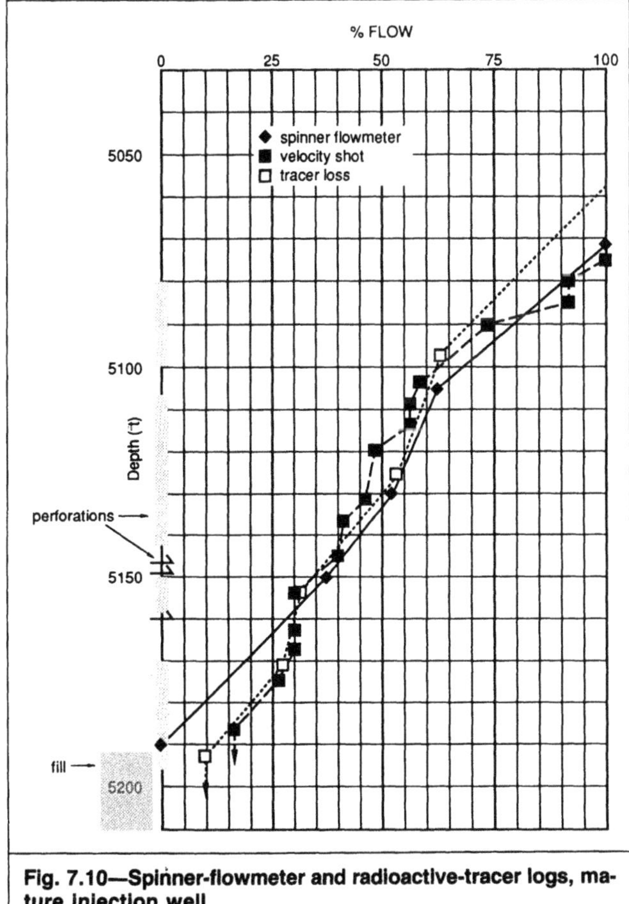

Fig. 7.10—Spinner-flowmeter and radioactive-tracer logs, mature injection well.

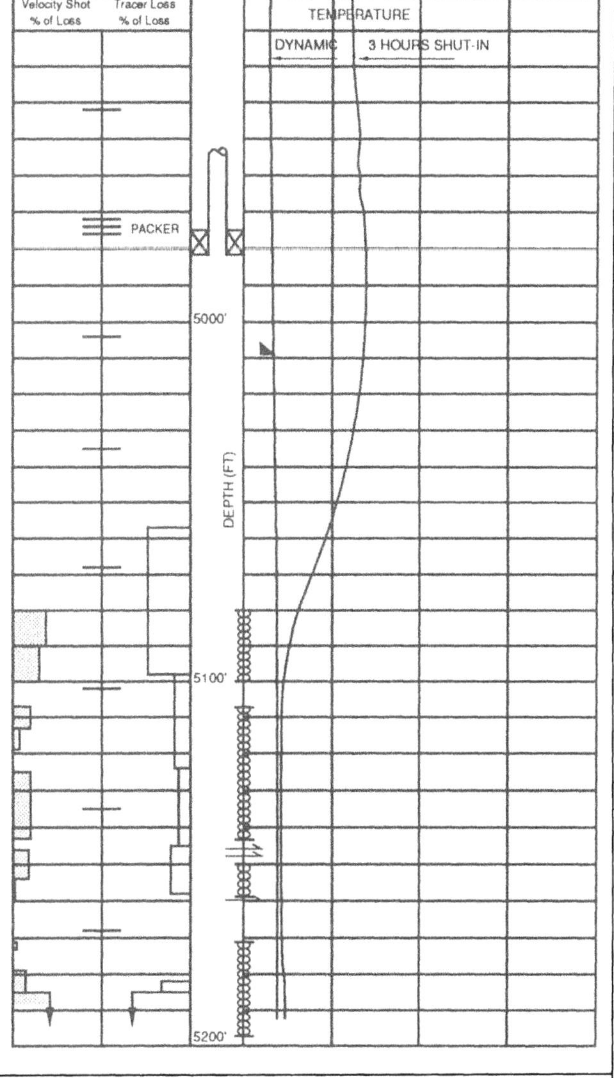

Fig. 7.11—Temperature log, mature injection well.

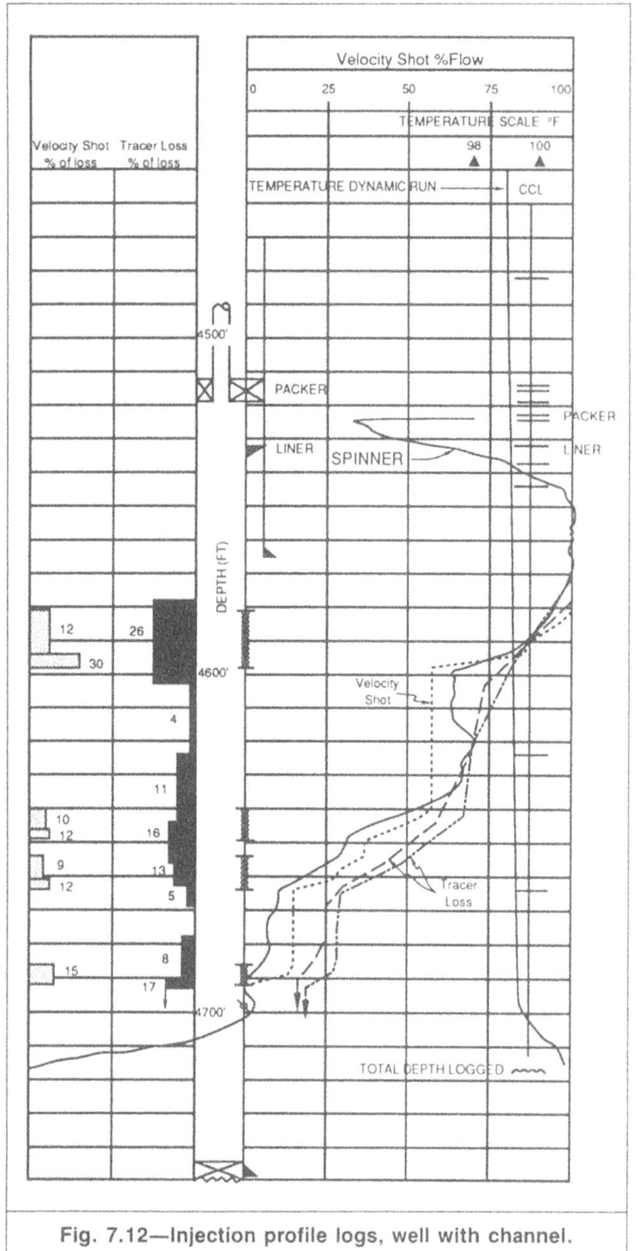

Fig. 7.12—Injection profile logs, well with channel.

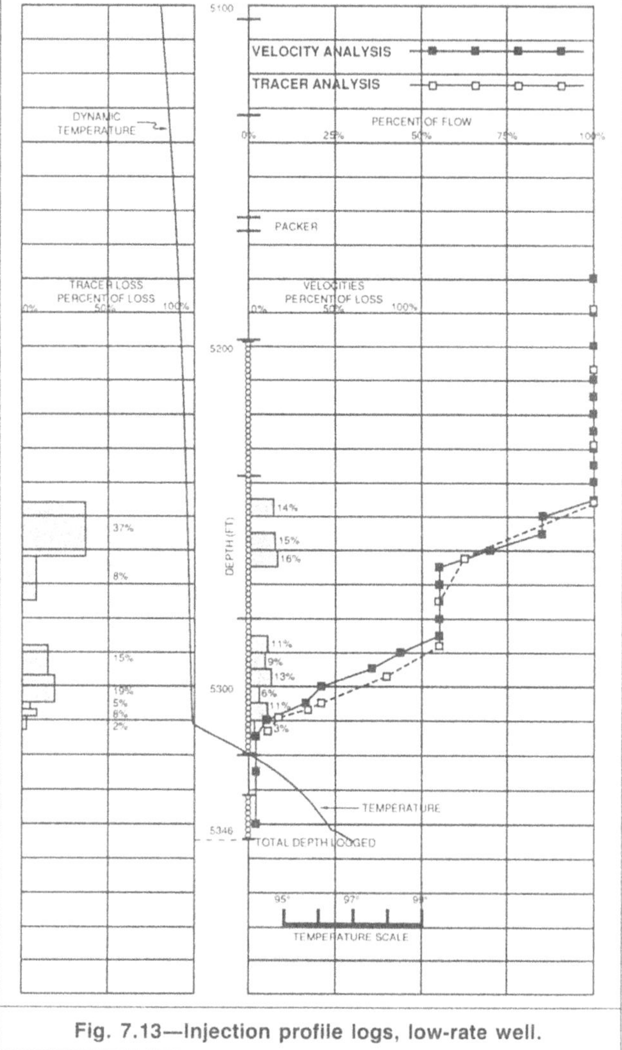

Fig. 7.13—Injection profile logs, low-rate well.

the tracer-loss log extends only a few feet below the perforations. The spinner-flowmeter log shown here is simply the raw spinner response from one pass in the well that is laid over the tracer results to show general agreement. Data taken were insufficient to interpret the spinner-flowmeter log properly. The flow below the lowest perforations could be occurring in the wellbore or through a channel. Carefully run velocity-shot or spinner-flowmeter logs in the bottom part of the well would determine whether the flow is inside or outside the casing.

Example 3—Low-Rate Injection Well. The well pictured in Fig. 7.13 had been on injection for several years, but always at a rate less than 200 B/D [32 m^3/d]. The velocity-shot and tracer-loss logs on this well agree as well as can ever be expected (within 15% throughout the well), particularly because the flow in much of the well was laminar, causing considerable tracer dispersion. Both these interpretations show a minor amount of flow continuing down to the lowest perforations. This is confirmed by the dynamic temperature log, which, because of the low injection rate, is similar to what would be expected in a young well. The sharp break to higher temperatures at 5,312 ft [1619 m] identifies this location as the bottom of the major injection interval. A second inflection at 5,342 ft [1628 m] suggests that the minor fluid injection from the bottom perforations enters the formation at this point.

SI Metric Conversion Factors

bbl	× 1.589 873	E−01	= m^3
ft	× 3.048*	E−01	= m

*Conversion factor is exact.

Chapter 8
Multiphase Flow in Pipes

8.1 Introduction

Production logging in a multiphase flow is much more difficult than logging in a single-phase flow because of the complexity of the multiphase flow. Multiphase flow presents two major difficulties in the measurement of flow-stream properties: the nonuniform distribution of the phases across the pipe and the variation of flow properties with time at any given location in the pipe. Thus, understanding the responses of production logging tools in multiphase flow and their interpretation requires an understanding of multiphase flow itself—e.g., the general structure of two- and three-phase flow, the relative motion of the phases, and other factors that will influence production logging. This chapter summarizes the fundamentals of multiphase flow. For more thorough treatment, refer to Refs. 1 through 5. Most discussions in this chapter are restricted to two-phase flow, with extensions to three-phase flow suggested.

8.2 The Holdup Phenomenon

Production log interpretation is generally based on the assumption that the logging tools measure average properties of the flow stream, such as average velocity or average density. Thus, the production logging measurements provide information about such in-situ flow conditions as the fraction of each phase present in the wellbore. Yet, in two-phase flow, the amount of pipe occupied by a phase is often different from its proportion of the total volumetric flow rate. An understanding of this phenomenon is central to understanding production logging interpretation in two-phase flow.

As an example of a typical two-phase flow situation, consider the upward flow of two phases, α and β, where Phase α is less dense than Phase β (Fig. 8.1). In upward two-phase flow, the lighter phase (α) typically will be moving faster than the denser phase (β), primarily because of buoyancy. Because of this fact, called the *holdup phenomenon*, the in-situ volume fraction of the denser phase will be greater than the input volume fraction of the denser phase—i.e., the denser phase is "held up" in the pipe relative to the lighter phase.

This relationship is quantified by defining a holdup parameter, y:

$$y_\beta=\frac{V_\beta}{V}, \quad \text{(8.1)}$$

where V_β=volume of denser phase in pipe segment and V= volume of pipe segment. If holdup does not change longitudinally,

$$y_\beta=\frac{A_\beta}{A}, \quad \text{(8.2)}$$

where A_β=pipe cross-sectional area occupied by Phase β and A=pipe cross-sectional area. y_β can also be defined in terms of a local holdup, $y_{\beta\ell}$:

$$y_\beta=\frac{1}{A}\int_0^A y_{\beta\ell}\,dA. \quad \text{(8.3)}$$

$y_{\beta\ell}$ is a time-averaged quantity—i.e., it is the fraction of the time a given location in the pipe is occupied by Phase β. The holdup of the lighter phase, y_α, is defined exactly like y_β:

$$y_\alpha=\frac{A_\alpha}{A}, \quad \text{(8.4)}$$

or, because the pipe is completely occupied by the two phases,

$$y_\alpha=1-y_\beta. \quad \text{(8.5)}$$

In gas/liquid flow, the holdup of the gas phase, y_α, is sometimes called the *void fraction*.

Another type of parameter used to describe two-phase flow is the input volume fraction, f:

$$f_\beta=\frac{q_\beta}{q_\alpha+q_\beta} \quad \text{(8.6)}$$

and

$$f_\alpha=1-f_\beta, \quad \text{(8.7)}$$

where q_α and q_β=volumetric flow rates of the two phases. f_α and f_β are also referred to as *no-slip holdups*.

In two-phase flow, the input volume fraction of a phase usually is not equal to the holdup:

$$f_\beta \neq y_\beta. \quad \text{(8.8)}$$

Slip velocity, v_s, is another measure of the holdup phenomenon that is commonly used in production log interpretation. Slip velocity is the difference between the average velocities of the two phases. Thus,

$$v_s=\bar{v}_\alpha-\bar{v}_\beta, \quad \text{(8.9)}$$

where $\bar{v}_\alpha$ and $\bar{v}_\beta$=average in-situ velocities of the two phases. Slip velocity is not a property independent of holdup, but simply another way to represent the holdup phenomenon. To show the relationship between holdup and slip velocity, we introduce the definition of superficial velocity, $v_{\alpha s}$ or $v_{\beta s}$:

$$v_{\alpha s}=\frac{q_\alpha}{A} \quad \text{(8.10)}$$

and

$$v_{\beta s}=\frac{q_\beta}{A}. \quad \text{(8.11)}$$

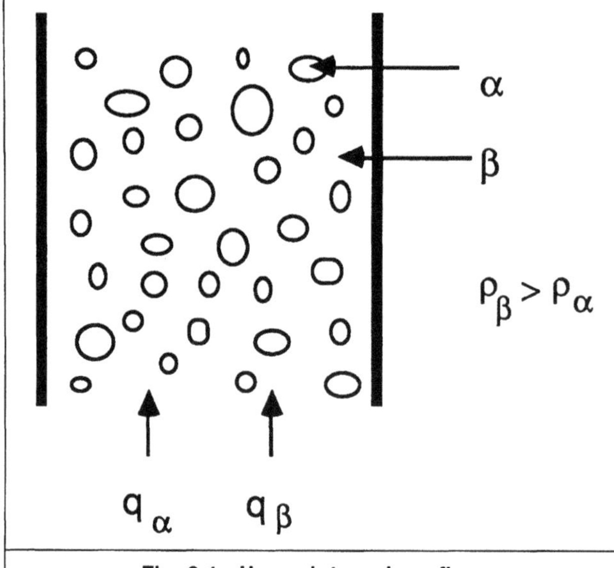

Fig. 8.1—Upward, two-phase flow.

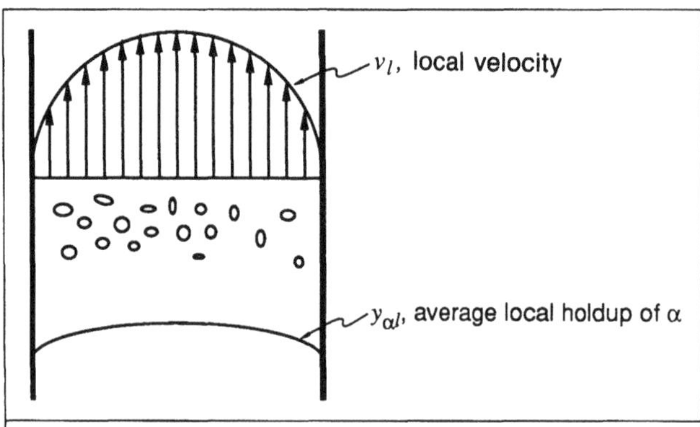

Fig. 8.2—Velocity and local holdup profiles in two-phase flow.

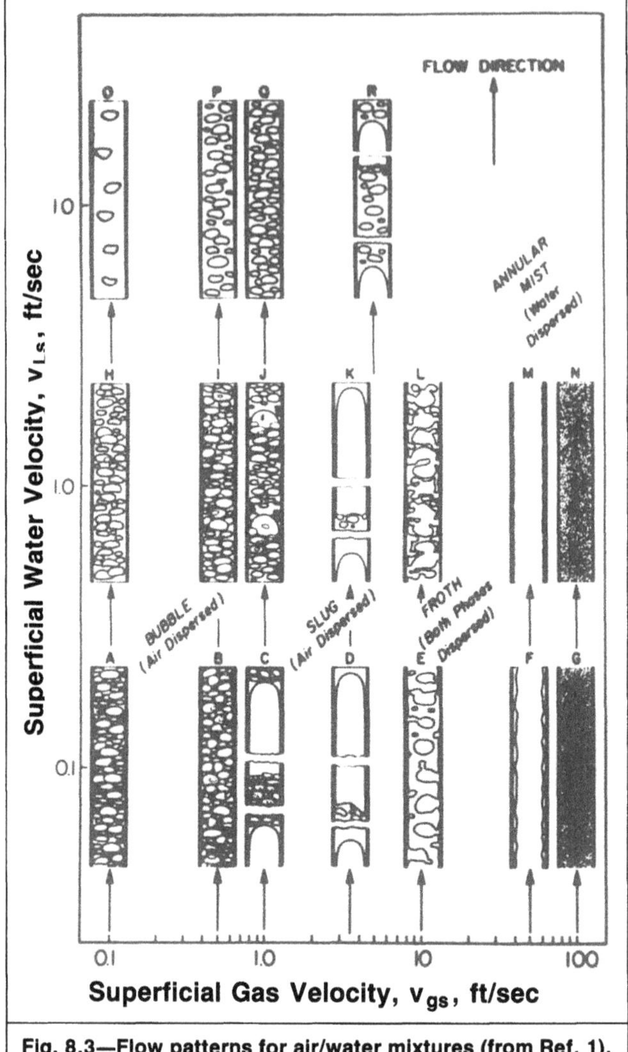

Fig. 8.3—Flow patterns for air/water mixtures (from Ref. 1).

The superficial velocity of a phase would be the average velocity of the phase if that phase filled the entire pipe—i.e., if it were single-phase flow. In two-phase flow, the superficial velocity is not a real velocity that physically occurs, but simply a convenient parameter.

The average in-situ velocities, $\bar{v}_\alpha$ and $\bar{v}_\beta$, are defined in terms of the superficial velocities and the holdup as

$$\bar{v}_\alpha=\frac{v_{\alpha s}}{y_\alpha} \quad (8.12)$$

and

$$\bar{v}_\beta=\frac{v_{\beta s}}{y_\beta}. \quad (8.13)$$

Substituting these expressions into the equation defining slip velocity (Eq. 8.9) yields

$$v_s=\frac{v_{\alpha s}}{y_\alpha}-\frac{v_{\beta s}}{y_\beta} \quad (8.14)$$

or

$$v_s=\frac{1}{A}\left(\frac{q_\alpha}{1-y_\beta}-\frac{q_\beta}{y_\beta}\right). \quad (8.15)$$

The holdup and the slip velocity concepts are perhaps the most straightforward ways to represent the holdup phenomenon. Other parameters, however, are sometimes more convenient in correlation work. One such parameter is the *holdup ratio, F_y*:

$$F_y=\frac{\bar{v}_\alpha}{\bar{v}_\beta}. \quad (8.16)$$

The holdup ratio is also called the *slip ratio.* By substituting for $\bar{v}_\alpha$ and $\bar{v}_\beta$, we can also show that

$$F_y=\frac{y_\beta/y_\alpha}{f_\beta/f_\alpha}. \quad (8.17)$$

Thus, the holdup ratio is the ratio of in-situ volume fractions of the two phases to the input volume fractions. The holdup ratio gives a clear indication of the degree of slip taking place. For $F_y=1$, no slip occurs between phases; for $F_y>1$, the lighter phase moves faster than the denser phase.

Three primary aspects of two-phase-flow physics contribute to the holdup phenomenon: (1) the relative local velocity difference between phases caused by gravitational effects, (2) the velocity profile across the pipe cross section, and (3) the local holdup profile of either phase.

The most important cause of the holdup phenomenon is the gravitational force that results from the difference in density between the two phases in the typical two-phase flow situation. This gravitational force causes a local relative velocity between the two phases that depends on many flow properties, including the geometry. For example, in two-phase flow consisting of bubbles of the lighter phase flowing within a continuous denser phase, the local relative velocity will depend on the bubble size. A limiting case of this type of flow is the rise of bubbles through a column of stagnant denser phase. In this instance, the bubbles will reach a terminal velocity when the buoyant forces on a bubble are balanced by the drag forces opposing the motion. For a single spherical bubble, this terminal velocity is

$$v_b=\sqrt{\frac{g\Delta\rho d_b}{\rho_\beta K_D}}, \quad (8.18)$$

where $\Delta\rho=\rho_\beta-\rho_\alpha$; d_b=bubble diameter; v_b=bubble terminal velocity; K_D=drag coefficient, $f(N_{Re})$; and g=acceleration of gravity. Thus, the bubble velocity depends on bubble size, density contrast between phases, and such dense-phase properties as viscosity and density.

In a typical two-phase flow situation, a wide range of bubble sizes or other flow geometries may influence the holdup phenomenon.

Two other factors, the velocity and local holdup profiles, interact to cause holdup. Both velocity and local holdup typically will vary as a function of radial position in vertical two-phase flow (Fig. 8.2). In this situation, slip would apparently be occurring because the local holdup of the lighter phase is higher in the center of the pipe, where the velocity is higher. Thus, even when there is no local relative velocity between phases, there is a holdup phenomenon caused by the interaction of the velocity and local holdup profiles. On the other hand, if either, or both, are flat, there will be no holdup effect (i.e., $y_\beta=f_\beta$). This is best understood by looking at the definition of holdup in terms of local holdup for the case where there is no local relative velocity between the phases (no buoyancy effect).

The volumetric flow rate of a phase can be found by integrating the product of local velocity and local holdup:

$$q_\alpha=\int_0^A v_\ell y_{\alpha\ell}dA \quad (8.19)$$

and

$$q_\beta=\int_0^A v_\ell y_{\beta\ell}dA. \quad (8.20)$$

y is

$$y_\alpha=\frac{1}{A}\int_0^A y_{\alpha\ell}dA \quad (8.21)$$

and

$$y_\beta=\frac{1}{A}\int_0^A y_{\beta\ell}dA. \quad (8.22)$$

Recalling the equations relating average in-situ velocity to holdup, we now have

$$\bar{v}_\alpha=\frac{q_\alpha}{Ay_\alpha} \quad (8.23)$$

and

$$\bar{v}_\beta=\frac{q_\beta}{Ay_\beta}, \quad (8.24)$$

or

$$\bar{v}_\alpha=\frac{\int_0^A y_{\alpha\ell}v_\ell dA}{\int_0^A y_{\alpha\ell}dA} \quad (8.25)$$

and

$$\bar{v}_\beta=\frac{\int_0^A y_{\beta\ell}v_\ell dA}{\int_0^A y_{\beta\ell}dA}. \quad (8.26)$$

If the velocity profile is flat, v_ℓ is a constant that can be taken out of the integrations, and Eqs. 8.25 and 8.26 reduce to

$$\bar{v}_\alpha=v_\ell \quad (8.27)$$

and

$$\bar{v}_\beta=v_\ell. \quad (8.28)$$

Thus, if the velocity profile is flat, the average velocities of the two phases are equal and no slip occurs. Similarly, if the concentration profile is flat, $y_{\alpha\ell}$ and $y_{\beta\ell}$ are constants, and

$$\bar{v}_\alpha=\bar{v}_\beta=\frac{1}{A}\int_0^A v_\ell dA. \quad (8.29)$$

Again there is no holdup phenomenon.

8.3 Two-Phase Flow Regimes

8.3.1 Flow-Regime Descriptions. Because holdup depends on the structure or geometry of two-phase flow, many researchers have tried to identify the various flow patterns or flow regimes that occur. The descriptions of these patterns are obviously somewhat arbitrary in nature and many different names have been used to label them. The two-phase flow literature now generally recognizes four flow patterns in gas/liquid, vertical, upward flow: bubble, slug, churn, and annular. These generally occur as a progression in the order listed, with increasing gas rate for a given liquid rate. Fig. 8.3 shows these flow patterns and the approximate regions in which they occur as functions of superficial velocities. The following brief descriptions of the flow patterns list the other names for the flow regimes in parentheses.

1. *Bubble (dispersed-bubble) flow* consists of dispersed bubbles of gas in a continuous liquid phase. (Note that most quantitative production logging interpretation methods for two-phase flow assume bubble flow.)

2. In *slug flow,* as the gas rate increases, the bubbles coalesce into larger bubbles, called Taylor bubbles, that eventually fill the entire pipe cross section. Between the large gas bubbles are slugs of liquid that contain smaller bubbles of gas entrained in the liquid.

3. With a further increase in gas rate, the larger gas bubbles become unstable and collapse, resulting in *churn (froth) flow,* a highly turbulent flow pattern with both phases dispersed. Churn flow is characterized by oscillatory, up-and-down motions of the liquid.

4. In *annular (annular-mist, mist) flow,* gas becomes the continuous phase at higher gas rates with liquid flowing in an annulus coating the surface of the pipe and with liquid droplets entrained in the gas phase.

Flow regime is of critical importance in two-phase flow because holdup depends on flow regime. For example, the data of Govier *et al.*[6] for air/water flow in a 1-in. [2.5-cm] -ID pipe show that the slip velocity generally increases, progressing from bubble to slug to churn to annular flow. Slip velocities ranged from a low of 0.03 ft/sec [0.01 m/s] in the bubble flow regime to as high as 93.1 ft/sec [28.4 m/s] in annular flow.

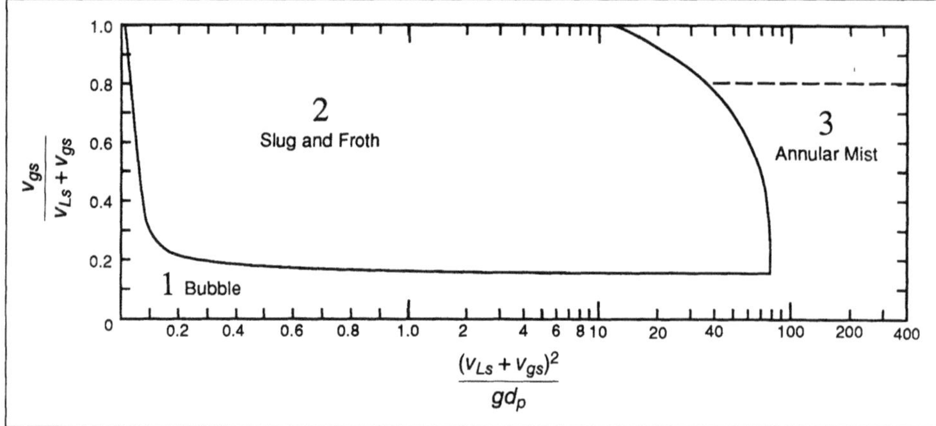

Fig. 8.4—Griffith-Wallis flow-regime map (from Ref. 7, courtesy The American Society of Mechanical Engineers).

8.3.2 Flow-Regime Maps. We have seen that holdup depends on flow regime and that flow regime depends on the velocity of each phase and other physical properties of the fluids. To quantify these relationships, several researchers have developed *flow-regime maps* that relate flow patterns to flow rates and fluid properties. Most of these maps are primarily empirical in nature; thus, their validity is limited to systems similar to those used in their development. Recent efforts have been directed at incorporating enough of the mechanics governing flow-regime transitions into flow-regime correlations to make them more general.

Fig. 8.4 shows one of the first flow-regime maps. Griffith and Wallis[7] defined three flow regimes—bubble, slug and froth (churn), and annular-mist—plotted on coordinates of

$$\frac{v_{gs}}{v_{Ls}+v_{gs}} \text{ vs. } \frac{(v_{gs}+v_{Ls})^2}{gd_p}.$$

Thus, this flow-pattern map considers only the flow rates of each phase and the pipe diameter and is independent of fluid properties.

Fig. 8.5 shows another commonly used flow pattern map, that of Duns and Ros.[8] Using a dimensional-analysis approach, Duns and Ros derived two dimensionless groups, the liquid and gas velocity numbers (N_{vL} and N_{vg}) to be used as coordinates. These groups are defined as

$$N_{vL}=v_{Ls}\sqrt[4]{\frac{\rho_L}{g\sigma}} \quad \text{(8.30)}$$

and

$$N_{vg}=v_{gs}\sqrt[4]{\frac{\rho_L}{g\sigma}}, \quad \text{(8.31)}$$

where ρ_L=liquid density, g=acceleration of gravity, and σ=interfacial tension (IFT) of the gas/liquid system. This flow pattern map accounts for some fluid properties; note, however, that for a given gas/liquid system, the superficial velocities of the phases are the only variables in the dimensionless groups.

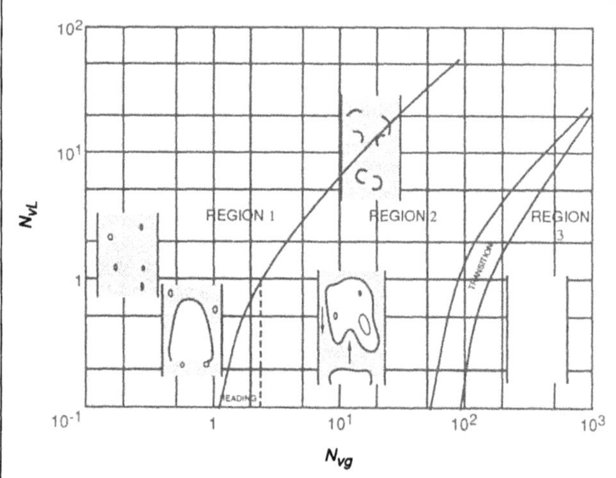

Fig. 8.5—Duns-Ros flow-regime map (after Ref. 8, courtesy World Petroleum Congresses).

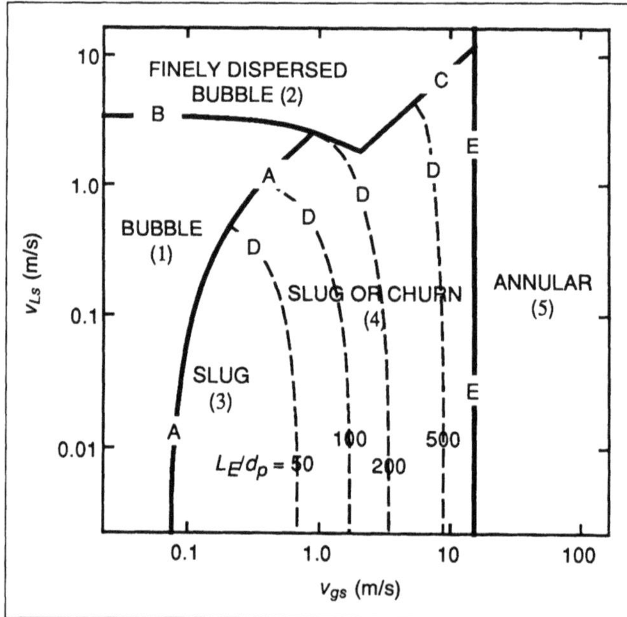

Fig. 8.6—Taitel-Dukler flow-regime map for air/water flow in a 2-in. pipe. (From Ref. 11, "Modelling Flow Pattern Transitions for Steady Upward Gas-Liquid Flow in Vertical Tubes," by Y. Taitel, D. Barnea, and A.E. Dukler, *AIChE J.*, Vol. 26, No. 6, 345–54, 1980. Reproduced by permission of the American Inst. of Chemical Engineers.)

Duns and Ros also defined three distinct regions on their map, but included a transition region where the flow changes from a liquid-continuous to a gas-continuous system. Region 1 contains bubble and low-velocity slug flow patterns, Region 2 is high-velocity slug and churn flow patterns, and Region 3 contains the annular-mist flow pattern.

Efforts to incorporate the fluid mechanics governing flow-regime transitions into flow-regime predictions have been led by Taitel and Dukler and their co-workers.[9-11] The Taitel-Dukler flow-regime map for upward gas/liquid vertical flow is based on an assumed mechanism for each transition. Unlike most other flow regime maps, the Taitel-Dukler map must be constructed for the particular pipe size and fluids of interest. Fig. 8.6 shows this type of flow-regime map for air/water flow in a 2-in. [5-cm] -diameter pipe. To construct this map, equations for the transition lines presented by Taitel *et al.*[11] are applied. The following paragraphs briefly describe the theory of these flow-regime transitions.

Line A—bubble-to-slug transition. At relatively low liquid velocities, slug flow occurs when bubbles are present in sufficient quantity to coalesce into large Taylor bubbles. Taitel *et al.* set this critical bubble concentration at 25% of the pipe volume. Thus, Line A defines the point at which the gas holdup reaches 0.25.

Line B—dispersed-bubble-to-slug transition. At sufficiently high liquid rates, turbulence in the liquid is high enough to break up the larger bubbles that form by coalescence, allowing for a dispersed bubble flow at gas holdups >0.25.

Line C—dispersed-bubble-to-slug transition. When the gas bubble concentration gets high enough in a dispersed bubble flow, the maximum packing density will be reached and coalescence into large bubbles will occur, even in the presence of turbulent liquid flow. Taitel *et al.* assumed this maximum gas holdup for dispersed bubble flow to be 0.52, defining Line C.

Line D—slug-to-churn transition. Perhaps the most innovative work of Taitel *et al.* was their conclusions regarding the transition from slug to churn flow as they postulated that churn flow is simply an entry-region phenomenon leading to slug flow. They observed that in all slug flows, the oscillatory liquid motion characteristic of churn flow occurs for some distance downstream from the pipe entrance before stable slug flow is established. Churn flow has been identified as a distinct flow regime because many two-phase-flow test sections are not long enough for slug flow to occur. The D lines on the flow-regime maps give the number of pipe diameters downstream from the pipe entrance required for slug flow to occur at given gas and liquid rates. Closer to the entrance, churn flow will be observed. This entry phenomenon is important in production logging because the measurements are often made close to the fluid-entry locations. Thus, in wellbores, churn flow probably occurs much more frequently than slug flow.

Line E—slug- or dispersed-bubble-to-annular transition. Annular flow will occur when the gas velocity is sufficient to lift liquid droplets, preventing them from falling down to form bridges leading to churn or slug flow. Line E gives the gas velocity assumed sufficient to carry liquid droplets.

In wellbores, where the fluids enter the pipe through small perforations, flow regimes may occur that are not observed in pipe flows, where the two phases enter the bottom of the pipe. Foams or emulsions may be formed because of the turbulence at the perforations, leading to flow regimes different from those typically described.

8.3.3 Oil/Water Flow Regimes. Unfortunately, little work has been done in describing the flow regimes that occur in an oil/water two-phase flow. Govier *et al.*[12] observed flow patterns with vertical oil/water flow (Fig. 8.7) similar to those found in gas/liquid flow, though an annular flow was not observed. On the other hand, Scott[13] and Cox[14] found flow regimes occurring in horizontal and near-horizontal oil/water flow much different from those typically observed in gas/liquid flow.

Zavareh *et al.*[15] studied oil/water flow regimes in a 7¼-in. [18.4-cm] -ID pipe for flow rates of each phase ranging from 300 to 5,000 B/D [48 to 795 m^3/d]. With the pipe vertical, bubble flow occurred for this entire flow-rate range. When the pipe was deviated

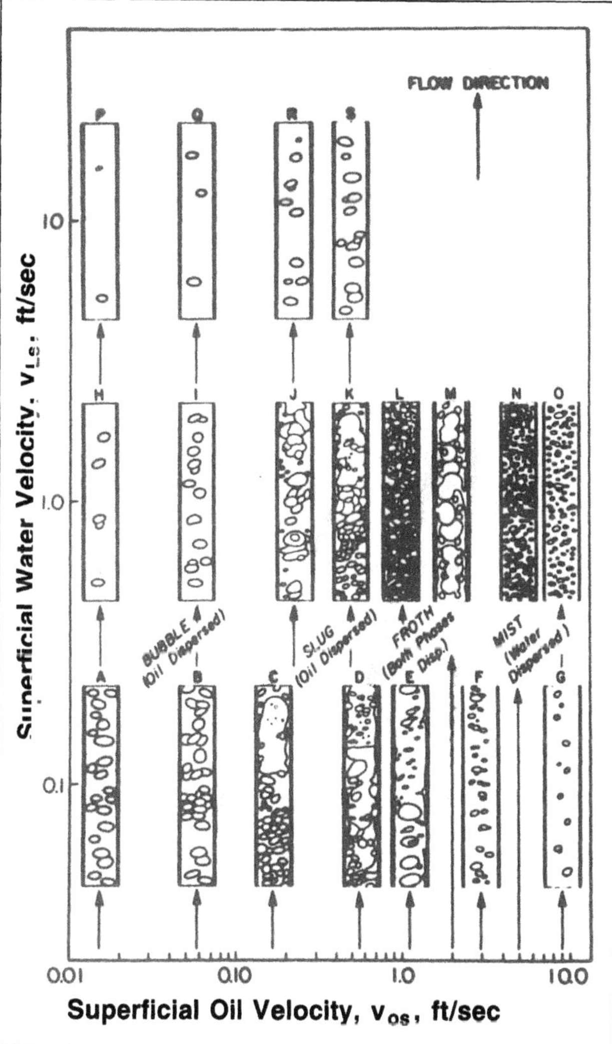

Fig. 8.7—Flow patterns for oil/water mixtures (from Ref. 1).

slightly from vertical, however, flow regimes characterized by a countercurrent flow of water along the lower side of the pipe occurred for most flow rates. Zavareh *et al.* found that the common flow-regime maps, which are based on gas/liquid flow, do not predict oil/water flow regimes reliably.

8.3.4 Implication of Flow Regimes for Production Logging. The flow regime existing in the wellbore is an important consideration in production logging because it significantly affects the performance of many tools. Conventional tools—e.g., spinner flowmeter and fluid-density and fluid-capacitance tools—are most reliable if the flow is relatively homogeneous and thus are best used in a bubble or dispersed-bubble flow. Slug or churn flows may be impossible to log with such devices; a flow-concentrating device may be required for reliable measurements.

Fig. 8.8 shows a Taitel-Dukler flow-regime map for a 7¼-in. [18.4-cm] -ID pipe (in the range of typical casing sizes). The boundaries show the regimes that can be encountered with liquid rates up to 5,000 B/D [795 m^3/d] and gas rates up to 367 Mcf/D [10.4×10^3 m^3/d]. Keep in mind that these rates must be the downhole flow rates to predict performance in the wellbore. These maximum rates would correspond to a GOR of 5,000 scf/bbl [900 std m^3/m^3] at a bottomhole pressure of 1,000 psia [6.9 MPa]. According to the Taitel-Dukler map, slug or churn flow would be predominant in this pipe size over this range of flow rates.

8.4 Two-Phase Pressure-Drop Behavior

Most studies of two-phase flow have been aimed at developing a means to predict the pressure drop for given flow rates of each

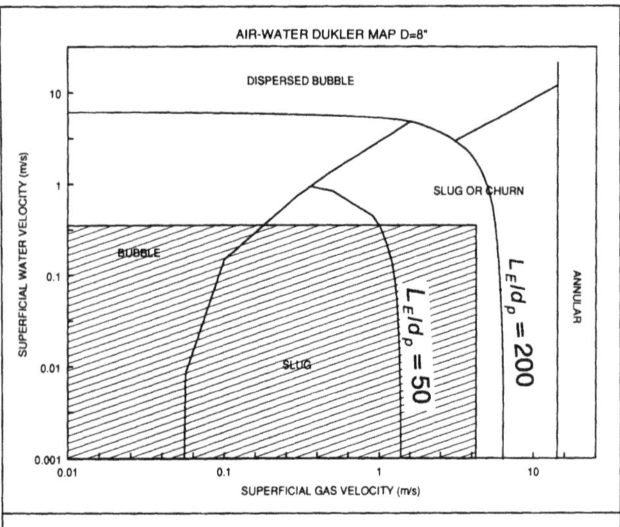

Fig. 8.8—Flow regimes predicted by Taitel-Dukler map in 7¼-in.-ID pipe.

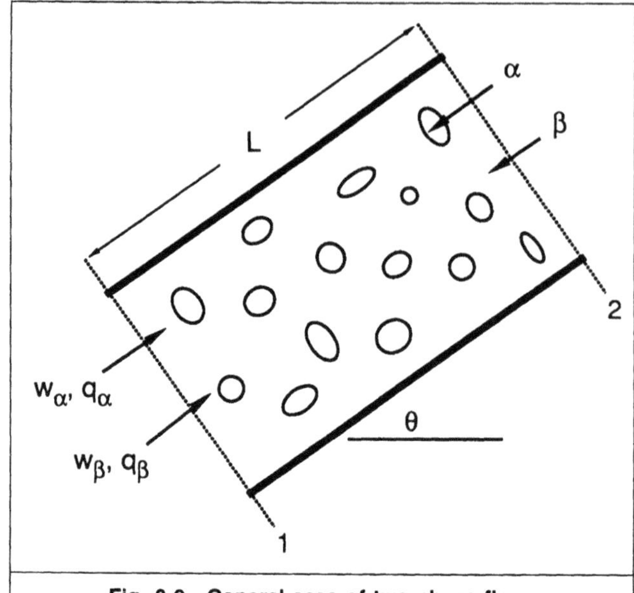

Fig. 8.9—General case of two-phase flow.

phase. In production logging, this is not usually our interest; however, the correlations used to predict two-phase pressure drops can, at least in theory, be inverted to predict flow rates from production logging measurements. Current two-phase pressure drop correlations are not generally accurate enough for this approach to yield valid results, but improvements to correlations could change this situation.

Two-phase flow correlations are based on a steady-state mechanical energy-balance equation. For a general two-phase flow such as that in Fig. 8.9, the mechanical energy balance[1] is

$$p_1-p_2=\left(\frac{w_\alpha+w_\beta}{q_\alpha+q_\beta}\right)gL\sin\theta$$

$$+\left[\frac{w_\alpha(v_{\alpha 2}^2-v_{\alpha 1}^2)}{2B_\alpha(q_\alpha+q_\beta)}+\frac{w_\beta(v_{\beta 2}^2-v_{\beta 1}^2)}{2B_\beta(q_\alpha+q_\beta)}\right]+\Delta p_f, \quad (8.32)$$

where w=mass flow rate, q=volumetric flow rate, v=velocity, p=pressure, L=pipe-segment length, θ=pipe inclination from horizontal, B=velocity profile correction factor, and Δp_f=frictional pressure drop. Subscripts α and β denote the two phases, and Subscripts 1 and 2 indicate positions along the pipeline. This equation can be written more simply as

$$\Delta p=\Delta p_{E_\Phi}+\Delta p_{E_K}+\Delta p_f, \quad (8.33)$$

indicating that the overall pressure drop is composed of potential energy, kinetic energy, and frictional terms. The potential energy term in Eq. 8.32 would be the hydrostatic head of the two-phase mixture if there were no slip between phases. To make the form of the equation more analogous to the usual expression for single-phase flow, Eq. 8.33 is often modified to

$$\Delta p=\Delta p_{HH}+\Delta p_{E_K}+\Delta p_f', \quad (8.34)$$

where Δp_{HH}=hydrostatic head of the two-phase flow stream, defined as

$$\Delta p_{HH}=g(\rho_\alpha y_\alpha+\rho_\beta y_\beta)L\sin\theta \quad (8.35)$$

or

$$\Delta p_{HH}=g\bar{\rho}L\sin\theta, \quad (8.36)$$

where $\bar{\rho}$=average in-situ density of flow stream. Two-phase flow correlations, then, use some technique based on theory, empirical correlations, or some combination of the two to calculate the terms Δp_{HH}, Δp_{E_K}, and $\Delta p_f'$.

For the conditions found in an oil well, the hydrostatic term often will be much larger than the kinetic-energy or frictional terms; thus, the heart of the correlation is a prediction of the holdup. Production logging instruments can sometimes measure the holdup and other average properties of the flow stream, such as average velocity. The manner in which a two-phase flow correlation can be inverted for use in production log interpretation will be illustrated with the Aziz *et al.*[16] correlation.

In this correlation, different calculation procedures are used for different flow regimes. Because production logging measurements are likely to be most accurate in bubble flow, we will examine that regime. For bubble flow, Aziz *et al.* calculate liquid holdup from

$$y_L=1-\frac{v_{gs}}{1.2v_M+v_{bs}}, \quad (8.37)$$

where v_M=mixture velocity, defined by

$$v_M=v_{gs}+v_{Ls}=\frac{q_g+q_L}{A}, \quad (8.38)$$

and v_{bs}=rise velocity of bubbles in a stagnant liquid, given by

$$v_{bs}=1.41\left[\frac{\sigma g(\rho_L-\rho_g)}{\rho_L{}^2}\right]^{1/4}. \quad (8.39)$$

In Eq. 8.39, the constant 1.41 is for units of ft/sec for v_{bs}, lbm/ft^3 for ρ, lbm/sec^2 for σ, and ft/sec^2 for g.

If the mixture velocity and the holdup have been obtained from production logging measurements, as is possible in a bubble flow, Eq. 8.37 can be solved for v_{gs} to yield

$$v_{gs}=(1-y_L)(1.2v_M+v_{bs}). \quad (8.40)$$

The superficial liquid velocity is then simply

$$v_{Ls}=v_M-v_{gs}, \quad (8.41)$$

and the volumetric flow rates are

$$q_g=v_{gs}A \quad (8.42)$$

and

$$q_L = v_{Ls} A. \quad (8.43)$$

Thus, the holdup correlation used to calculate pressure drop can be applied to production log interpretation. The accuracy of this technique, however, has not been tested and may not be high.

8.5 Effect of Pipe Inclination on Two-Phase Flow

The already complex flow geometry of two-phase flow becomes even more complex when the wellbore is inclined from vertical. Three major changes take place compared with vertical flow at comparable rates—the flow regime may change, the holdup behavior will likely be different, and the flow stream will no longer be symmetric about the pipe axis. All these effects complicate our efforts to make production logging measurements in deviated wells.

When the pipe is inclined, the lighter phase will naturally concentrate on the upper side of the pipe. Recall that slug or churn flow results when a sufficient concentration of the gas phase is reached to allow coalescence into large Taylor bubbles. In the inclined pipe, this concentration can be reached at lower gas flow rates because of gravity segregation of the phases. Experimental evidence confirms that slug or churn flow is more prevalent in an inclined pipe than in a vertical pipe at the same flow rates. Figs. 8.10 and 8.11 are Duns-and-Ros-type flow-regime maps for vertical and inclined flow (45°) with experimental data indicated by the lettering on the maps. The regions described as both phases continuous (Region B) or alternating phases continuous (Region A) correspond to slug or churn flow. Notice that in the inclined-flow case, these flow regimes occur for a much larger range of flow rates than in the vertical-flow case.

Because holdup depends strongly on flow regime, it is not surprising that holdup behavior also changes as a pipe is inclined from vertical. Fig. 8.12 shows the liquid holdup measured for air/water flow for a wide range of deviation angles for both upward and downward flow. In this plot, a positive angle indicates upward flow, with +90° being upward vertical flow. As the pipe is deviated from vertical, the liquid holdup increases until a maximum is reached at around 45° deviation, then decreases. The primary implication of this phenomenon for production logging is that a holdup correlation for vertical flow will be in error when used to interpret a log from a deviated well.

Perhaps the most troublesome effect of well deviation on production logging is the nonuniform distribution of the phases across the pipe. In this situation, a small-diameter production logging tool that samples only a small portion of the pipe cross section has little chance of measuring average properties of the flow stream. As an example, Fig. 8.13 shows velocities measured with a spinner flowmeter at the bottom, in the middle, and at the top of a pipe deviated at various angles in an oil/water flow. With the pipe vertical, the spinner yields similar results at any position in the pipe, but for deviations as small as 5°, significant differences are observed in the velocity field across the pipe. At angles ≥15°, the net flow on the lower side of the pipe is actually downward because water flows down the lower side of the pipe and oil flows rapidly upward along the upper side of the pipe.

8.6 Flow From Perforations in a Two-Phase Well

Because production logging measurements are generally made near fluid entries into the wellbore, entry effects that are not normally considered in studies of two-phase flow are likely to be significant. One such effect is whether slug or churn flow will occur, as discussed in Sec. 8.3.2. The influence of dense-phase fluid entries on the flow *upstream* from the entry location can also significantly affect production log interpretation. For example, if oil is flowing up a well and water enters the well from perforations above the oil entry location, it is possible that some of the water entering the well will fall down through the oil column, resulting in water being present in a flow stream in which the only phase having a net upward movement is the oil. If water was measured in the lower part of such a well, all current quantitative interpretation procedures would agree that the water was produced from a lower part of the well.

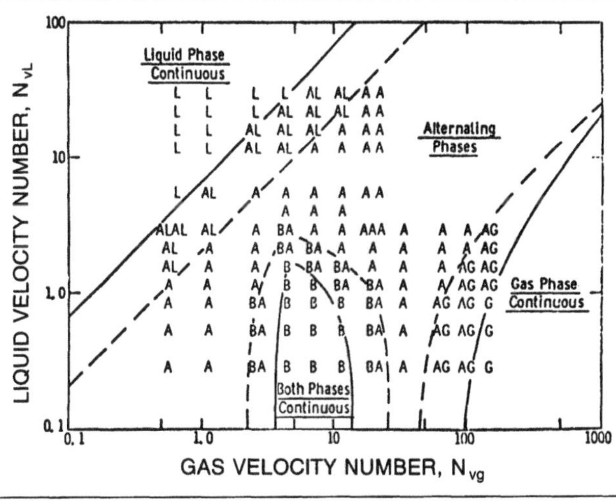

Fig. 8.10—Flow-regime map for vertical flow (from Ref. 17).

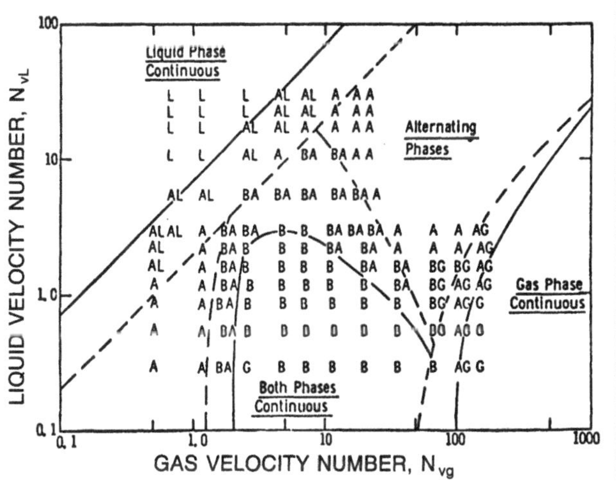

Fig. 8.11—Flow-regime map for upward flow, 45° inclination (from Ref. 17).

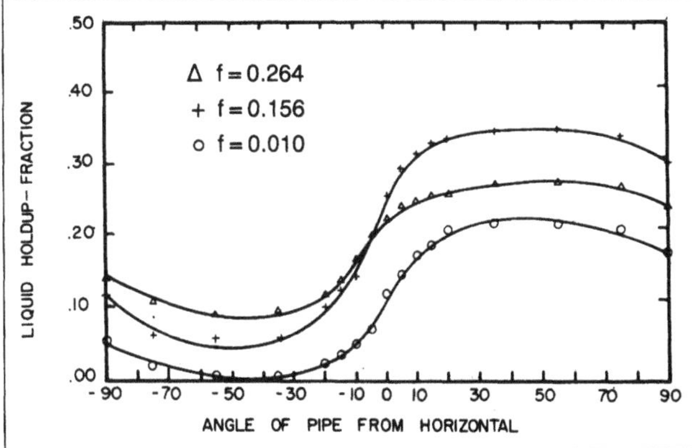

Fig. 8.12—Dependence of holdup on pipe inclination (from Ref. 18).

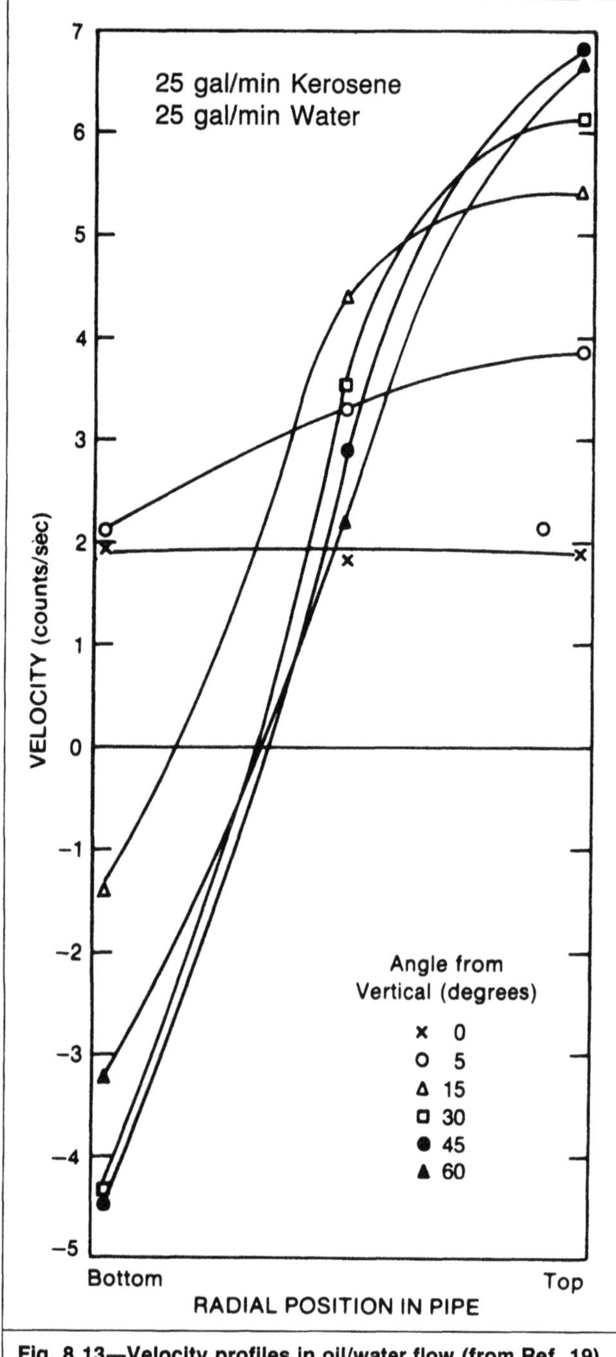

Fig. 8.13—Velocity profiles in oil/water flow (from Ref. 19).

Zhu and Hill[20] studied the flow of water from perforations into gas or oil flowing in a 7¼-in. [18.4-cm] -ID pipe. For the gas/water case, they found that some water fell through the gas stream below the perforations for a wide range of gas and liquid flow rates. Thus, it is likely that in a gas well with water production, some water will be present throughout the well, no matter where the water entry is located. In oil/water flow in a vertical pipe, with an oil rate of 5,000 B/D [795 m^3/d] from below the perforations, and 50 B/D of water entering from the perforations, all water was swept upward. For any lower oil rate, any higher water rate, or any pipe deviation from vertical, some water was present in the oil stream below the perforations.

8.7 Three-Phase Flow

Many oil wells may be producing three distinct phases—oil, water, and gas—at the bottomhole conditions where production logging measurements are made. Because very little work has been done on three-phase flow, it is generally assumed that the two liquids can be treated as one liquid phase with average properties of the two liquids, so that the problem can be treated as two-phase gas/liquid flow. In some cases, the turbulence caused by the gas flow will be sufficient essentially to emulsify the liquids, so this assumption is valid. It is not known at this time when this will hold.

Nomenclature

A = pipe cross-sectional area, ft^2 [m^2], L^2
A_α = cross-sectional area of pipe occupied by Phase α, ft^2 [m^2], L^2
A_β = cross-sectional area of pipe occupied by Phase β, ft^2 [m^2], L^2
B_α = velocity-profile correction factor for Phase α
B_β = velocity-profile correction factor for Phase β
d_b = bubble diameter, ft [m], L
d_p = pipe diameter, ft [m], L
f_α = input volume fraction of Phase α
f_β = input volume fraction of Phase β
F_y = holdup ratio
g = acceleration of gravity, ft/sec^2 [m/s^2], L/t^2
K_d = drag coefficient
L = distance between Positions 1 and 2, ft [m], L
L_E = distance from entrance of pipe, ft [m], L
N_{Re} = Reynolds number
N_{vg} = gas velocity number
N_{vL} = liquid velocity number
p_1 = pressure at Position 1, psi [kPa], m/Lt^2
p_2 = pressure at Position 2, psi [kPa], m/Lt^2
Δp = overall pressure drop, psi [kPa], m/Lt^2
Δp_{E_K} = kinetic energy pressure drop, psi [kPa], m/Lt^2
Δp_{E_Φ} = potential energy pressure drop, psi [kPa], m/Lt^2
Δp_f = frictional pressure drop, psi [kPa], m/Lt^2
$\Delta p_f'$ = modified frictional pressure drop, psi [kPa], m/Lt^2
Δp_{HH} = hydrostatic head, psi [kPa], m/Lt^2
q_g = gas volumetric flow rate, ft^3/sec [m^3/s], L^3/t
q_L = liquid volumetric flow rate, ft^3/sec [m^3/s], L^3/t
q_α = volumetric flow rate of Phase α, ft^3/sec [m^3/s], L^3/t
q_β = volumetric flow rate of Phase β, ft^3/sec [m^3/s], L^3/t
v_b = bubble terminal velocity, ft/sec [m/s], L/t
v_{bs} = bubble rise velocity, ft/sec [m/s], L/t
v_{gs} = superficial gas velocity, ft/sec [m/s], L/t
v_ℓ = local velocity, ft/sec [m/s], L/t
v_{Ls} = superficial liquid velocity, ft/sec [m/s], L/t
v_M = mixture velocity, ft/sec [m/s], L/t
v_s = slip velocity, ft/sec [m/s], L/t
$v_{\alpha s}$ = superficial velocity of Phase α, ft/sec [m/s], L/t
$v_{\alpha 1}$ = velocity of Phase α at Position 1, ft/sec [m/s], L/t
$v_{\alpha 2}$ = velocity of Phase α at Position 2, ft/sec [m/s], L/t
$\bar{v}_\alpha$ = average in-situ velocity of Phase α, ft/sec [m/s], L/t
$v_{\beta s}$ = superficial velocity of Phase β, ft/sec [m/s], L/t
$v_{\beta 1}$ = velocity of Phase β at Position 1, ft/sec [m/s], L/t
$v_{\beta 2}$ = velocity of Phase β at Position 2, ft/sec [m/s], L/t
$\bar{v}_\beta$ = average in-situ velocity of Phase β, ft/sec [m/s], L/t
V = volume of pipe segment, ft^3 [m^3], L^3
w_α = mass flow rate of Phase α, lbm/sec [kg/s], m/t
w_β = mass flow rate of Phase β, lbm/sec [kg/s], m/t
y_L = liquid holdup
y_α = holdup (in-situ fraction) of Phase α
$y_{\alpha\ell}$ = local holdup of Phase α
y_β = holdup (in-situ fraction) of Phase β
$y_{\beta\ell}$ = local holdup of Phase β
θ = pipe inclination from horizontal, degrees
ρ_g = gas density, lbm/ft^3 [kg/m^3], m/L^3
ρ_L = liquid density, lbm/ft^3 [kg/m^3], m/L^3

ρ_α = density of Phase α, lbm/ft^3 [kg/m^3], m/L^3
ρ_β = density of Phase β, lbm/ft^3 [kg/m^3], m/L^3
$\bar{\rho}$ = average in-situ density, lbm/ft^3 [kg/m^3], m/L^3
$\Delta\rho$ = density difference between phases, lbm/ft^3 [kg/m^3], m/L^3
σ = IFT, dynes/in. [N/m], m/t^2

References

1. Govier, G.W. and Aziz, K.: *The Flow of Complex Mixtures in Pipes*, Robert E. Krieger Publishing Co., Huntington, NY (1977).
2. Hewitt, G.F.: *Measurement of Two Phase Flow Parameters*, Academic Press, London (1978).
3. Wallis, G.B.: *One Dimensional Two Phase Flow*, McGraw-Hill Publishing Co., New York City (1969).
4. Brill, J.P. and Beggs, H.D.: *Two-Phase Flow in Pipes*, U. of Tulsa, Tulsa, OK (1978).
5. Persen, L.N.: *The Mechanics of Two Phase Flow*, Inst. for Mechanics, Norwegian Inst. of Technology, Trondheim, Norway (1986).
6. Govier, G.W., Radford, B.A., and Dunn, J.W.S.: *Cdn. J. Chem. Eng.* (1957) **35,** 58.
7. Griffith, P. and Wallis, G.B.: "Two-Phase Slug Flow," *J. Heat Transfer* (Aug. 1961) 307-20; *Trans.*, ASME, Ser. C, **83.**
8. Duns, H. Jr. and Ros, N.C.J.: "Vertical Flow of Gas and Liquid Mixtures in Wells," *Proc.*, Sixth World Pet. Congress, Frankfurt (1963) **2,** Paper 22.
9. Taitel, Y. and Dukler, A.E.: "A Model for Predicting Flow Regime Transition in Horizontal and Near Horizontal Gas-Liquid Flow," *AIChE J.* (Jan. 1976) **22,** 47-55.
10. Barnea, D. *et al.*: "Flow Pattern Transition for Gas-Liquid Flow in Horizontal and Inclined Pipes: Comparison of Experimental Data with Theory," *Intl. J. Multiphase Flow* (June 1980) **6,** No. 3, 217-26.
11. Taitel, Y., Barnea, D., and Dukler, A.E.: "Modelling Flow Pattern Transitions for Steady Upward Gas-Liquid Flow in Vertical Tubes," *AIChE J.* (May 1980) **26,** No. 6, 345-54.
12. Govier, G.W., Sullivan, G.A., and Wood, R.K.: "The Upward Vertical Flow of Oil-Water Mixtures," *Cdn. J. Chem. Eng.* (April 1961) **39,** 67.
13. Scott, G.M.: "A Study of Two-Phase Liquid-Liquid Flow at Variable Inclinations," MS thesis, U. of Texas, Austin (Aug. 1985).
14. Cox, A.L. Jr.: "A Study of Horizontal and Downhill Two-Phase Oil-Water Flow," MS thesis, U. of Texas, Austin (Aug. 1985).
15. Zavareh, F., Hill, A.D., and Podio, A.L.: "Flow Regimes in Vertical and Inclined Oil/Water Flow in Pipes," paper SPE 18215 presented at the 1988 SPE Annual Technical Conference and Exhibition, Houston, Oct. 2-5.
16. Aziz, K., Govier, G.W., and Fogarasi, M.: "Pressure Drop in Wells Producing Oil and Gas," *J. Cdn. Pet. Tech.* (July-Sept. 1972) **11,** 38-45.
17. Gould, T.L., Tek, M.R., and Katz, D.L.: "Two-Phase Flow through Vertical, Inclined, or Curved Pipe," *JPT* (Aug. 1974) 915-26; *Trans.*, AIME, **257.**
18. Beggs, H.D. and Brill, J.P.: "A Study of Two-Phase Flow in Inclined Pipes," *JPT* (May 1973) 607-17; *Trans.*, AIME, **255.**
19. Hill, A.D. and Oolman, T.: "Production Logging Tool Behavior in Two-Phase Inclined Flow," *JPT* (Oct. 1982) 2432-40.
20. Zhu, D. and Hill, A.D.: "The Effect of Flow from Perforations on Two-Phase Flow: Implications for Production Logging," paper SPE 18207 presented at the 1988 SPE Annual Technical Conference and Exhibition, Houston, Oct. 2-5.

SI Metric Conversion Factors

ft	× 3.048*	E−01	=	m
gal	× 3.785 412	E−03	=	m^3
in.	× 2.54*	E+00	=	cm

*Conversion factor is exact.

Chapter 9
Production Logging in Multiphase Flow

9.1 Introduction

Production logging in multiphase flow is generally much more difficult than logging in single-phase flow because the flow system is often complex. The objective of production logging measurements is to determine the fluid production rates from various intervals in a well; in multiphase flow, however, the objective is also to determine the type of fluid produced. Thus, a profile of oil, gas, and water production is the desired result of the production log. As in single-phase logging, the flow profile is inferred from measurements of flow-stream properties in the wellbore at various depth locations.

Production logging in multiphase flow is difficult because the measurements are statistical in nature and because the interpretation methods require uncertain assumptions. Logging-tool responses are often "noisy" because of the stochastic nature of multiphase flow—e.g., even in a so-called steady-state flow, such flow properties as average velocity and holdup will fluctuate with time at any pipe location. In addition to time variations, there are often variations in flow properties across the pipe cross section, particularly if the well is deviated, that further complicate the production logging measurement. Once measurements are obtained, assumptions about the multiphase flow, usually concerning the slip velocity between the phases, must be made to interpret the logs quantitatively.

Because of these complexities, a qualitative picture of the flow profile is often all that can be obtained in a multiphase well with any certainty. In fact, the most reliable interpretation in a multiphase well will often be from a simple temperature log.[1] Thus, the engineer interpreting logs in multiphase wells should apply the interpretation procedures carefully, examining the expected accuracy of each log to determine the accuracy of the interpreted flow profile.

In many cases, particularly in deviated wells, a standard interpretation of production logs yields a completely false picture of the flow profile. Smolen[2] illustrated this difficulty clearly for the case of a spinner-flowmeter log (Fig. 9.1). A two-pass interpretation of the spinner response shows an apparent downflow in the lower part of the well. There is actually no net downflow; the spinner reversal is caused by a downward circulation of the denser phase along the lower side of the pipe. To avoid such erroneous interpretations, the log interpreter must be aware of the unusual flow conditions that can occur in a multiphase well and must use all available well and reservoir information to corroborate the production logging diagnosis. For the case shown in Fig. 9.1, a simple calculation of the reservoir pressures that would have to exist for the apparent downflow actually to occur would probably show the crossflow to be physically impossible.

Four basic measurements are made to determine flow profiles in production wells—fluid velocity, fluid density, water fraction, and temperature. Velocity is measured with spinner flowmeters, radioactive-tracer-logging techniques, and flow-concentrating flowmeters. Fluid density logs are based on gamma ray absorption or pressure-differential measurements. Water fraction is measured by determining the capacitance or dielectric constant of the multiphase mixture. Temperature logging in production wells does not differ greatly from that in injection wells and was discussed in Chap. 4.

This chapter presents the basic theory underlying all interpretation procedures for multiphase flow and discusses operational procedures in production wells, which are relevant to production logging in any production well, whether the flow is single- or multiphase. It also describes the primary logs used in multiphase flow, grouped in categories of fluid-velocity and fluid-identification logs. Production log interpretation, both quantitative and qualitative, is then discussed, followed by a brief treatment of three-phase flow. This chapter also presents examples of qualitative interpretation of temperature logs in multiphase flow and how they relate to other production logs. Finally, guidelines for logging in multiphase flow are presented. Most of the discussion of production logging in multiphase flow concerns two-phase flow of oil and water or gas and liquid because two-phase flow is more easily analyzed.

9.2 Production Log Interpretation in Multiphase Flow

The fundamental problem of multiphase-flow log interpretation can be illustrated by considering two-phase-flow interpretation. To determine the flow profile of each phase, two logs are required—one to measure the average velocity of the flow stream and one to measure the amount of each phase present at the measurement location. As an example, consider the simplified problem of oil/water vertical flow in Fig. 9.2. The objective of the production log is to determine the volumetric flow rates of the two phases, q_o and q_w. Let us assume that the logging tools measure a fluid velocity and a fluid density. We will further assume that the velocity measured is the true in-situ average velocity (the mixture velocity) and that the density measured is the in-situ average density. (This is a best-case estimate with current logging tools.) Thus,

$$\bar{\rho}=y_o\rho_o+y_w\rho_w \quad \text{(9.1)}$$

and

$$v_M=\frac{q_o+q_w}{A_w}=\frac{q_t}{A_w}, \quad \text{(9.2)}$$

where $\bar{\rho}$ and v_M=measured density and velocity, respectively, ρ_o and ρ_w=oil and water densities, respectively, and y_o and y_w=oil and water holdups, respectively. Because the oil and water holdups must sum to 1, the water holdup can be determined from the density measurement because Eq. 9.1 can be rearranged to yield

$$y_w=\frac{\bar{\rho}-\rho_o}{\rho_w-\rho_o}. \quad \text{(9.3)}$$

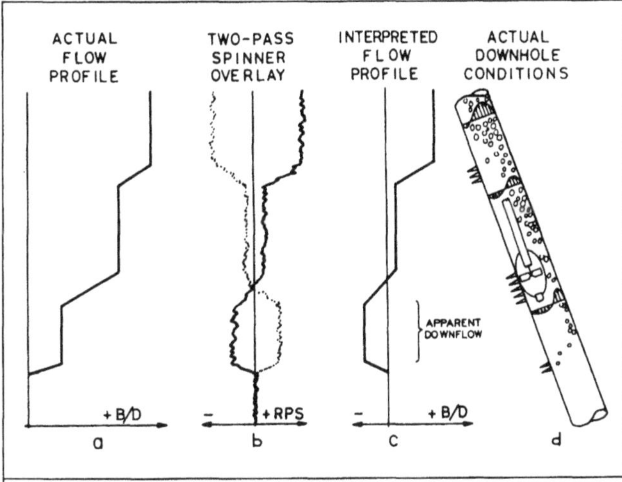

Fig. 9.1—Effect of well deviation on spinner-flowmeter interpretation in multiphase flow (from Ref. 2, courtesy Society of Professional Well Log Analysts).

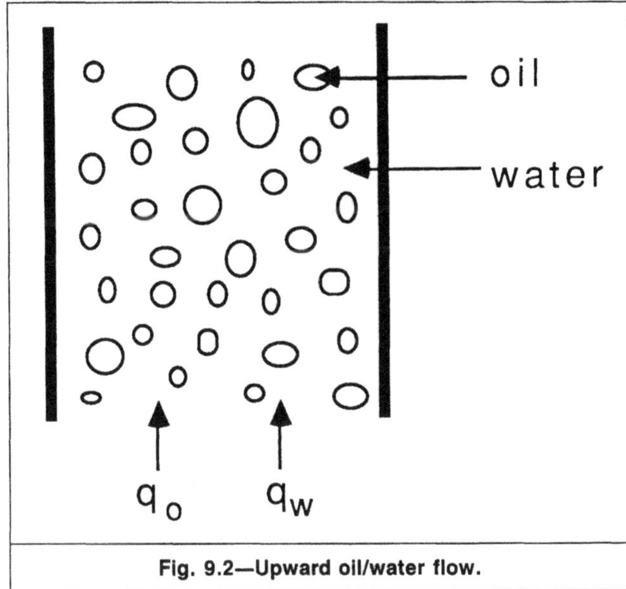

Fig. 9.2—Upward oil/water flow.

Thus, the velocity measurement determines the total volumetric flow rate, q_t, and the density measurement determines the holdup. The question now is whether the volumetric flow rates of each phase can be extracted from the knowledge of the total volumetric flow rate and the holdup. The mixture velocity is related to the average in-situ velocities of the phases, $\bar{v}_o$ and $\bar{v}_w$, by the holdup as

$$v_M=\bar{v}_o y_o+\bar{v}_w y_w. \quad (9.4)$$

Eliminating the oil holdup yields

$$v_M=\bar{v}_w y_w+\bar{v}_o(1-y_w). \quad (9.5)$$

Recall that y_w is known from the density measurement. Solving for $\bar{v}_w$ gives

$$\bar{v}_w=\frac{v_M-\bar{v}_o(1-y_w)}{y_w}. \quad (9.6)$$

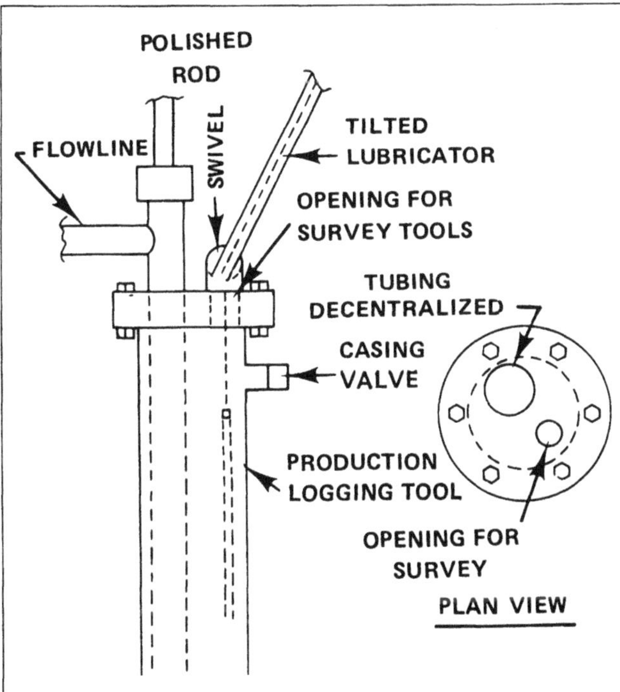

Fig. 9.3—Wellhead arrangement to log in casing/tubing annulus (from Ref. 3).

We are still not able to calculate the water velocity from the production logging measurement because $\bar{v}_o$ is not known in Eq. 9.6. However, $\bar{v}_o$ can be eliminated in terms of the slip velocity, v_s, because

$$v_s=\bar{v}_o-\bar{v}_w. \quad (9.7)$$

Thus, substituting for $\bar{v}_o$ in Eq. 9.6 yields

$$\bar{v}_w=v_M-v_s(1-y_w). \quad (9.8)$$

Similarly, it can be shown that

$$\bar{v}_o=v_M+v_s y_w. \quad (9.9)$$

Thus, to determine the volumetric flow rates of each phase, the slip velocity must be known. Because no current production logging tool directly measures slip velocity, the fundamental problem in quantitative interpretation of production logs in multiphase flow is the need to estimate the slip velocity independently. Many techniques have been used to estimate slip velocity, including empirical guidelines based on laboratory measurements, inference from surface flow conditions, and two-phase holdup correlations. Because of the complexity of multiphase flow, some uncertainty is always introduced, no matter what procedure is used.

9.3 Operational Procedures in Production Wells

When production logs are run in producing wells, a few potential problems that do not occur in injection wells must be considered. For example, there may be difficulty in getting the tools in the well, the lift mechanism may affect the tool responses, and the fluids may be carrying sand or other detritus that will affect the measurements.

An added concern in production-well logging that is not a typical problem in injection-well logging is simply getting the logging tools in the well. This problem depends on the lift mechanism used in the well. In a naturally flowing or a gas-lift well, there generally will be no impediments in the tubing to obstruct the logging tool. In a well being pumped, however, the tubing is occupied by the pump (and by sucker rods in a rod-pumped well), so a wireline tool cannot be used. A production logging tool can be run in the annulus between the tubing and casing, but only at the risk of the wireline wrapping around the tubing so that the logging tool cannot be retrieved. Hammack *et al.*[3] illustrate the surface arrangement for logging through the casing-tubing annulus (Fig. 9.3) They used a small tool string (7/8-in. [0.9-cm] OD), and reported problems with the tool wrapping around the tubing in 8% of more than 400 logging jobs.

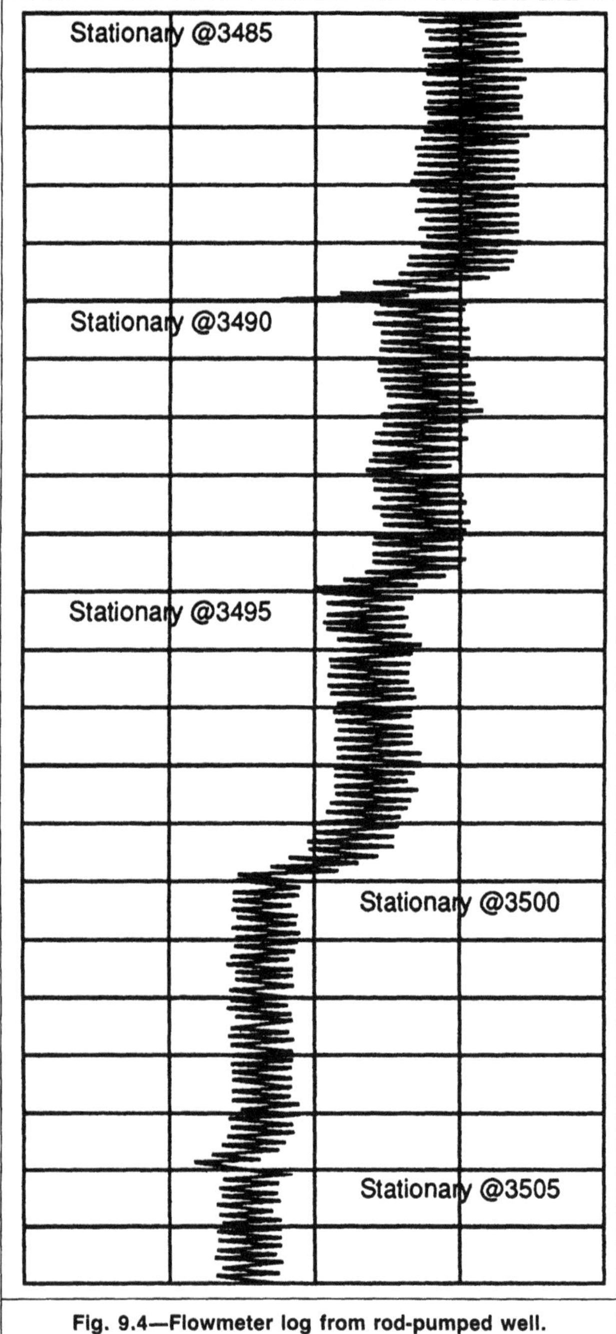

Fig. 9.4—Flowmeter log from rod-pumped well.

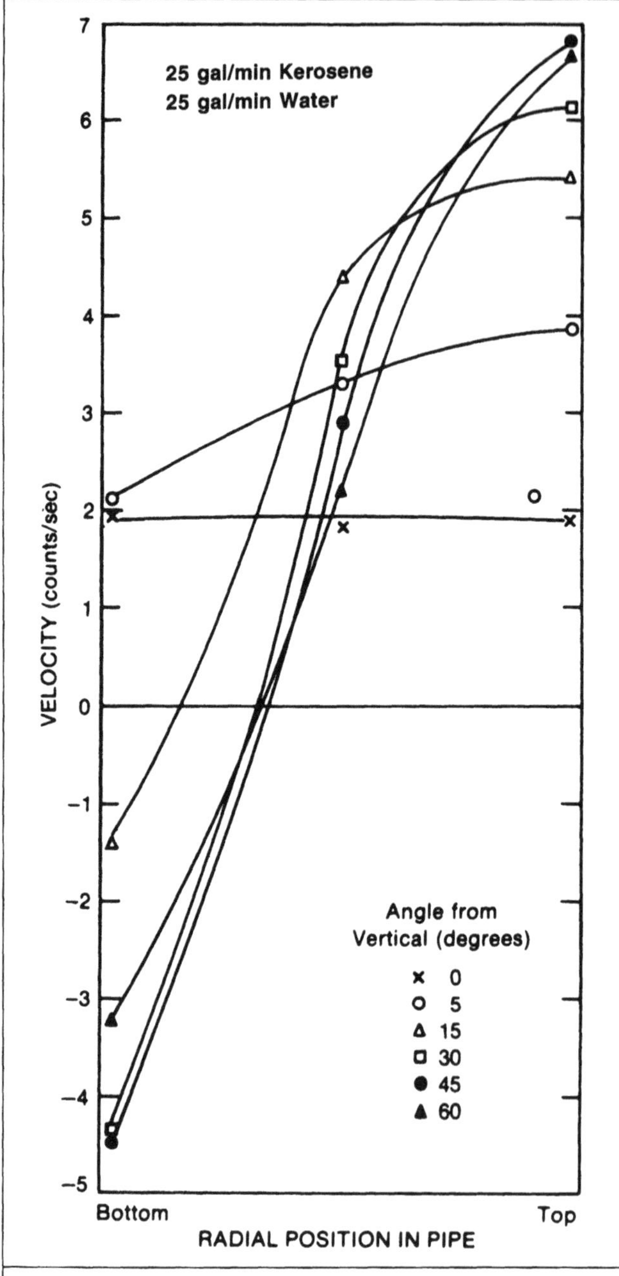

Fig. 9.5—Velocity profiles in oil/water two-phase flow (from Ref. 4).

If the casing is sufficiently large, the problem of the wireline wrapping around the tubing can be avoided by running a second tubing string in the well so that the logging tool can be run in the empty tubing string. The operator must decide whether the expected logging results are worth the added expense of the second tubing string.

Another problem encountered in logging rod-pumped wells is the cyclic nature of the flow rate. Because the pump does not provide a constant flow rate, considerable noise can be introduced into the production logs. Fig. 9.4 shows the response of a flowmeter log in a rod-pumped well. As this log shows, a significant portion of the signal is noise resulting from the cyclic nature of the flow. The log interpreter must choose a consistent basis upon which to read such a log, choosing either the midpoint or the maximum of the signal in each region to interpret the log.

A final consideration in production-well logging is the cleanliness of the fluids. Sand production can alter the response or even damage certain production logging tools, primarily flowmeters. If sand production is a possibility, the logging operator should carefully monitor any flowmeter logs to watch for diminished or erratic responses.

9.4 Fluid-Velocity Measurements

Three common logging methods are used to measure fluid velocity in multiphase wells—spinner-flowmeter, flow-concentrating-flowmeter, and radioactive-tracer logging. This section discusses the characteristics of these techniques in multiphase flow.

9.4.1 Spinner Flowmeters. The same spinner flowmeters used in single-phase-flow logging are used to measure velocity in multiphase flow. The spinner flowmeter's efficacy is greatly diminished in multiphase flow, however, because of the complexity of the flow. One major difficulty is that a spinner flowmeter measures a localized velocity that does not necessarily represent the average flow velocity. Because the velocity profile in multiphase flow is much less predictable than that in single-phase flow, the small sampling area of a spinner flowmeter can be a serious detriment. In a 6-in. [15-cm] -ID casing, a standard $1^{11}/_{16}$-in. [4.3-cm] spinner samples no more than 8% of the flow stream; one of the larger spinners currently in use, with $3\frac{1}{2}$-in. [8.9-cm] blades, samples only one-third of the pipe cross section. Thus, though it generally must be assumed for interpretation that the velocity being measured is the average

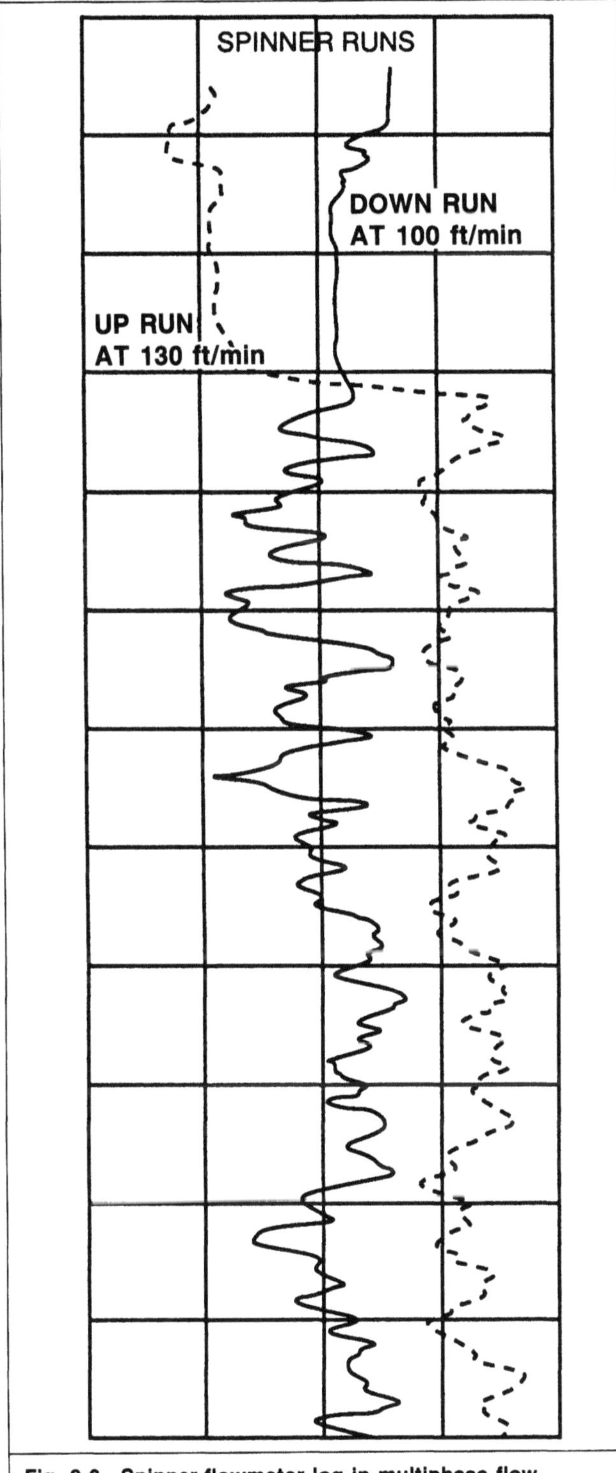

Fig. 9.6—Spinner-flowmeter log in multiphase flow.

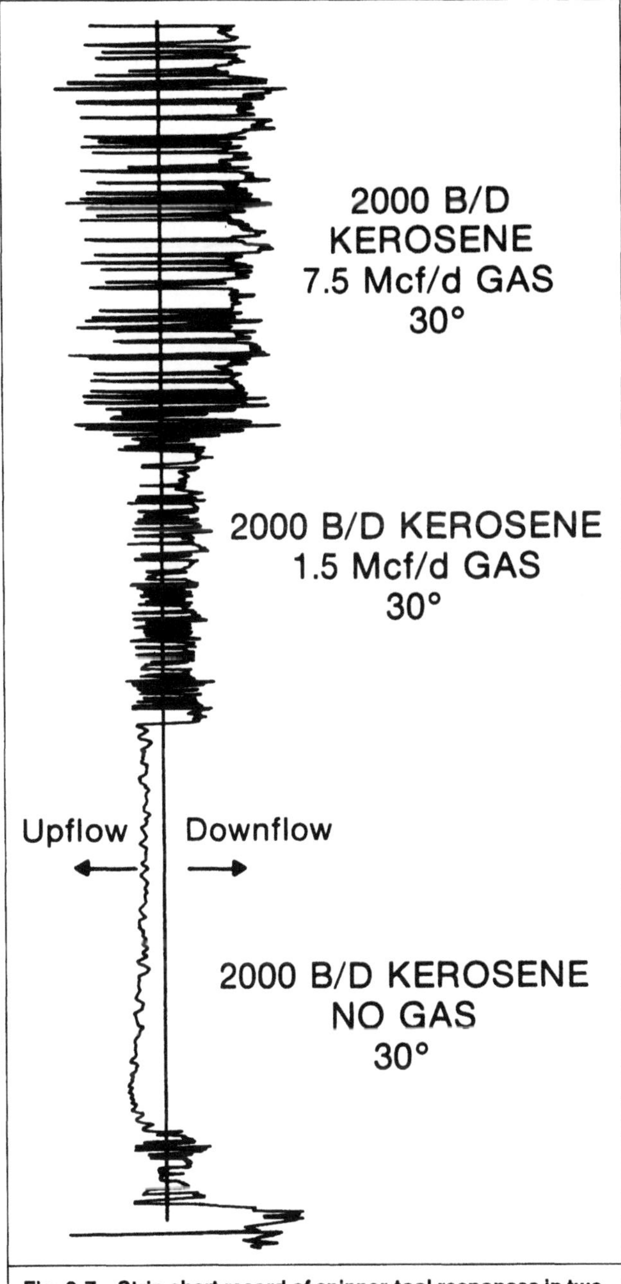

Fig. 9.7—Strip chart record of spinner-tool responses in two-phase inclined flow (from Ref. 4).

flow velocity, this is often not the case with spinner flowmeters in two-phase flow.

The response of a spinner flowmeter in inclined flow illustrates the error caused by sample size. Fig. 9.5 shows the velocity measured at the top, in the center, and at the bottom of a pipe for oil/water flow at various inclinations. As the figure shows, the measured velocity (and even the direction of flow) can vary greatly across the pipe.

Another problem encountered when spinner flowmeters are used in two-phase flow is the high level of noise in the signal. In two-phase flow, particularly in the slug and churn flow regimes, the flow often is intermittent and the fluid velocity varies greatly with time. Thus, a spinner flowmeter yields a response that may be uninterpretable, even if the flowmeter is accurately measuring velocity at all times. Fig. 9.6 shows a portion of a spinner-flowmeter log run at two cable speeds in a multiphase-flow production well. In this log, the level of noise is such that no reasonable interpretation can be made. Fig. 9.7 illustrates the fluctuation of velocity at a point and its result on spinner response. With the spinner stationary, the noisiness of the signal increases dramatically as gas is introduced to the flow stream. This example is an extreme case because the pipe is inclined 30°.

Spinner flowmeters are often poor logs in inclined wells, sometimes giving very misleading results. As Fig. 9.5 shows, when the wellbore is inclined, there is often a downward stream of the denser phase on the lower side of the pipe. If the flowmeter is placed in this region, the log interpreter may conclude that there is a net downward flow present that does not really exist. This can be minimized by always centralizing the spinner flowmeter. Fig. 9.8 shows that for given oil and gas rates, a centralized spinner gives a reasonable response to the flow up to a deviation of about 15° from vertical, while a decentralized spinner measures a downward flow in this upward two-phase flow with only a slight deviation of the wellbore.

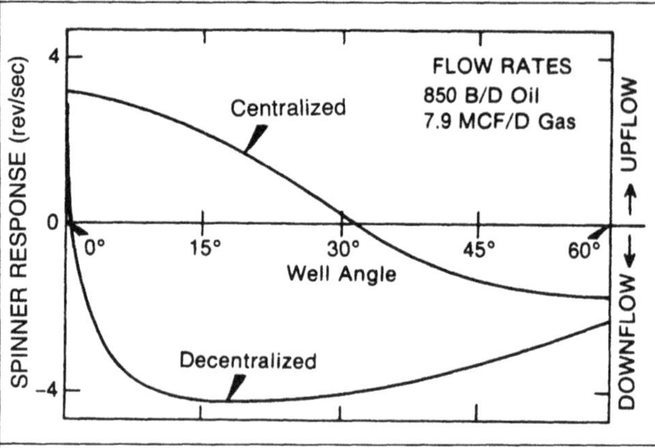

Fig. 9.8—Spinner response as a function of pipe inclination in two-phase flow (from Ref. 4).

Anderson *et al.*[5] illustrate how a spinner flowmeter will sometimes indicate an apparent downward flow in a multiphase production well. Upward and downward passes of a spinner flowmeter are overlaid in Fig. 9.9. The intersection of the response curves indicates a change in flow direction. Above this point the flow is upward; below this point the flow *measured* is downward. When standard interpretation methods are applied, the flow profile (Fig. 9.10) indicates more than 800 BWPD [1272 m^3/d water] flowing down from Zone 2 to Zones 3 and 4. The spinner flowmeter undoubtedly sensed a downward flow stream. The question to be answered is whether a net downward flow exists or whether the spinner is simply measuring a downward flow along the lower side of the pipe when the net flow is upward.

Consideration of other logs and reservoir conditions proves that no net downflow exists. The temperature log alone (Fig. 9.9) is convincing evidence that the net flow throughout the well is upward. The decrease in temperature at the bottom of Perforated Interval C is characteristic behavior for a cooler fluid entering a warmer stream flowing up the wellbore from lower zones. If the fluid entering the wellbore at Perforated Interval C were flowing downward, the wellbore below this entry point would be cooled, and no cool anomaly would occur at the entry location.

Considering other reservoir conditions, it is not likely that a net downflow exists. The average permeability reported for this reservoir was less than 5 md, and all zones were miscibly flooded. Because the pressure in the wellbore would differ by <50 psi [<345 kPa] between Zones 2 and 3, there would have to be a difference of several thousand psi in the reservoir pressures of these zones for there to be flow of 1,650 B/D [262 m^3/d] *into* the well in Zone 2 and a simultaneous flow of 740 B/D [118 m^3/d] *out of* the well in Zone 3. It is difficult to surmise how such a large pressure difference could ever be established in commingled zones less than 100 ft [30 m] apart.

Because of the complexities mentioned, spinner flowmeters are often not adequate for use in multiphase flow. When the flow is homogeneous enough that the behavior is similar to single-phase flow, a spinner flowmeter can provide accurate results. This would be the case in three situations.

1. *Annular-mist flow.* In a high-rate gas well with some liquid production, a spinner flowmeter may give an accurate measurement of the gas velocity in the center of the pipe. The spinner flowmeter would probably not respond to the liquid film flowing along the pipe wall.

2. *Emulsion flow.* If oil and water are emulsified, they will appear as a single phase to a spinner flowmeter but with a higher effective viscosity than either of the individual phases. This will occur primarily at high flow rates and will depend on the surface chemistry of the particular oil and water.

3. *Dispersed-bubble flow.* At sufficiently high rates of the denser phase, the less dense phase will be dispersed as small bubbles moving at the same average velocity as the continuous phase. This is the dispersed-bubble flow regime described by Taitel *et al.*[6] In

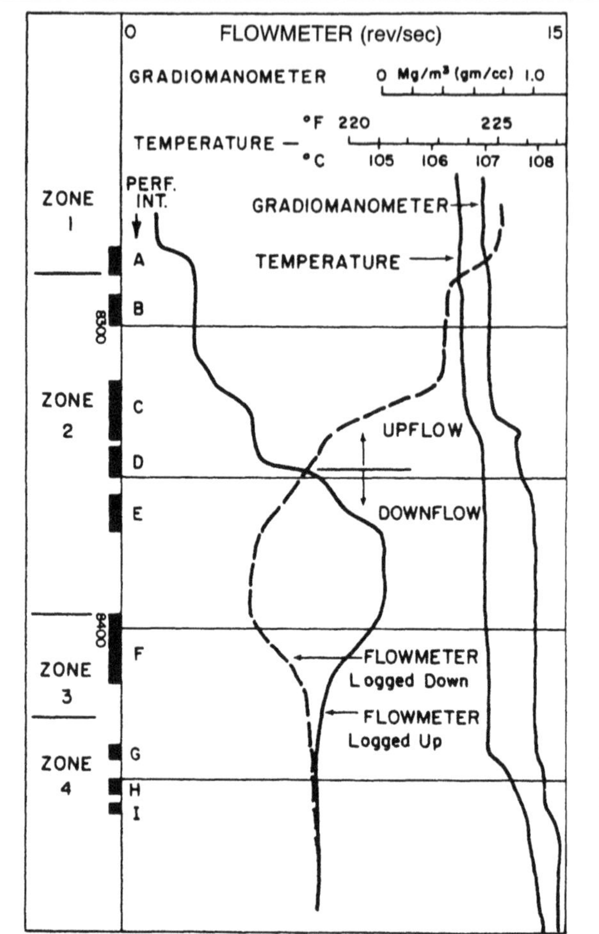

Fig. 9.9—Spinner-flowmeter log showing apparent downflow (from Ref. 5).

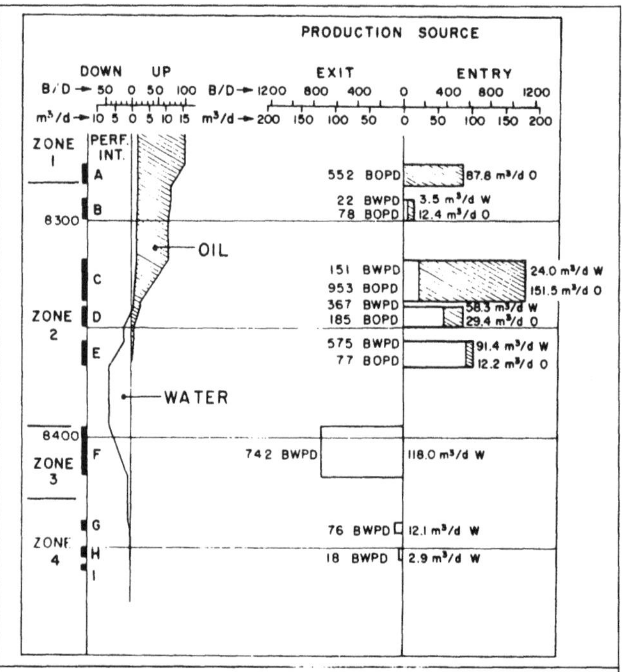

Fig. 9.10—Interpreted flow profile using standard spinner interpretation (from Ref. 5).

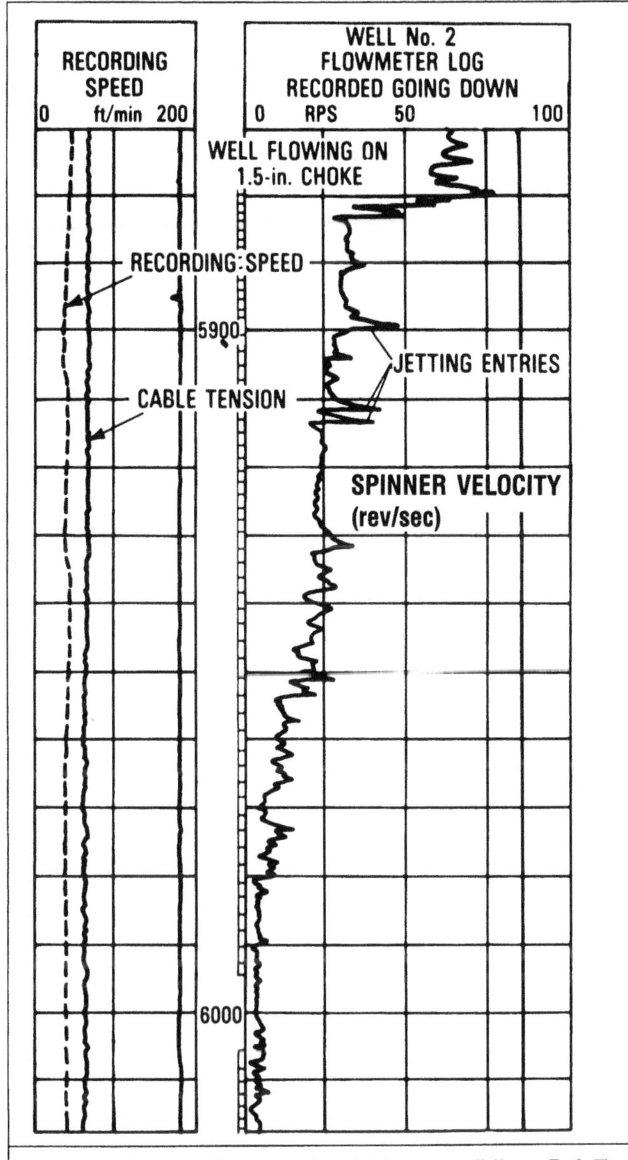

Fig. 9.11—Spinner-flowmeter log, high-rate well (from Ref. 7).

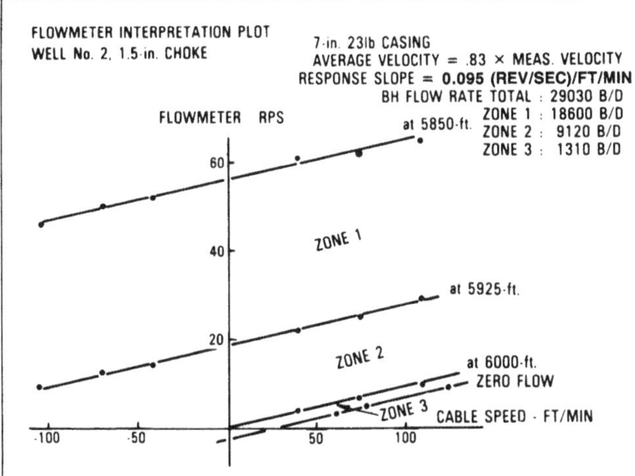

Fig. 9.12—In-situ calibration plot, spinner flowmeter in high-rate well (from Ref. 7).

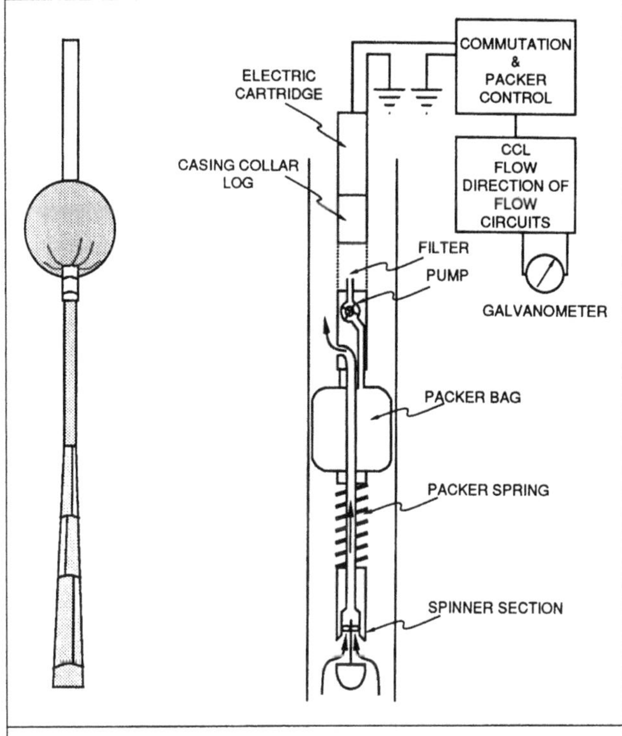

Fig. 9.13—Schlumberger packer flowmeter (from Ref. 9, courtesy Schlumberger).

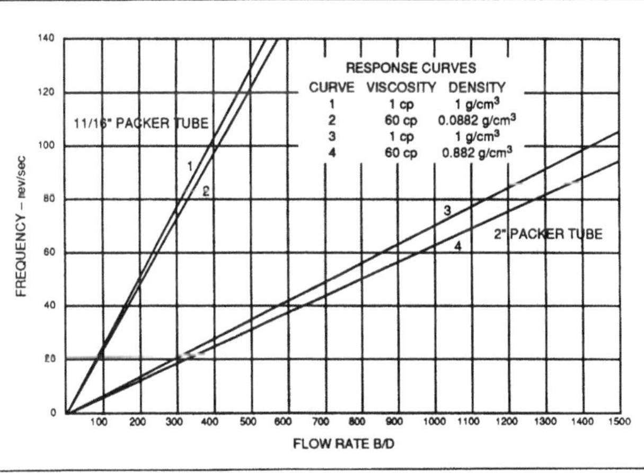

Fig. 9.14—Packer-flowmeter response in single-phase flow (from Ref. 9, courtesy Schlumberger).

this flow regime, a spinner flowmeter should respond much as it does in a single-phase flow.

Spinner flowmeters generally will be more reliable in high-velocity flow streams in multiphase flow. No absolute guidelines are currently available to predict which flow rates are sufficient for accurate use of spinner flowmeters. When a spinner flowmeter is run in a multiphase well, the standard single-phase interpretation methods are often unreliable. It is often best to make a few passes first with the spinner flowmeter in the well to get a general picture of the flow behavior, and then to make a series of stationary measurements. With the spinner stationary, the response can be averaged over time, eliminating some of the error caused by flow fluctuations.

Example—Spinner Flowmeter in High-Rate, Two-Phase Well. Weiss *et al.*[7] presented a case history of a spinner flowmeter run in a multiphase well in which the rates were sufficiently high for the flow to behave like a single-phase flow. This oil and gas well was producing at a downhole rate of 29,030 B/D [4615 m^3/d] in 7-in. [18-cm] casing. From a density log, it was determined that 40% of the production was gas. Fig. 9.11 shows one downward

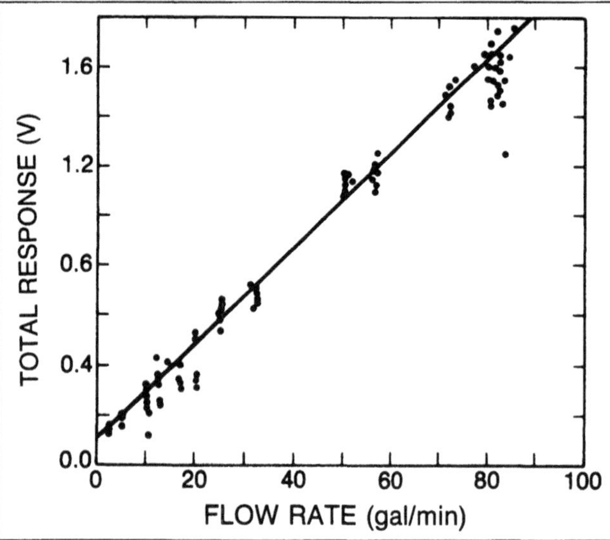

Fig. 9.15—Packer-flowmeter responses in one-, two-, and three-phase flows (from Ref. 4).

pass of the spinner at an unspecified cable speed. Though the spinner response is fairly noisy, there are regions where the response is reasonably smooth, such as from 5,915 to 5,930 ft [1803 to 1807 m]. Multiple downward passes were made in the well (Fig. 9.12 shows the in-situ calibration plot). A reasonably linear response was obtained at each station. No upward passes were made, so the threshold velocity could not be determined. In a high-rate well such as this one, however, the threshold velocity would be a very minor correction.

9.4.2 Flow-Concentrating Flowmeters. The best devices currently available to measure velocity in multiphase flow when volumetric flow rates are relatively low are tools that force all flow to pass through a small chamber containing a turbine flowmeter. When the flow is forced into a much-smaller-diameter flow path inside the tool, the fluid velocities are increased enough to make the flow stream relatively homogeneous; thus, the turbine meter responds linearly to total velocity. The packer flowmeter and the basket flowmeter are two types of flow-concentrating flowmeters that have been used. The basket flowmeter is currently the most commonly used.

Connolly[8] described two packer flowmeters in use in 1965—the Birdwell-Humble flowmeter and the Schlumberger packer flowmeter. Today, only the Schlumberger packer flowmeter remains in very limited use in isolated cases, probably because of operational difficulties often encountered.

Fig. 9.13 shows the Schlumberger packer flowmeter. Flow is forced through the measurement section by an inflatable bag or packer. When the tool is at the desired location for a measurement, a pump in the tool fills the packer with wellbore fluids, sealing the wellbore and forcing all flow through the turbine meter. Because the fluids are forced through a small opening, they are well mixed and their behavior is similar to that of fluids in single-phase flow.

Packer-flowmeter measurements must be made stationary. Because the packer bag is made of a rubber material that can tear, measurements are generally made only in regions of smooth pipe—i.e., between perforated intervals. This reduces the depth-resolution capability of the packer flowmeter. Problems are also sometimes encountered when the deflated packer bag tears as the tool is moved. Because of this, some wells with rough walls cannot be logged with a packer flowmeter. The fragility of the packer bag is probably why its use is very limited.

The packer flowmeter yields a fairly linear response to total flow rate or average velocity in multiphase flow. A Schlumberger Ltd.[9] calibration chart (Fig. 9.14) shows that there is some sensitivity to fluid viscosity, as is expected with any turbine flowmeter. In tests in multiphase flow, the packer flowmeter also exhibited a linear response to total flow rate, as shown in Fig. 9.15. In this plot, the data that deviated most significantly from the linear trend were those for high gas flow rates.

Fig. 9.16—Atlas Wireline Services basket flowmeter (courtesy Atlas Wireline Services, Western Atlas Intl. Inc.).

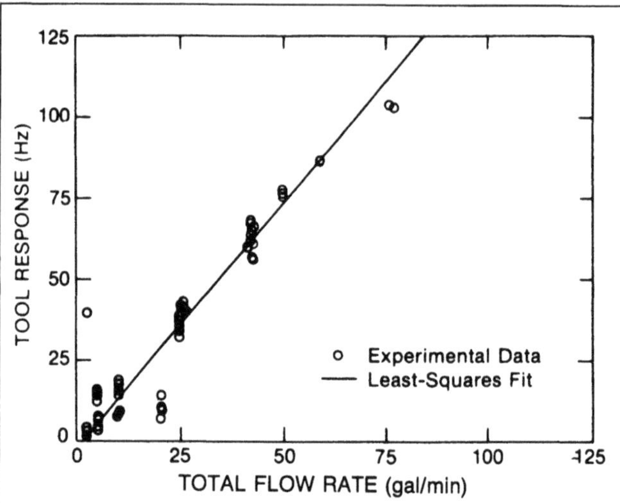

Fig. 9.17—Basket-flowmeter responses in one- and two-phase flows (from Ref. 4).

The packer flowmeter has a maximum rate limitation. Schlumberger Ltd.[10] recommended a maximum rate of about 1,900 B/D [302 m^3/d] for the $2\frac{1}{8}$-in. [5.4-cm] packer flowmeter; however, the tool has been used successfully at rates as high as 3,000 B/D [477 m^3/d]. One primary factor contributing to the flow-rate limitation is the force acting on the packer flowmeter resulting from the pressure drop across the packer. As the flow rate increases,

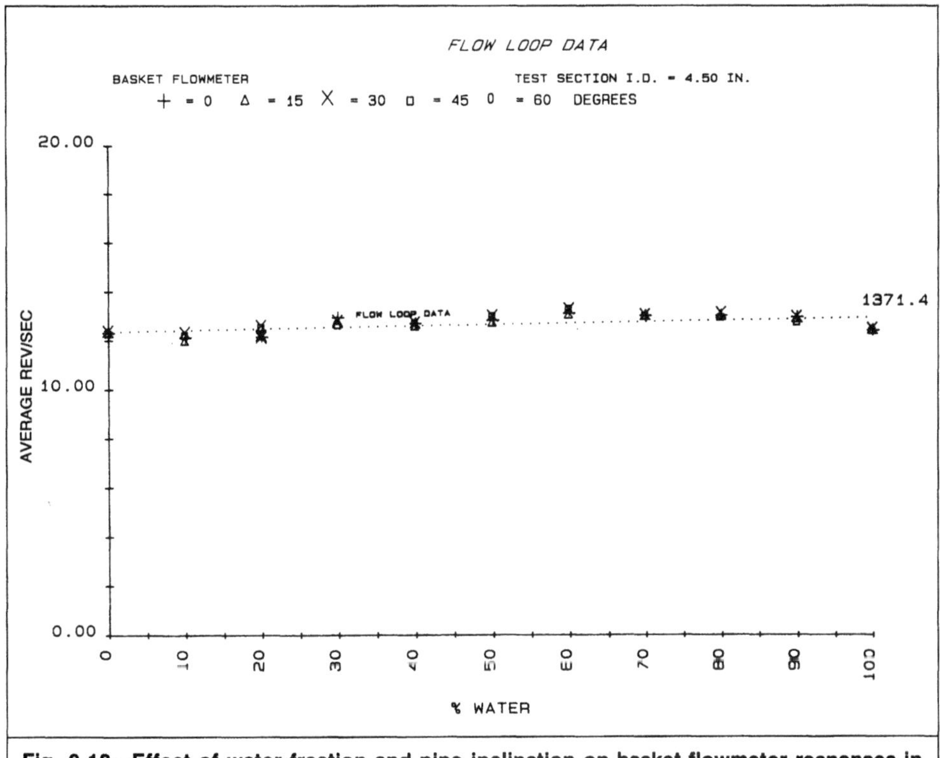

Fig. 9.18—Effect of water fraction and pipe inclination on basket-flowmeter responses in oil/water flow (from Ref. 11, courtesy Society of Professional Well Log Analysts).

this force can become high enough that the packer cannot be held in place.

The basket flowmeter works on the same principle as the packer flowmeter, differing only in the mechanism used to force the flow through the tool. The basket flowmeter has a metal funnel that can be expanded or contracted to capture the flow. The funnel is composed of overlapping metal petals attached to centralizer struts that can be expanded or contracted with a motor activated from the surface. Fig. 9.16 shows an Atlas Wireline Services basket flowmeter.

The basket-flowmeter readings should be obtained when the tool is stationary; the basket is usually closed to move from station to station. This tool is rugged enough that measurements can be made in perforated intervals. In some cases, a basket-flowmeter log is run by opening the basket at the lowest location to be logged, and then pulling the tool back up the well with the basket open, stopping whenever a reading is to be taken.

The basket-flowmeter response is similar to that of the packer flowmeter—fairly linear with total flow rate. Fig. 9.17 shows the responses of a basket flowmeter in single- and two-phase flow. Other tests have shown that this response is not sensitive to water fraction or pipe inclination in oil/water flow,[11] as Fig. 9.18 illustrates. The basket flowmeter may show a somewhat diminished response compared with the packer flowmeter, particularly in gas/liquid flow, because of fluid leakage through the basket. This leakage also limits the basket flowmeter to rates under about 3,000 B/D [477 m^3/d], with this upper limit depending on the tool and casing size. Above this rate, the tool response is essentially flat because the added flow passes through the basket rather than being forced through the turbine meter in the tool.

A flow-concentrating flowmeter that combines features of a basket flowmeter and a packer flowmeter is the Schlumberger inflatable diverter tool[12] (Fig. 9.19). This device uses a fabric diverter with an inflatable ring inside a metal cage. When the tool is positioned at a location to make a measurement, the metal cage is opened and the inflatable ring is filled with a fluid contained in the tool to form a seal with the casing wall.

The response of the inflatable diverter tool to total flow rate in oil/water flow has been shown to be quite linear in flow-loop tests (Fig. 9.20). As with any flow-concentrating flowmeter, there is an upper limit on the flow rate that can be measured; above this rate, the tool will be damaged or pushed up the hole. Piers *et al.*[12] suggest an upper limit of about 2,100 B/D [334 m^3/d] for the inflatable diverter tool.

When a flow-concentrating-flowmeter log is interpreted, it is assumed that the tool output is linear with average in-situ velocity, or

$$f = mv_M. \quad (9.10)$$

Because the average velocity is just the total volumetric flow rate divided by the cross-sectional area of the pipe, the response is thus linear with total volumetric flow rate. This neglects the effect of a threshold velocity, which should be small for a flow-concentrating flowmeter.

9.4.3. Radioactive-Tracer Logging. Radioactive-tracer logging (velocity-shot method) can be used to measure velocity in production wells. Because radioactive material is produced to the surface, these techniques are seldom used in production wells. However, by taking such precautions as storing the produced fluid until the level of radioactivity has decayed to a safe level, radioactive-tracer logs can be used in production wells.

A radioactive-tracer log is potentially valuable in multiphase flow because, in most instances, the tracer will travel at the velocity of the continuous phase, no matter which phase is miscible with the tracer. As an example, consider an aqueous tracer in an oil/water two-phase flow. If the flow regime consists of oil bubbles dispersed in a continuous water stream, the tracer will mix with the water and travel at the velocity of the water. If the flow is water droplets entrained in a continuous oil phase, however, the aqueous tracer will most likely travel at the velocity of the oil because there will be insufficient time for the small droplets of tracer to coalesce with the water bubbles before the tracer passes the gamma ray detectors. This will depend to a large extent on how the tracer is initially dispersed. If the tracer bubbles are small, they will likely travel at the continuous-phase velocity. The larger the tracer bubbles, the more they will be slowed relative to the oil phase because of the density difference between the oil and the tracer.

The advantage offered by the tracer log is its ability to measure directly the velocity of just one of the phases. This, coupled with

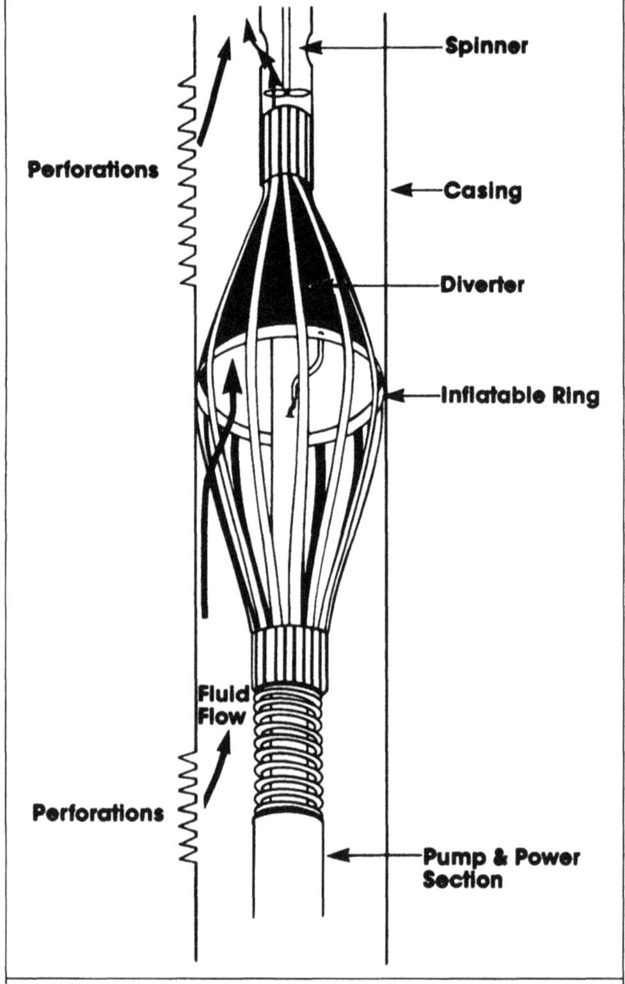

Fig. 9.19—Schematic of inflatable diverter tool (from Ref. 12).

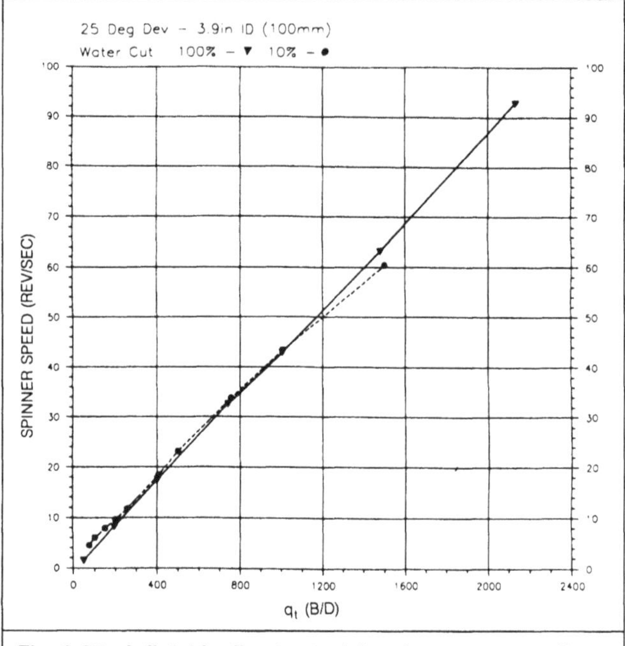

Fig. 9.20—Inflatable diverter tool flow-loop response (from Ref. 12).

measurements of total velocity and holdup, provides a means of determining the volumetric flow rates of both phases in a two-phase flow without making any of the assumptions usually required in the interpretation of fluid-velocity and holdup measurements.

Consider an oil/water flow in which the water is the continuous phase. A radioactive-tracer log would measure the average in-situ water velocity, $\bar{v}_w$. If the holdup were also measured (with a density or capacitance log), the volumetric flow rate of water could be directly calculated as

$$q_w = \bar{v}_w y_w A_w. \quad \text{(9.11)}$$

If the average total velocity, v_M, was also measured, such as with a flow-concentrating flowmeter, the total volumetric flow rate is

$$q_t = v_M A_w, \quad \text{(9.12)}$$

and the volumetric flow rate of oil is

$$q_o = q_t - q_w. \quad \text{(9.13)}$$

It is also possible to use a tracer in a multiphase well in a manner analogous to that of a tracer-loss log. This would be accomplished by ejecting a slug of tracer into a low-flow-rate region in the lower part of a well and monitoring the dilution of the tracer slug as it moves up the wellbore past fluid-entry locations.

In addition to the problem of producing radioactivity to the surface, a drawback of using radioactive tracers in a multiphase stream is that dispersion of the tracer is likely to be more pronounced than in a single-phase flow because of the additional turbulence in the multiphase flow stream. In such flow regimes as slug or churn flow, where there is periodic fallback of the denser phase, the tracer may be dispersed so rapidly that velocity-shot measurements may be impossible. In such cases as deviated wells, where flow is upward on the upper side of the pipe and downward on the lower side of the pipe, some tracer will likely be carried in each direction, quickly diluting the tracer slug. In a gas/liquid flow, tracer may be stripped into the gas phase, which also tends to disperse the tracer slug rapidly. To minimize tracer-dispersion effects, the detector spacing should be kept as small as possible in a multiphase flow.

9.5 Fluid-Identification Logs

In a two-phase flow, a log that can measure the in-situ fraction of the wellbore occupied by each phase is needed. Two types of meas-

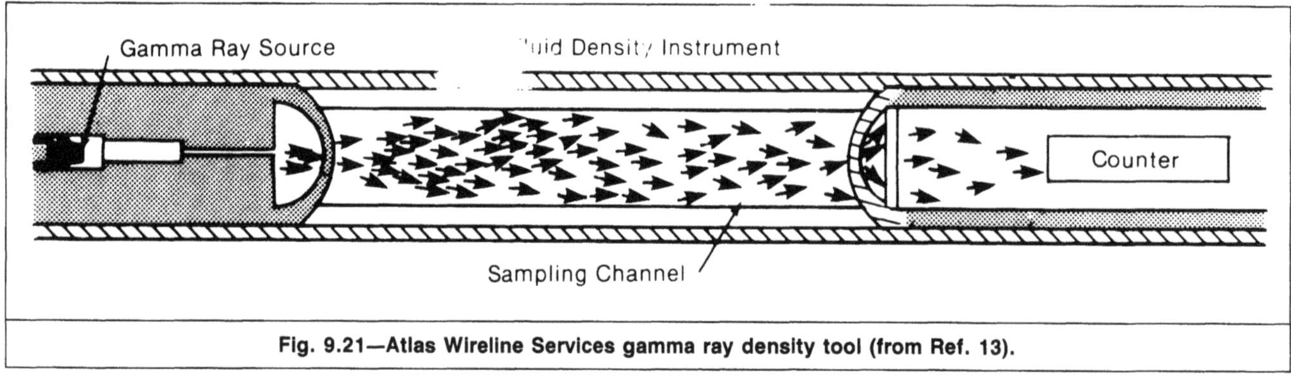

Fig. 9.21—Atlas Wireline Services gamma ray density tool (from Ref. 13).

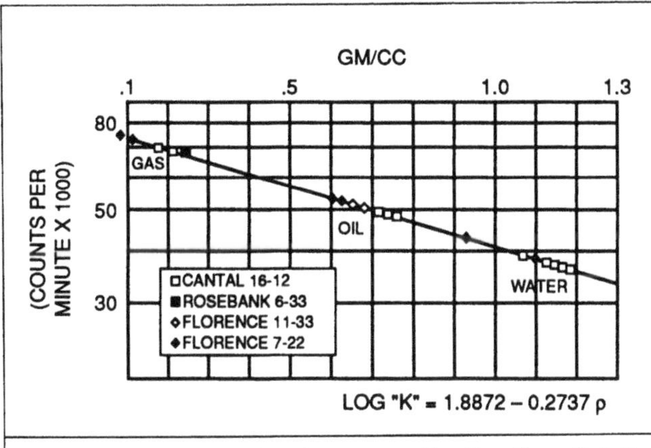

Fig. 9.22—Gamma ray densitometer calibration (from Ref. 8, courtesy Society of Professional Well Log Analysts).

urements are commonly used for this purpose: fluid-density and fluid-electrical-capacitance.

9.5.1 Fluid-Density Logs. Two types of tools are used to measure fluid density—those based on gamma ray absorption and those based on a pressure-drop measurement.

The most common type of fluid-density tool is the gamma ray densitometer. This tool is based on the fact that gamma ray absorbance is inversely proportional to the density of the material through which the gamma rays are passing. The device consists of a gamma ray source (a collimated beam, used to prevent backscatter from the casing wall), an open space through which wellbore fluids can pass, and a gamma ray detector. Fig. 9.21 shows a schematic of Atlas Wireline Services' density tool. The tool response decreases logarithmically as the fluid density increases so that

$$\log N_{GR}=a-b\rho_f, \quad (9.14)$$

where N_{GR}=count rate, a and b=constants, and ρ_f=fluid density. Fig. 9.22 shows a typical response of a gamma ray densitometer to single-phase fluids.

A gamma ray densitometer has several inherent limitations, including the statistical nature of the measurement, small sampling size, and low sensitivity in oil/water flow.

As in any nuclear measurement, there will be statistical fluctuations in gamma ray densitometer readings because the gamma ray source does not emit radiation at a constant rate. This effect can be minimized by using a strong gamma ray source and by averaging the tool response over a finite time period. Thus, stationary measurements are advantageous with a gamma ray densitometer so that the response can be averaged to eliminate statistical fluctuations.

Another major problem with the gamma ray tool is the small sampling size. As with a spinner flowmeter, a gamma ray densitometer senses density of only a small portion of the flow stream; thus, the density measured may not represent the average fluid density. If the flow regime is highly nonuniform, as in inclined flow, a gamma ray densitometer may yield a reading that has no relation to average density.

As with any density tool, sensitivity is low in an oil/water stream in which there is a small difference between the densities of the oil and the water. A density tool will most clearly distinguish between gas and liquids.

The primary density tool based on a measurement of pressure drop is the Schlumberger Gradiomanometer™. The Gradiomanometer determines average fluid density by measuring the pressure differential across 2 ft [0.6 m] of wellbore. Recall that the pressure gradient in multiphase flow can be written as

$$\Delta p=\Delta p_{HH}+\Delta p_f+\Delta p_{E_K}, \quad (9.15)$$

where Δp_{HH}=hydrostatic head of the flow stream, Δp_f=frictional pressure gradient, and Δp_{E_K}=pressure gradient resulting from ki-

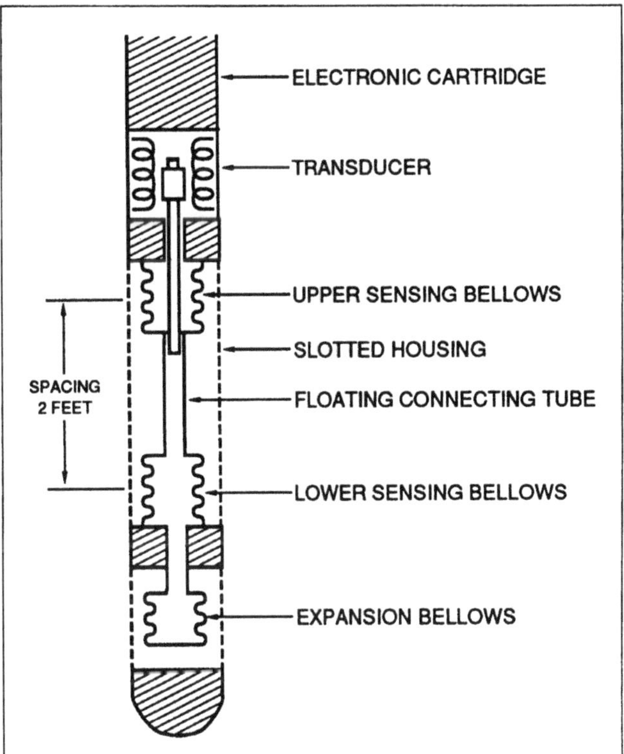

Fig. 9.23—Schlumberger Gradiomanometer (from Ref. 9, courtesy Schlumberger).

netic energy change. Assuming that Δp_{E_K} is negligible and that Δp_f is small compared to Δp_{HH}, then

$$\Delta p=g\bar{\rho}\cos\theta, \quad (9.16)$$

where $\bar{\rho}$=average in-situ density and θ=deviation from vertical. Thus, for these conditions, $\bar{\rho}$ can be determined from a pressure-gradient measurement in the wellbore. The holdup in two-phase flow can then be calculated according to Eq. 9.3.

With the Gradiomanometer, the pressure drop is measured as the difference in pressure exerted on two bellows spaced 2 ft [0.6 m] apart (Fig. 9.23). To distinguish between oil and water reliably, the pressure-drop measurement must be quite accurate. For example, assuming a water density of 1 g/cm^3 and an oil density of 0.8 g/cm^3, the pressure drop over 2 ft [0.6 m] for 100% water will be 0.866 psi [5.97 kPa], while the pressure drop in 100% oil will be 0.693 psi [4.78 kPa]. Thus, the maximum difference in pressure that would be measured would be 0.173 psi [1.19 kPa]. To measure holdup with 10% accuracy, the pressure-drop sensing device must be able to measure a change of only 0.017 psi [0.117 kPa] accurately.

The Gradiomanometer is calibrated to read fluid density directly in g/cm^3. A second curve, called an amplified Gradiomanometer curve, is also usually recorded. The amplified Gradiomanometer curve is simply an electronic amplification of the Gradiomanometer curve so that changes are more pronounced. The amplified Gradiomanometer typically will have five times the sensitivity of the Gradiomanometer reading so that a shift of one chart division on the Gradiomanometer will appear as a shift of five chart divisions on the amplified Gradiomanometer. Fig. 9.24 shows a typical Gradiomanometer log. The amplified Gradiomanometer curve is not normally zeroed, so it must be compared with the Gradiomanometer curve at some point to determine absolute values.

If the frictional pressure drop is significant compared to the gravitational term, the fluid density calculated from Eq. 9.16 will, of course, be in error. Using ρ_{Gr} to represent the apparent density from the Gradiomanometer reading, we can write

$$g\rho_{Gr}=g\bar{\rho}+\Delta p_f \quad (9.17)$$

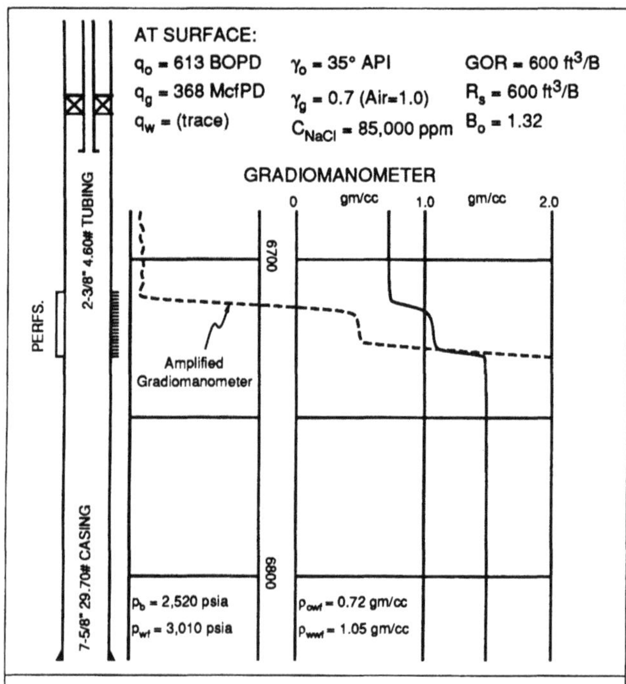

Fig. 9.24—Gradiomanometer log (from Ref. 10, courtesy Schlumberger).

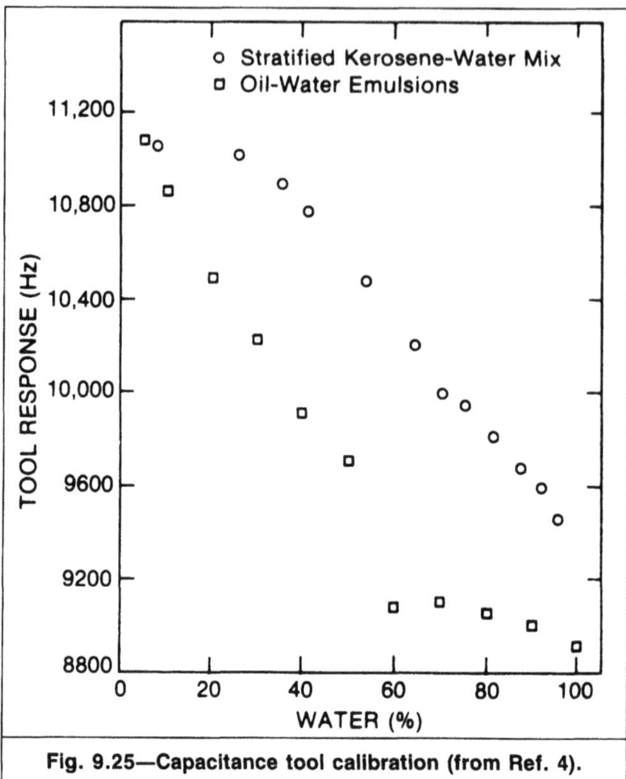

Fig. 9.25—Capacitance tool calibration (from Ref. 4).

or

$$\bar{\rho}=\frac{\rho_{Gr}}{1+F}, \quad (9.18)$$

where F=correction factor to account for frictional pressure losses. Schlumberger has empirically determined F values for flow rates greater than 2,000 B/D [318 m³/d].

The Gradiomanometer readings must be corrected for well deviation if the well is not vertical. The actual density in a deviated well is calculated from the Gradiomanometer readings by

$$\bar{\rho}=\frac{\rho_{Gr}}{\cos\theta}, \quad (9.19)$$

where $\bar{\rho}$ =actual in-situ density, ρ_{Gr}=Gradiomanometer density reading, and θ=deviation from vertical.

9.5.2 Capacitance Logs. To overcome the problems inherent in the ability of fluid density tools to distinguish between oil and water, another class of tools was developed to measure water fraction more accurately in multiphase flow. The devices are based on an electrical-capacitance measurement and are sometimes referred to as holdup meters or water-cut meters.

Capacitance tools are essentially coaxial capacitors. By application of a voltage potential between a central electrode and the outside of the logging tool, the capacitance of the device is determined. Because the measured capacitance is a function of the dielectric constant of the fluids in the sample chamber, the capacitance tool provides a measurement of dielectric constant. Liquid hydrocarbons have dielectric constants on the order of 2 to 6, while water has a dielectric constant of about 80. Thus, a dielectric-constant measurement distinguishes easily between hydrocarbons and water. Because the dielectric constants of gases are near one, the capacitance log does not discriminate well between oil and gas.

The advantage, then, of a capacitance-tool measurement is the wider separation between the responses to oil and water, thus providing a better means of measuring water holdup. Furthermore, in three-phase flow, a capacitance measurement in conjunction with a density measurement can provide an estimate of the holdup of each phase.

Capacitance tools are not without their drawbacks, the most serious of which are their nonlinear responses to water fraction. When water is the continuous phase in a multiphase flow, a capacitance tool has little sensitivity because a continuous electrical path exists through the water. Fig. 9.25 shows a calibration of a capacitance tool using mixtures of water and a manometer oil with a specific gravity of one so that stable mixtures were obtained. The tool sensitivity is good up to about 50%. Above this water fraction, water becomes the external or continuous phase and tool sensitivity is greatly diminished. The water fraction at which the change from oil external to water external occurs will depend on the particular oil and water in the system. It must also be remembered that the in-situ fraction of water present (the water holdup) will generally be larger than the input water fraction (the water cut.) For example, an in-situ water fraction of 40% may occur when only 20% of the total volumetric flow rate is water. Thus, a capacitance tool will measure water holdup most reliably when relatively small amounts of water are present.

The capacitance tool shares the problem of small sampling size with the density and spinner-flowmeter logs. A capacitance log may indicate 100% water, particularly in inclined wells, even though a significant oil stream is present because the oil is flowing along the upper side of the pipe and is undetected by the capacitance tool.

Carlson and Roesner[11] reported a series of tests of a capacitance log in oil/water flow in a flow loop that illustrates the behavior of this tool. Kerosene and fresh water were the fluids used, and the capacitance tool was placed just above a basket flowmeter that was kept closed in some tests and opened in others. Fig. 9.26 shows the capacitance-tool response as a function of input water fraction for deviations of 18 and 36° from vertical for a total volumetric flow rate of 1,029 B/D [164 m³/d]. The normalized tool response was 6.7 in 100% oil and 0 in 100% water. In this test, the capacitance response dropped dramatically as soon as water was introduced, and then gradually diminished with increasing input water fraction. Several effects combined to reduce the capacitance-tool sensitivity in this test. At this relatively low rate in an inclined pipe, the flow regime will be a stratified flow, with most of the oil flowing along the upper side of the pipe. The water holdup will be relatively high, and water will be the dominant phase in the center of the pipe, where the tool is located. Because the capacitance tool is not very sensitive to water fraction when water is the continuous

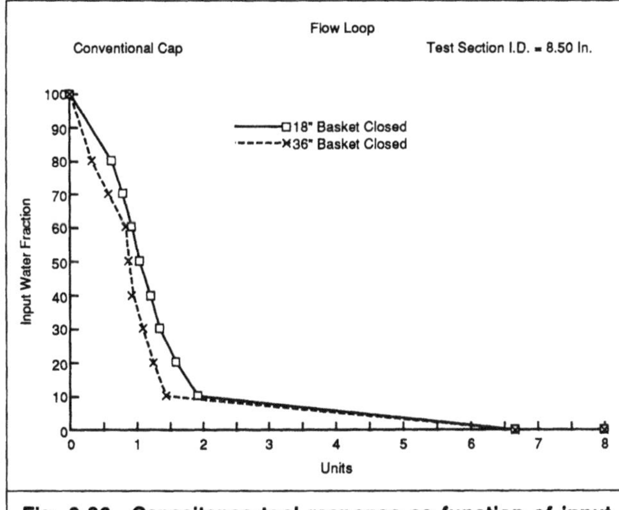

Fig. 9.26—Capacitance tool response as function of input water fraction (from Ref. 11, courtesy Society of Professional Well Log Analysts).

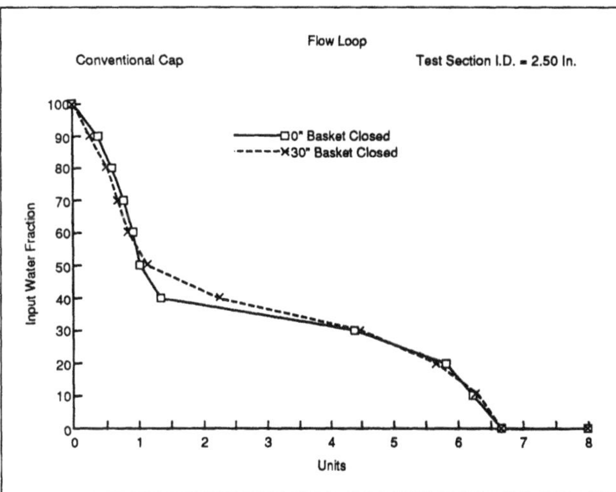

Fig. 9.27—Capacitance tool response to water fraction in smaller pipe size (from Ref. 11, courtesy Society of Professional Well Log Analysts).

phase, sensitivity is low over the entire range of input water fractions in this test.

Increasing the velocity will tend to mix the oil and water more uniformly, increasing the sensitivity of the capacitance tool. Figs. 9.27 and 9.28 illustrate this effect. At a higher total flow rate in a smaller pipe (Fig. 9.27), it appears that oil is the external phase up to 40 to 50% input water fraction, as indicated by the greater sensitivity of the capacitance tool. At a significantly higher flow rate, 6,657 B/D [1058 m^3/d] in the 6.5-in. [16.5-cm] -ID pipe (Fig. 9.28), sensitivity is again greater up to an input water fraction of 50% compared with the previous tests at 1,029 B/D [164 m^3/d].

Even when water is the external phase, the capacitance tool has some sensitivity to water fraction, though it is much lower than when oil is the external phase. To apply a capacitance tool properly, it must be recognized that the response to water fraction is nonlinear, with a discontinuity when the external phase changes from oil to water. With a calibration run with the particular oil and water from the well to be logged, the capacitance tool can be interpreted to yield a reasonable estimate of in-situ water fraction or holdup. Of course, well deviation that causes a nonuniform concentration profile across the pipe can still diminish the accuracy of the log.

9.5.3 Combination Flow-Concentrating Tools. A flow-concentrating flowmeter is often a significant improvement over a continuous spinner flowmeter in a multiphase flow because it tends to

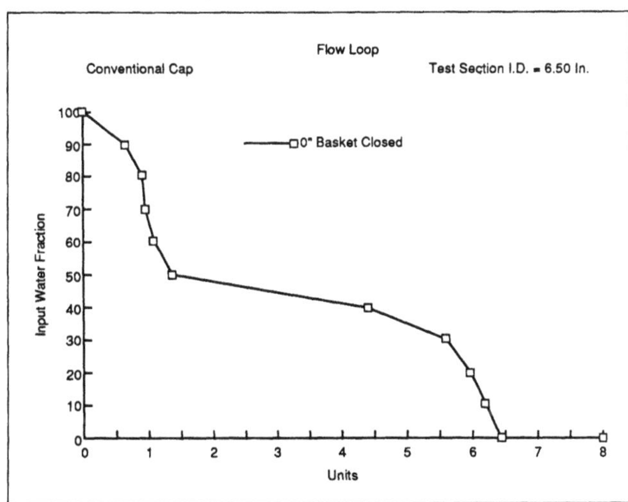

Fig. 9.28—Capacitance tool response to water fraction at higher flow rate (from Ref. 11, courtesy Society of Professional Well Log Analysts).

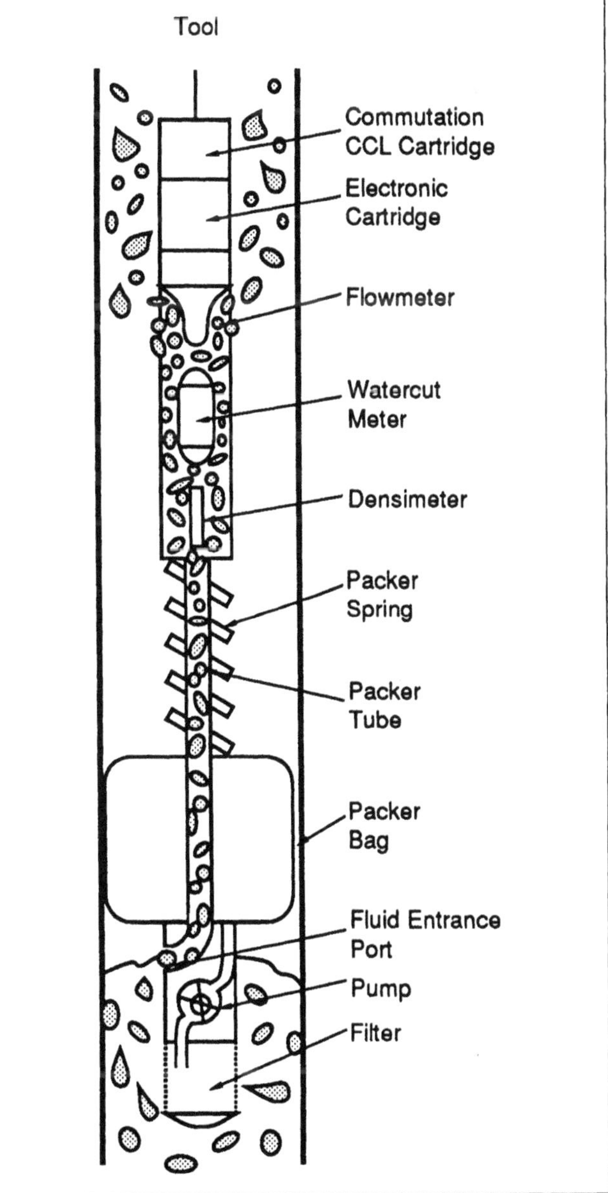

Fig. 9.29—Schlumberger packer-flowmeter-fluid analyzer (from Ref. 9, courtesy Schlumberger).

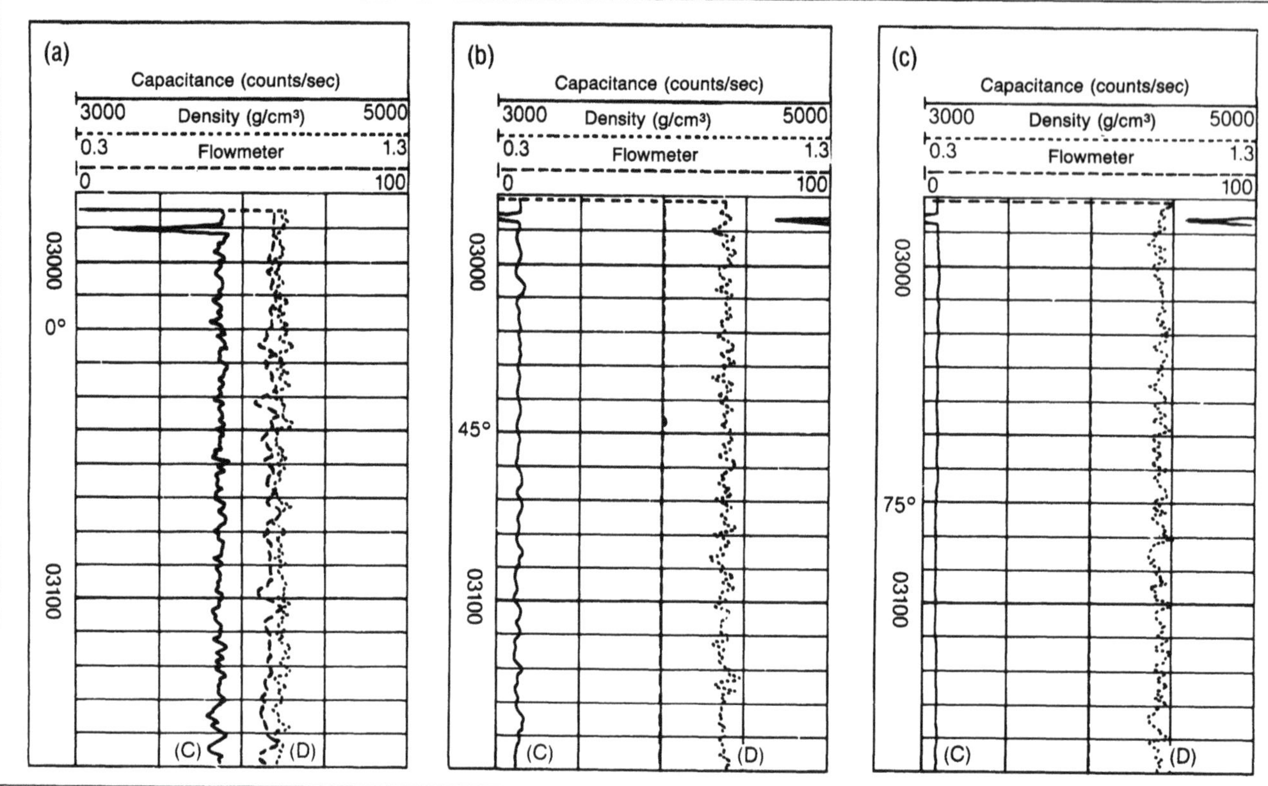

Fig. 9.30—Sensitivity of capacitance and density responses to pipe inclination (3,500 B/D, 50% oil, basket closed) (from Ref. 14, courtesy Society of Professional Well Log Analysts).

homogenize the flow sufficiently for a representative measure of average flow velocity to be made. The same principle has been extended to density and capacitance measurements by adding these devices to a flow-concentrating instrument while the flow stream is still captured in the tool. The advantages of this approach are that it will usually eliminate the problem of small sampling area that is so common to density and capacitance measurements and that the flow stream is usually mixed sufficiently that the water fraction measured will be the input fraction (no-slip holdup). Of course, such instruments are limited to fairly low flow rates, as are the flow-concentrating flowmeters.

Fig. 9.29 shows one of the first logging tools of this type, the Schlumberger Packer Flowmeter-Fluid Analyzer.[9] Density, capacitance, and spinner-flowmeter instruments measure flow properties of the captured fluid stream. Like the packer flowmeter, this instrument is no longer generally available.

More recently, Atlas Wireline Services introduced the basket dielectric logging instrument,[11] a basket flowmeter with a capacitance tool added to measure the captured flow. Carlson and Roesner[11] and Carlson *et al.*[14] also reported improved capacitance and density log results, particularly in inclined flow, when a capacitance or density tool was used above an opened basket flowmeter. The density and capacitance measurements are made after the wellbore fluids exit from the basket tool; however, it appears that the fluids remain sufficiently mixed for a short distance above the basket tool to yield improved density and capacitance measurements. Figs. 9.30 and 9.31 illustrate the difference this approach can make. The density and capacitance tool responses in Fig. 9.30 were obtained by measuring the unmixed (basket closed) well stream for deviation angles of 0, 45, and 75° from vertical. A comparison of Figs. 9.30a and 9.30b shows that significant changes in capacitance and density responses occurred, even though the flow rates of oil and water did not change. This results from the increased water holdup and less-uniform distribution of water across the pipe as the wellbore is inclined. The responses in Fig. 9.31 were obtained with a basket flowmeter opened just below the density and capacitance tools. In this case, the density and capacitance tool responses show little sensitivity to well deviation because of the mixing of the flow stream caused by the basket flowmeter.

9.6 Quantitative Analysis of Multiphase-Flow Logs

Quantitative interpretation of logs in multiphase flow consists of techniques that yield values of volumetric flow rate of each phase at several depth locations in the wellbore as compared with qualitative-interpretation methods that provide a general indication of the location of oil, water, or gas entries. In this section, the analysis methods are restricted to two-phase flow; the extension to three-phase flow is presented later. Quantitative analysis of two-phase-flow logs is based on a measurement of holdup (a density or capacitance log) and a flowmeter log that yields the average velocity of the flow stream. The basic analysis method was presented by Curtis[15,16] and by Nicolas and Witterholt[17] and will be illustrated here for oil/water two-phase flow. Identical interpretation procedures can be applied to gas/liquid flow by substituting gas for oil and liquid for water in the equations.

Consider oil/water flow in which average velocity, v_M, is measured with a flow-concentrating flowmeter, and average density, $\bar{\rho}$, is measured by a density tool. $\bar{\rho}$ is assumed to be the true average density of the wellbore flow stream. As shown earlier (Eqs. 9.3, 9.8, and 9.9), holdup and velocity of each phase can be calculated with

$$y_w=\frac{\bar{\rho}-\rho_o}{\rho_w-\rho_o},$$

$$\bar{v}_o=v_s y_w+v_M,$$

and

$$\bar{v}_w=v_M-(1-y_w)v_s.$$

The volumetric flow rates of each phase, q_o and q_w, are related to the velocities by Eq. 9.11,

$$q_o=\bar{v}_o A_w y_o$$

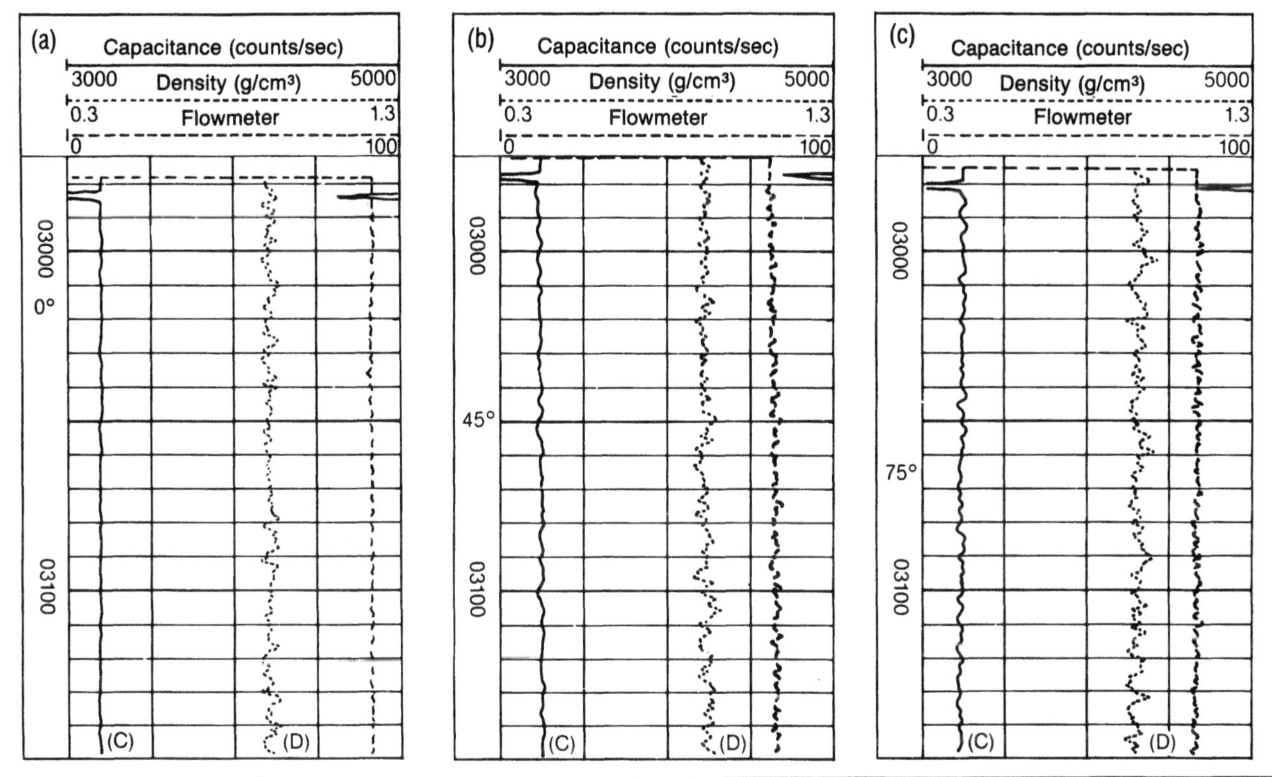

Fig. 9.31—Capacitance and density responses above open basket flowmeter (3,500 B/D, 50% oil) (from Ref. 14, courtesy Society of Professional Well Log Analysts).

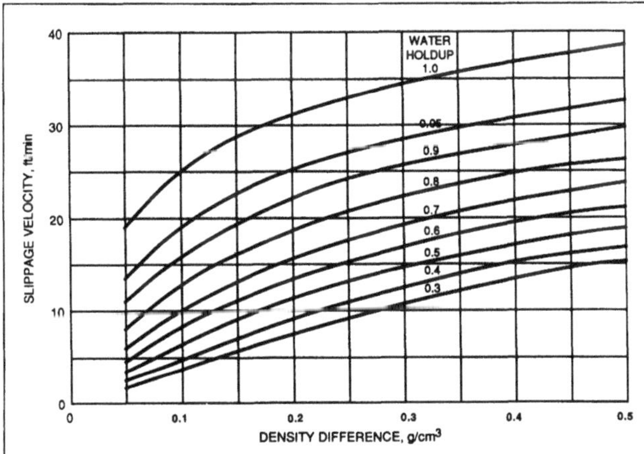

Fig. 9.32—Slip velocity as a function of density difference in oil/water bubble flow (from Ref. 9, courtesy Schlumberger).

and

$$q_w = \bar{v}_w A_w y_w,$$

so that

$$q_o = A_w(1-y_w)(v_s y_w + v_M) \quad \text{(9.20)}$$

and

$$q_w = A_w y_w [v_M - (1-y_w) v_s]. \quad \text{(9.21)}$$

To analyze a log, then, some means must be used to estimate the slip velocity. The following sections describe three approaches—estimating slip velocity from laboratory data, calculating slip velocity from log responses above all perforations, and calculating slip velocity with a two-phase-flow correlation.

9.6.1 Slip Velocity From Laboratory Data. Several empirical methods have been used to estimate slip velocity from laboratory data. Curtis[16] presented slip-velocity values depending only on the density difference between the phases. On the basis of laboratory data taken for vertical bubble flow in oil/water and gas/water flow and field experience, the following slip velocities were recommended:

v_s = 10 ft/min [3 m/min] for oil bubbles in water, $\rho_w - \rho_o < 0.15$ g/cm^3,
v_s = 20 ft/min [6 m/min] for oil bubbles in water, 0.15 g/cm$^3 < \rho_w - \rho_o < 0.35$ g/cm^3, and
v_s = 60 ft/min [18 m/min] for gas bubbles in oil or water.

Considering the complicated dependence of holdup and thus slip velocity on flow regime, flow rate, and fluid properties, this approach must be considered very approximate.

To improve this approach, Schlumberger Ltd.[9] presented a chart (Fig. 9.32) based on laboratory oil/water bubble flow experiments to estimate slip velocity as a function of both density difference and water holdup.

On the basis of a theoretical analysis of the rise velocity of gas bubbles in a stagnant liquid and experiments with oil/water bubble flow, Nicolas and Witterholt[17] suggested empirical equations relating slip velocity to fluid properties and holdup. Their model is

$$v_s = (y_w)^n v_{\lim} \text{ with } 1/2 < n < 2, \quad \text{(9.22)}$$

where

$$v_{\lim} = C[g\sigma\Delta\rho/\rho_L^2]^{1/4} \text{ with } 1.53 < C < 1.61. \quad \text{(9.23)}$$

In these equations, $v_{\lim}$ = bubble rise velocity in a stagnant liquid, g = acceleration of gravity, σ = interfacial tension (IFT), ρ_L = density of the denser phase, and $\Delta\rho$ = density difference between the phases. Nicolas and Witterholt suggest a value of 1 for n.

Any of the empirical approaches for estimating slip velocity must be considered very approximate. Laboratory measurements with

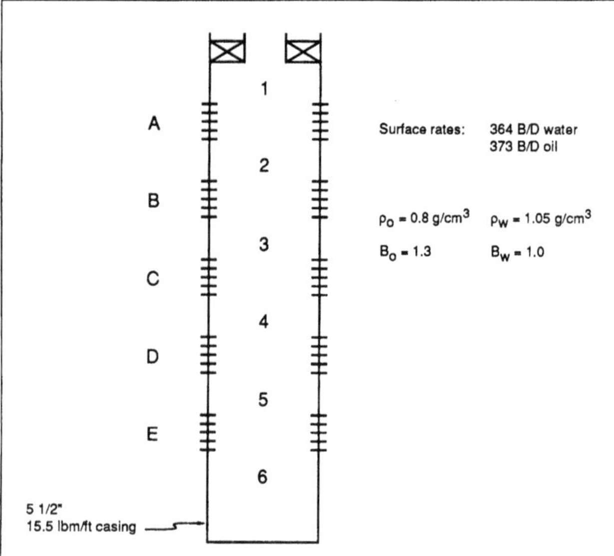

Fig. 9.33—Well producing oil and water (from Ref. 15, courtesy Schlumberger).

the particular oil and water from a well being logged would provide good estimates of slip velocity, but that is obviously an expensive proposition.

9.6.2 Slip Velocity From Log Responses Above All Perforations. Another procedure Curtis[15,16] suggested is to calculate slip velocity from one location in the well and assume that the slip velocity remains constant throughout the well. If the flow rates of oil and water at the surface and the formation volume factors (FVF's) are known and if no fluid leakage occurs in the tubing, the volumetric flow rates of oil and water above all production zones can be calculated from surface rates. With a logging measurement of average density and flowmeter response above all perforations, the slip velocity is calculated from

$$v_s=\frac{q_{o100}}{A_w(1-y_w)}-\frac{q_{w100}}{A_w y_w}, \quad \text{(9.24)}$$

where $q_{o100}=q_{os}B_o$ and $q_{w100}=q_{ws}B_w$.

The water holdup is calculated from the density measurement according to Eq. 9.3. The flowmeter is also calibrated from the measurement made above all production zones because, at this location,

$$q_{t100}=q_{o100}+q_{w100} \quad \text{(9.25)}$$

and

$$\bar{v}_{t100}=\frac{q_{t100}}{A_w}. \quad \text{(9.26)}$$

The flowmeter-response slope is then calculated as

$$m=\frac{\bar{v}_{t100}}{f_{100}}, \quad \text{(9.27)}$$

where the subscript 100 designates the measurement location above all production. The flowmeter response is assumed linear with flow rate, with no threshold velocity, so that at any other location in the well the fluid velocity is given by

$$v_f=mf. \quad \text{(9.28)}$$

If the wellbore cross-sectional area is constant, Eqs. 9.27 and 9.28 can be written in terms of volumetric flow rate:

$$m=q_{t100}/f_{100} \quad \text{(9.29)}$$

and

$$q=mf. \quad \text{(9.30)}$$

TABLE 9.1—PACKER FLOWMETER AND GRADIOMANOMETER DATA

Station	Packer Flowmeter (rev/sec)	Gradiomanometer (g/cm³)
1	60	0.95
2	36	1.00
3	23	0.99
4	12.4	1.03
5	3.5	1.05
6	0	1.18

9.6.3 Slip Velocity From Two-Phase-Flow Pressure-Drop Correlation. Chap. 8 showed that holdup and slip velocity are not independent, but simply two different ways to express the relative velocities between phases in two-phase flow. Central to most two-phase-flow correlations is a prediction of holdup; thus, slip velocity can be estimated with these correlations.

The use of a two-phase-flow correlation to calculate the volumetric flow rates of each phase on the basis of measurements of holdup and average velocity was illustrated in Chap. 8 for the Aziz *et al.* (Ref. 16 in Chap. 8) correlation for bubble flow.[18] From this correlation, Eq. 8.40 was derived for the superficial gas velocity as

$$v_{gs}=(1-y_L)(1.2v_M+v_{bs}),$$

where y_L=liquid holdup, v_M=mixture velocity ($v_{gs}+v_{Ls}$) and v_{bs} =rise velocity of a gas bubble in a stagnant liquid, given by

$$v_{bs}=1.41\left[\frac{g\sigma(\rho_L-\rho_g)}{\rho_L^2}\right]^{1/4}.$$

Eq. 9.20, written for a gas/liquid system, relates volumetric flow rate to slip velocity as

$$q_g=A_w(1-y_L)(v_s y_L+v_M). \quad \text{(9.31)}$$

Because superficial gas velocity is just volumetric flow rate divided by cross-sectional area, Eq. 9.31 can be written as

$$v_{gs}=(1-y_L)(v_s y_L+v_M). \quad \text{(9.32)}$$

Equating the right sides of Eqs. 8.40 and 9.32 yields

$$v_s=\frac{0.2v_M+v_{bs}}{y_L}. \quad \text{(9.33)}$$

For a given gas/liquid system, v_{bs} can be calculated from the properties of the fluids; slip velocity can then be obtained from measurements of total flow rate and holdup. It is interesting to compare this prediction of slip velocity with Nicolas and Witterholt's[17] empirical correlation (Eqs. 9.22 and 9.23). The $v_{\lim}$ in the Nicolas and Witterholt correlation is the same as v_{bs} in the Aziz *et al.* correlation, except for a slight difference in the constant. Thus, Nicolas and Witterholt found that slip velocity depends on the product of bubble-rise velocity and liquid holdup, while Aziz *et al.* predict that slip velocity depends on bubble-rise velocity divided by holdup. This is another example of how little two-phase-flow behavior is understood.

Example—Two-Phase-Flow Log Analysis: Slip Velocity From Flow Above All Perforations. Curtis[15] illustrates the two-phase-flow log analysis method in which slip velocity is determined from

TABLE 9.2—LOG ANALYSIS RESULTS

Station	q_t (B/D)	y_w	q_o (B/D)	q_w (B/D)
1	849	0.60	485	364
2	509	0.80	199	310
3	325	0.76	188	137
4	175	0.92	59	116
5	50	1.00	0	50
6	0	1.00+	0	0

the log responses above production. Five productive zones were perforated in the well shown in Fig. 9.33. A packer flowmeter and a Gradiomanometer were run in the well, with measurements taken at the six stations shown between each perforated interval (see Table 9.1 for data).

From the surface flow rates, the flow rates at Station 1 are calculated:

$$q_{o100}=(q_{os})B_o=(373 \text{ B/D})(1.3)=485 \text{ B/D } [77 \text{ m}^3/\text{d}],$$

$$q_{w100}=q_{ws}=364 \text{ B/D } [58 \text{ m}^3/\text{d}],$$

and

$$q_{t100}=q_{o100}+q_{w100}=849 \text{ B/D } [135 \text{ m}^3/\text{d}].$$

The flowmeter response is then obtained as

$$m=\frac{q_{t100}}{f_{100}}=\frac{849 \text{ B/D}}{60 \text{ rev/sec}}=14.15\frac{\text{B/D}}{\text{rev/sec}} \quad [2.25 \text{ (m}^3\text{/d)/(rev/sec)}].$$

The volumetric flow rate at each station can now be calculated with Eq. 9.30. Also, the holdup at each station is calculated with Eq. 9.3. Table 9.2 shows the results of these calculations.

Next, the slip velocity must be calculated at Station 1. First, the cross-sectional area is calculated for the annular space between the tool and the casing. For a $1^{11}/_{16}$-in. [4.3-cm] tool,

$$A_w=\frac{\pi}{4}(d_{ci}^2-d_T^2)$$

$$=\frac{\pi}{4}\left[\left(\frac{4.95}{12}\right)^2-\left(\frac{1.6875}{12}\right)^2\right]=0.1181 \text{ ft}^2 \ [0.01 \text{ m}^2].$$

Note that the annular area between the tool and the casing wall is used in this analysis because the density measurement was made with a Gradiomanometer. If a gamma ray densitometer is run, it is probably more appropriate to use the full wellbore cross-sectional area in this calculation because of the open construction of a gamma ray densitometer.

Converting the volumetric flow rates to cubic feet per minute yields

$$q_o=\frac{(485 \text{ B/D})(5.615 \text{ ft}^3\text{/bbl})}{1{,}440 \text{ min/D}}=1.89 \text{ ft}^3\text{/min } [0.05 \text{ m}^3\text{/min}]$$

TABLE 9.3—SENSITIVITY OF LOG INTERPRETATION TO DENSITY MEASUREMENT

ρ (g/cm³)	Zone B Production (B/D) Oil	Water	WOR
1.00	11	173	15.7
0.99	45	139	3.1

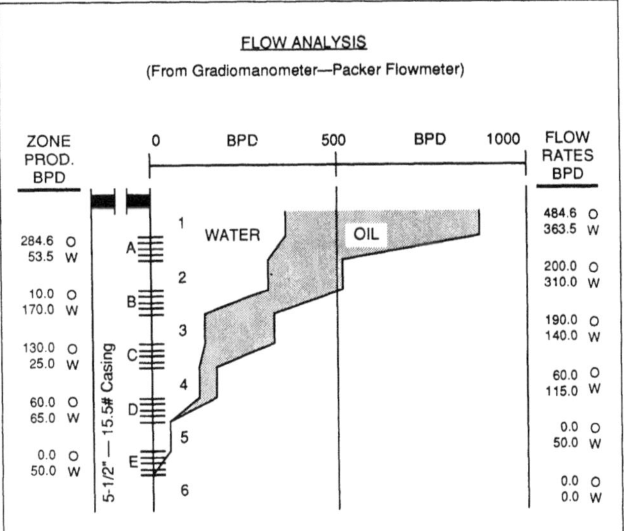

Fig. 9.34—Interpreted flow profile (from Ref. 15, courtesy Schlumberger).

and

$$q_w=1.42 \text{ ft}^3\text{/min } [0.04 \text{ m}^3\text{/min}].$$

With Eq. 9.24, the slip velocity is

$$v_s=\frac{1.89 \text{ ft}^3\text{/min}}{(0.1181 \text{ ft}^2)(1-0.6)}-\frac{1.42 \text{ ft}^3\text{/min}}{(0.1181)(0.6)}$$

$$=20 \text{ ft/min } [6 \text{ m/min}].$$

With this value, the volumetric flow rates of oil and water at the other stations are calculated from Eqs. 9.20 and 9.21. For example, at Station 2,

$$q_o=A_w(1-y_w)(v_s y_w+v_M),$$

$$q_{o_2}=(0.1181 \text{ ft}^2)(1-0.80)[(20 \text{ ft/min})(0.80)+16.8 \text{ ft/min}]$$

$$=0.775 \text{ ft}^3\text{/min}=199 \text{ B/D } [31.6 \text{ m}^3\text{/d}],$$

and

$$q_{w_2}=q_{t_2}-q_{o_2}=509-199=310 \text{ B/D } [49.3 \text{ m}^3\text{/d}].$$

Table 9.2 gives the final results obtained at each station. The interpreted log is usually presented as a plot of the flow rate of one phase and the total flow rate as a function of depth (e.g., Fig. 9.34).

One thing that must be remembered when two-phase-flow logs are analyzed is that the results are sensitive to small errors in the measurements. In the example analysis just shown, the results are very sensitive to the density measurements because of the strong dependence of all the calculations on holdup and the sensitivity of holdup to average density. For example, consider the results that would be obtained if the density measured at Station 2 had been 0.99 instead of 1.0. With this minor change, the holdup at Station 2 would be 0.76 instead of 0.80 and the flow rates would be 233 BOPD [37 m³/d oil] and 276 BWPD [44 m³/d water]. Table 9.3 shows the effect of this change on the interpreted production from Zone B. In the first case, Zone B is interpreted to be primarily a water zone, while in the second case, significant oil production appears to occur.

Because of this sensitivity, the log interpreter must estimate the accuracy of each tool response carefully and consider the range of possible flow-rate interpretations. Some idea of accuracy can be obtained by tool calibrations on the surface and by logging measurements in locations where the fluid composition and rate are known (e.g., in static water below all perforations). Another indication of the errors involved with the logging measurements is the noise level of the signal. For example, if a density tool measurement taken with a stationary tool varies with time over a range of 0.02 g/cm^3, the log interpreter knows that this error must be considered in the interpretation. Because of tool nonlinearities, it is a good practice to calibrate any density tool in situ with the well shut in.

9.7 Quantitative Analysis in Three-Phase Flow

To analyze three-phase flow, an additional logging measurement is needed to determine the in-situ fractions of each phase. In some cases, this third measurement can be obtained with a water fraction or a capacitance log. In this instance, the water holdup is obtained directly from the capacitance log because oil and gas behave similarly on the capacitance log. In three-phase flow,

$$y_w+y_o+y_g=1. \quad (9.34)$$

Again assuming that measurements of average density, $\bar{\rho}$, and average velocity, $\bar{v}$, are available, the oil holdup is calculated from

$$y_o=\frac{(\bar{\rho}-\rho_g)-y_w(\rho_w-\rho_g)}{\rho_o-\rho_g}. \quad (9.35)$$

The gas holdup is then

$$y_g=1-y_o-y_w. \quad (9.36)$$

The holdup of each phase is thus determined from the capacitance and density measurements. To conclude the analysis, it is assumed that no slip occurs between the oil and water. Thus,

$$\bar{v}_w=\bar{v}_o=\bar{v}_L \quad (9.37)$$

and

$$v_s=\bar{v}_g-\bar{v}_L. \quad (9.38)$$

This problem is now identical to the problem in the two-phase-flow case. The slip velocity can be determined from laboratory correlations or from logging measurements above all production zones by correcting surface flow rates to downhole conditions.

9.8 Qualitative Production Log Interpretation in Multiphase Flow

The discussions of production logging tool behavior in multiphase flow and of the interpretation methods have shown that many uncertainties exist when quantitative interpretation of production logs is attempted in multiphase flow. If the interpretation procedures are applied without consideration of the many factors that can adversely affect the log responses, a completely erroneous interpretation of the flow profile may result. Thus, it is often better to use multiphase production logs in a qualitative way. The log interpreter must examine all logs critically, emphasizing those best suited to the flow conditions and using other information about the well or reservoir to substantiate the log interpretation.

In multiphase-flow production logging, sufficient information is often unavailable to analyze the flow rates of each phase quantitatively. For example, a flow-concentrating flowmeter may have been run that yields good data, but a density or capacitance log was not run or the results are poor. Conversely, a good density log may be available, but flowmeter data are erratic. In these situations, the reliable logs are used quantitatively and any other logs will be examined to see whether they provide any evidence to refine the log interpretation further.

If a usable density or capacitance log has been obtained but no flowmeter log is available, Curtis'[15] log-analysis method designed for use with the Gradiomanometer can be used. In this method, oil and water flow rates are determined at discrete stations by assuming that each productive zone produces either oil or water, not both. This fixes the flow rate of one phase on either side of a production zone, allowing the flow rate of the other phase to be calculated from the density measurements. If the density increases with depth between stations, it is assumed that only oil was produced from the zone to account for the lower average density above the zone. Similarly, a density decrease moving down across a zone is attributed solely to water production. This analysis begins with calculation of the slip velocity from the value of holdup above all productive zones, where the flow rates of oil and water are known from surface rates. At each station, the flow rate of one of the phases is calculated (assuming that the other phase has a constant rate across the zone.) These rates are given by

$$q_o=A_w(1-y_w)v_s+\frac{1-y_w}{y_w}q_w \quad (9.39)$$

or

$$q_w=\frac{y_w}{1-y_w}(q_o)-v_sA_wy_w. \quad (9.40)$$

This analysis procedure can be illustrated with the density measurements from the example log in Table 9.1. At Station 1, the slip velocity is calculated to be 20 ft/min [6 m/min]. Between Stations 1 and 2, the density increases, so it is assumed that Zone A is producing only oil. At Station 2, the flow rate of water is interpreted to be the same as at Station 1, 364 B/D [58 m^3/d]. The oil flow rate, from Eq. 9.39, is

$$q_o=(0.1181\ \text{ft}^2)(1-0.8)(20\ \text{ft/min})\frac{(1{,}440\ \text{min/D})}{(5.615\ \text{ft}^3/\text{bbl})}$$

$$+\frac{(1-0.8)}{0.8}364\ \text{B/D}=212\ \text{B/D}\ [34\ \text{m}^3/\text{d}].$$

According to this analysis, Zone A produced 273 BOPD [43 m^3/d oil] and no water, as opposed to 286 BOPD [45 m^3/d oil] and 54 BWPD [8.6 m^3/d water] when the flowmeter information was used. The larger the density change across a zone, the more accurate the semiquantitative analysis based only on a density log would be.

Analysis of the flow profile based solely on a holdup measurement should be used only as a last resort and should be supported by other information, such as a temperature log. A major drawback to the method is that the most important information sought—the location of major oil, gas, and water entries—must be assumed before the analysis begins. The value of the incremental information obtained after this crucial assumption is questionable.

When a good flowmeter log is obtained but no reliable holdup log (density or capacitance) is available, the only quantitative analysis that can be made is a total flow profile. Sometimes, inferences about the oil and water production from zones can be made on the basis of total flow rates from the wells. For example, if a zone is producing 50% of the total flow rate according to the flowmeter log and the surface water cut is 35%, the zone is obviously producing some oil. In some instances, a density or capacitance log can yield supplemental information to a good flowmeter log. If a density or capacitance log indicates a significant change in holdup in a region where the flowmeter shows a large fluid entry, one can generally infer the phase of the entry, depending on the direction of the holdup change.

Many times, the most accurate interpretation of flow conditions in a multiphase well can be made with a temperature log[1] because it is not greatly affected by the composition of the flow stream in the wellbore. The temperature log will not yield precise information about the flow profile, but will often locate water- or gas-production zones. The following examples illustrate qualitative interpretation of production logs in multiphase flow.

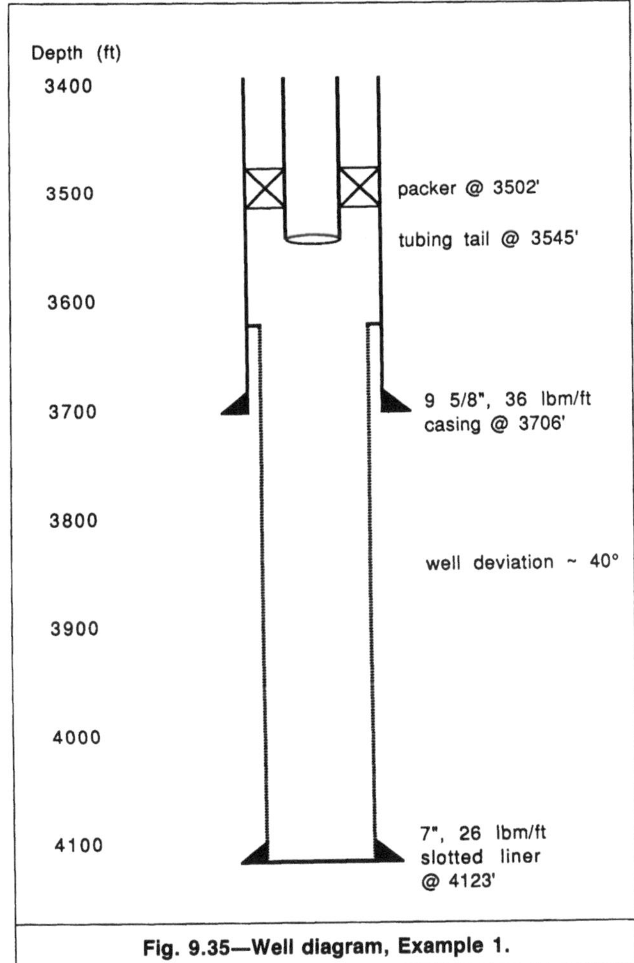

Fig. 9.35—Well diagram, Example 1.

Example 1—Basket Flowmeter and Gamma Ray Densitometer in Oil/Water Flow. The deviated well in Fig. 9.35 was producing 480 B/D [76 m^3/d] of liquid with 40% water cut from a slotted liner. The primary objective of the production logging was to identify the water-entry location. A spinner flowmeter had been run previously, but the data were very erratic, as expected at such a low flow rate in a deviated well. A basket flowmeter was run with a temperature log and a gamma ray densitometer to provide a better flow log. Fig. 9.36 shows data from the basket flowmeter, temperature, and density logs. The solid bars along the depth track indicate the locations of permeable zones.

Look first at the basket-flowmeter response. The response above all zones is 36 rev/sec, so this value corresponds to 100% flow. The fraction of total flow at any other location is obtained by dividing the basket-flowmeter response at that location by 36 rev/sec. Look now at the lower part of the well. The lowest zone is producing about 55% of the total flow because the basket-flowmeter response is 20 rev/sec at 3,994 ft [1217 m]. Fig. 9.37 shows the flow profile obtained with this analysis procedure.

The density log should be used, where possible, to indicate the type of fluid entering from the various flow-entry locations identified with the basket flowmeter. In Fig. 9.36, the dashed lines indicate the densities of the oil (0.9 g/cm^3) and water (1.02 g/cm^3) in this well. From the noisiness of the density-tool response, it is apparent that no quantitative information about holdup can be obtained, particularly with such a small difference in oil and water densities. Opposite the lowest zone, the density log fluctuates about the density of water, apparently indicating that this zone produces only water. However, the flowmeter shows that the lowest zone produces 55% of the total flow and that the well produces only 40% water at the surface. Because the FVF for oil is generally higher than that of water, the downhole oil flow rate will be proportionately higher, meaning that at downhole conditions, less than 40% of the total flow is water. Thus, despite the density-log response,

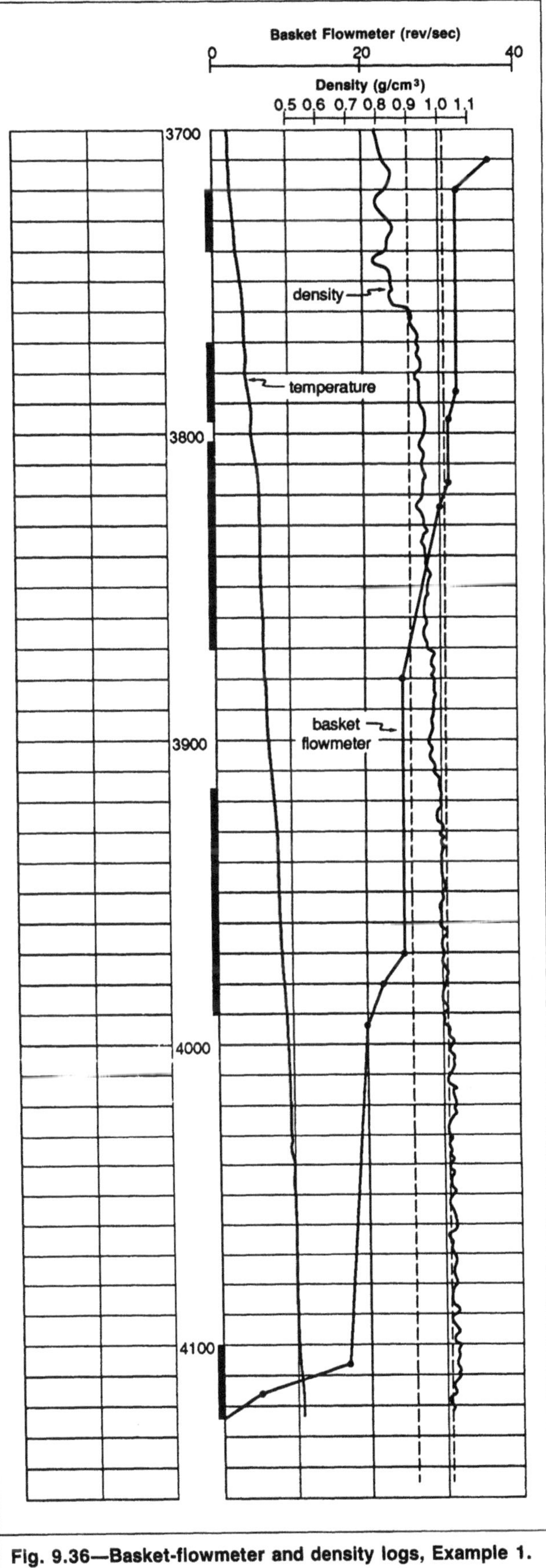

Fig. 9.36—Basket-flowmeter and density logs, Example 1.

the lowest zone cannot possibly be producing only water; some of its production must be oil. The density-tool response is caused by the oil flowing along the upper side of the pipe in the deviated well so that it is not passing through the tool's sample chamber.

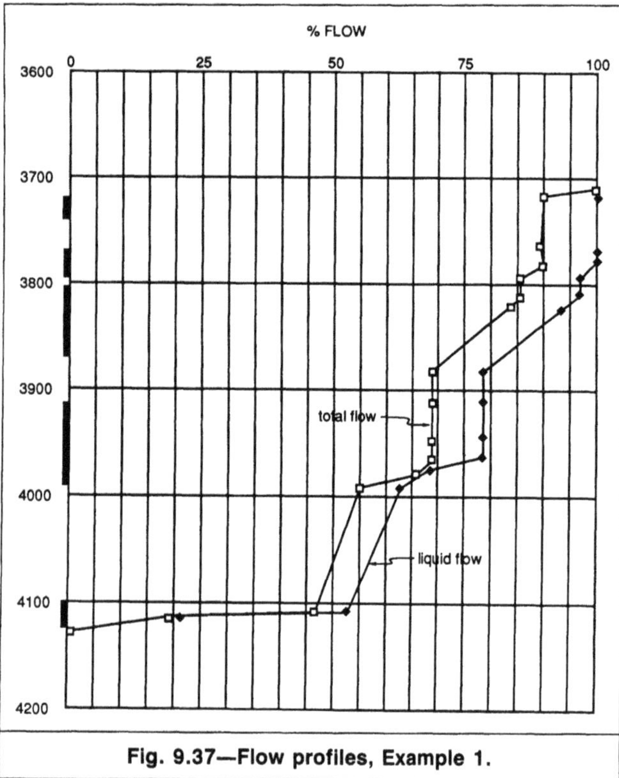

Fig. 9.37—Flow profiles, Example 1.

In the next zone up the well (3,916 to 3,990 ft [1194 to 1216 m]), there is a gradual shift to lower density as the tool moves through this zone. This would suggest that the production measured from this zone is oil. At about 3,910 ft [1192 m], a larger shift in density occurs, yet the basket flowmeter indicates no fluid entry at this depth. The decrease in density observed at this depth probably results from a small restriction in the wellbore or a small change in well deviation—i.e., some physical change that directs more of the oil stream to the center of the pipe where it can be detected by the densitometer.

At about 3,760 ft [1146 m], the fluid density drops below the oil density. This must result from gas in the flow stream, either solution gas that flashed from the oil or free gas that entered the wellbore. The increase in total flow rate above this location probably results from this gas flow. Thus, the flow response above all the zones, at 3,710 ft [1131 m], is influenced by the flow of this gas as well as by the oil and water flow. To get a more accurate profile of the liquid flow into the well, it would be preferable to assume that the basket-flowmeter response just below the first gas detected is the response to total liquid flow. Thus, 32 rev/sec would be the response to total liquid flow, yielding a slightly different flow profile for the liquid than for total flow (see Fig. 9.37).

The temperature log strongly suggests that the major water entry in this well is in the bottom zone. The break to a cooler temperature that occurs at about 4,110 ft [1253 m] is characteristic of a fluid entry. Because the produced water contains most of the thermal capacity of the produced fluids, it is likely that this entry consists of a significant amount of water. The temperature log does not rule out the possibility that some oil is also being produced from the lowest zone.

To summarize the interpretation for this well, the basket flowmeter gives a good indication of the overall flow profile. Because of the amount of fluid produced, the bottom zone must be producing some oil, even though the density log indicates only water opposite this zone. The temperature log indicates that most of the water production is from the lowest zone. The density log suggests that oil is produced from 3,970 to 3,990 ft [1210 to 1216 m], though this is not conclusive. Perhaps the most important information supplied by this log is that there are large sections of supposedly productive zones that are producing no fluids.

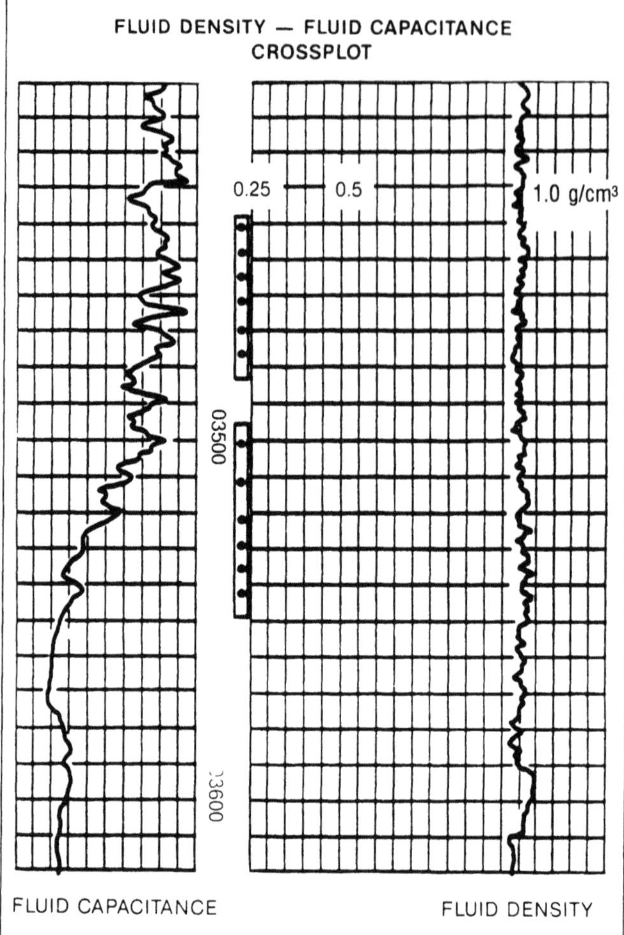

Fig. 9.38—Density and capacitance logs, Example 2 (from Ref. 14, courtesy Society of Professional Well Log Analysts).

Example 2—Capacitance and Density Logs in Oil/Water Flow. Carlson *et al.* [14] illustrate the improved sensitivity of the capacitance log compared with a gamma ray densitometer in an oil/water flow. This well was producing 500 BWPD [79.5 m^3/d water] and 350 B/D [56 m^3/d] of heavy oil in 7-in. [18-cm] casing. The water was nearly fresh with a density of 1.0 g/cm^3, and the oil had a surface density of 0.92 g/cm^3, so there was little density contrast for the density tool to distinguish. With these fluids, it is not surprising that the density log (Fig. 9.38) fluctuates around 1.0 g/cm^3, the density of the water. There is a very gradual shift to lower density moving up across the lower perforated interval, but it is not significant enough to identify the oil production regions conclusively. The capacitance log, shown in the left track of the log, shows a more positive response to oil production from the lower zone. A shift to the right on the capacitance log is caused by hydrocarbon, so the large shift to the right across the lower zone identifies it as the major oil production zone in this well. The water production in this well was apparently from lower zones not shown on the log.

Example 3—Temperature, Spinner-Flowmeter, and Density Logs in Gas/Water Flow. Carlson and Barnette [1] present temperature, spinner-flowmeter, and gamma-ray-density logs around one zone in a gas well that has production from lower zones (Fig. 9.39). The well was producing 1 MMscf/D [2864×10^3 std m^3/d] of gas and 300 BWPD [48 m^3/d water]. Both the temperature and density logs give a qualitative indication of significant gas production from the perforations shown—the temperature decreases across the zone because of Joule-Thomson cooling and the density decreases from ~ 0.63 to ~ 0.53 g/cm^3 moving up across the zone. The 0.63-g/cm^3 density below this zone indicates that gas and water are being produced from lower zones. The spinner-flowmeter log shown

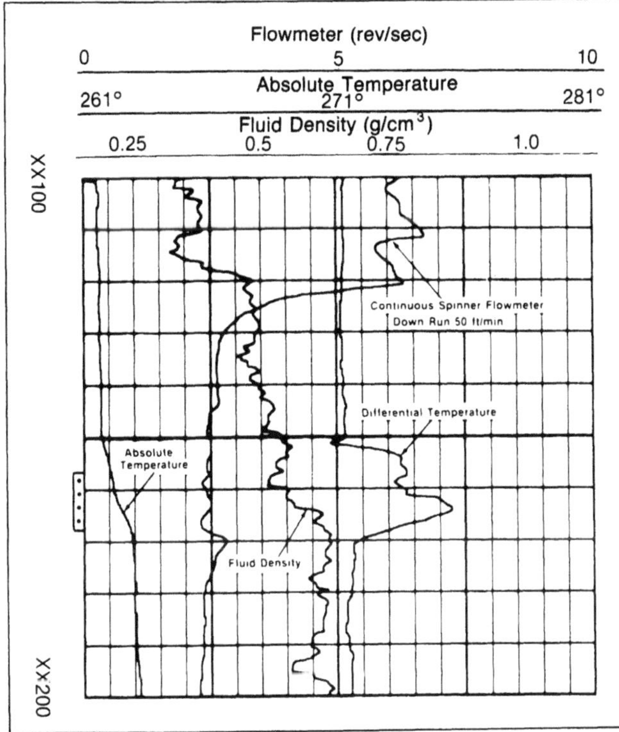

Fig. 9.39—Temperature, density, and spinner-flowmeter logs, Example 3 (from Ref. 1, courtesy Society of Professional Well Log Analysts).

was a down pass at 50 ft/min [15 m/min]. Notice that it shows little or no response to the gas production from the perforations shown. This is not unusual in a two-phase flow and, in this case, the spinner flowmeter should be ignored.

Example 4—Basket-Flowmeter, Gamma-Ray-Density, and Temperature Logs in Highly Deviated Oil/Water Flow. Carlson *et al.* [19] illustrate that, at times, log responses in difficult logging environments can be best explained by simulating the flow conditions in a surface flow loop. The well in Fig. 9.40 was deviated 70° from vertical and produced 500 BOPD [79.5 m^3/d oil] and 20 BWPD [3.2 m^3/d water] through a gravel-pack completion with a 4-in. [10-cm] -ID liner. The objective of the production logging job was to identify the major oil-producing intervals. The basket-flowmeter log was unfortunately not shown; however, it was reported that above the lowest perforations, the basket-flowmeter response was 0.9 rev/sec. To calibrate this response, kerosene and water were run in a pipe of similar size inclined at 70° in a surface flow loop.

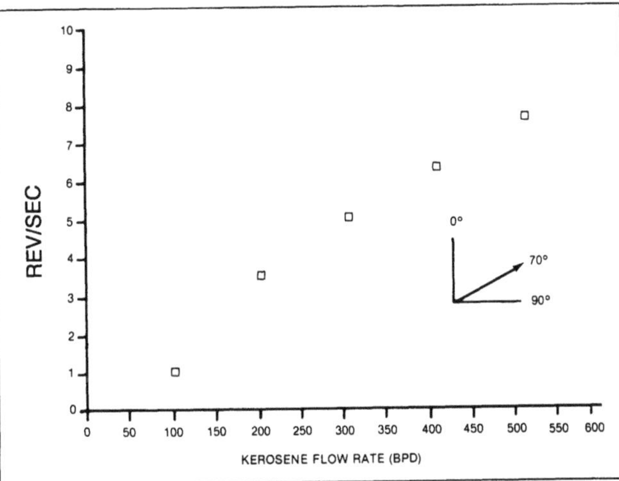

Fig. 9.41—Flow-loop calibration, Example 4 (from Ref. 19, courtesy Society of Professional Well Log Analysts).

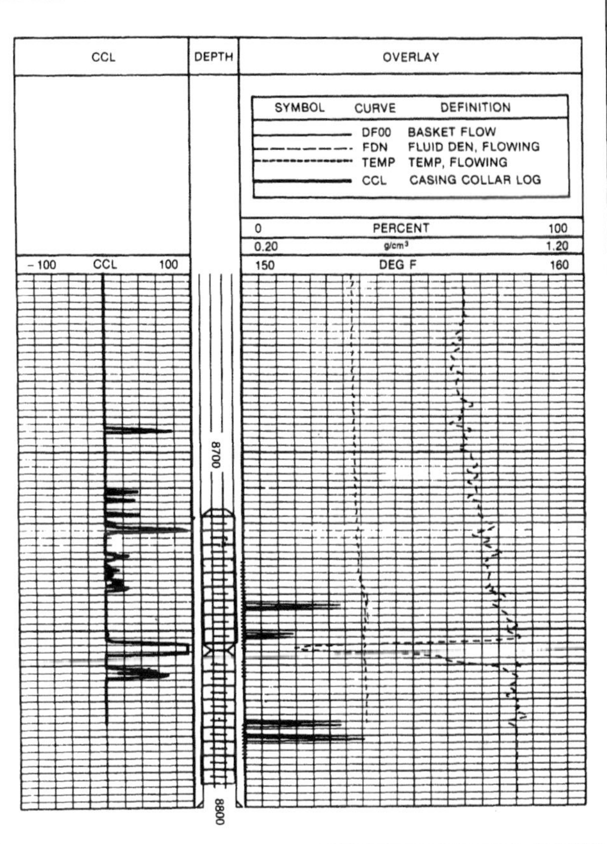

Fig. 9.40—Density and temperature logs, Example 4 (from Ref. 19, courtesy Society of Professional Well Log Analysts).

From this calibration (Fig. 9.41), the log response of 0.9 rev/sec corresponds to about 80 BLPD [13 m^3/d liquid] production. Because the total water production from the well was only 20 B/D [3.2 m^3/d], the lowest zone must be producing at least 60 BOPD [9.5 m^3/d oil], despite the fact that the density log reads the density of water across the entire lower zone. Observations in the flow loop also explain the density-log response. At these low flow rates and high deviation angle, the oil flows along the upper side of the pipe and completely bypasses the centralized density tool.

The temperature log indicates that the majority of the oil production from the uppermost perforated interval is occurring from about 8,734 to 8,740 ft [2662 to 2664 m] because a shift to cooler temperatures at this location is caused by mixing of the cooler entering fluid with warmer fluid produced below. The density log also shows a decrease across this interval; however, the density response continues to decrease above this location and never indicates the top of the production zone. The basket-flowmeter responses around this region were not reported. There is a good chance that a flow-concentrating flowmeter would not have conclusively identified the fluid-entry location in this gravel-pack completion because the fluid can bypass the tool by flowing through the gravel pack.

9.9 Guidelines for Running and Interpreting Production Logs in Multiphase Flow

Multiphase flow presents the most difficult conditions faced in production logging. In these conditions, the logger must recognize that tool responses may not represent average properties of the flow stream and that considerable error may be introduced by the interpretation procedures. The following guidelines summarize the most important factors to consider when multiphase-flow production logs are run and interpreted.

1. Multiphase flow may occur in the region of the wellbore being logged in any production well, even if multiphase-flow conditions are not clearly indicated by surface conditions. For example, a well

producing only gas at the surface may contain gas and condensate or gas and water at bottomhole conditions. Thus, the logger must be prepared for the prospect of multiphase flow in any production well.

2. Under many multiphase-flow conditions, flow properties will fluctuate with time at any pipe location. This leads to "noisy" log responses that can be difficult to interpret. Stationary measurements are particularly necessary in this type of flow so that log responses can be averaged with time.

3. As in single-phase-flow logging, stable, known flow rates are needed for proper logging measurements and interpretation. Fluid properties at downhole conditions of each phase are also required for interpretation of multiphase-flow logs.

4. Running a suite of logs with the well shut in can be very helpful if operations permit. A shut-in temperature log in multiphase flows has the same advantages as in single-phase flows. A shut-in well is also helpful in calibrating density and capacitance logs because the phases will separate with the well shut in, allowing measurements to be made on the individual phases.

5. Well deviation will often profoundly affect production logs in multiphase flow, particularly at low flow rates. Inclinations as small as 2° from vertical lead to a nonuniform distribution of the phases across the pipe cross section, making the responses of spinner-flowmeter, gamma-ray-densitometer, and capacitance logs inaccurate.

6. A temperature log is not greatly affected by multiphase-flow conditions in the wellbore. For this reason, though it can usually be used only in a qualitative way, the temperature log is often the most reliable log in multiphase flow.

7. Spinner-flowmeter logs are not reliable in multiphase flow unless the flow rates are sufficiently high to make the flow similar to a single-phase flow. Flow-concentrating flowmeters provide the best velocity measurements at applicable flow rates (generally less than about 3,000 to 4,000 B/D [477 to 636 m^3/d]).

8. In gravel-pack completions, flow-concentrating flowmeters may yield erroneous results because of fluid bypassing the tool by flowing through the gravel pack.

9. Radioactive-tracer logs are not often used in production wells by some operators because radioactive fluids will be produced to the surface, though their use varies significantly from company to company and region to region. Radioactive-tracer logs offer a possible means to measure the velocity of the continuous phase directly.

10. Density tools are best suited for distinguishing between gas and liquid because of the larger density contrast than that between oil and water. Gamma ray densitometers are hampered by small sample size and the statistical natures of the measurement and of the flow stream. Stationary measurements are advantageous with any density tool to minimize the effect of statistical fluctuations.

11. A capacitance tool is often better than a density tool for distinguishing between oil and water. However, keep in mind that capacitance-tool response is not a linear function of water fraction and that capacitance tool sensitivity is greatly diminished when water is the continuous phase.

12. Quantitative analysis of production logs in multiphase flow requires independent estimation of the slip velocity. Slip velocity can be obtained from laboratory data, from log response above all production zones, or from two-phase-flow correlations. Whatever source is used, the estimation of slip velocity lends considerable uncertainty to any quantitative interpretation of the flow profile.

13. When logs are interpreted in three-phase flow, it is generally assumed that there is no slip between oil and water so that the flow can be treated like gas/liquid two-phase flow. The validity of this assumption is untested.

14. Log-interpretation results in multiphase flows are very sensitive to the holdup measurement (density or capacitance log). The log interpreter should estimate the possible range of these measurements when the accuracy of the interpreted flow profile is considered.

15. Because of the many uncertainties, the more logs run the better, with a combination of temperature, flow-concentrating-flowmeter, density, and capacitance logs preferred. This provides a means of recognizing erroneous or misleading log responses by checking each log for consistency with the other logs.

16. Any other available information about the well or reservoir (reservoir pressure, behavior of offset wells, etc.) should be used to check the validity of the production log interpretation. Unusual conditions indicated by the logs—e.g., apparent downflows—often result from the nonuniform nature of the multiphase-flow stream. Independent information can sometimes distinguish between "real" conditions and artifacts resulting from misleading production logging responses.

Nomenclature

a = constant
A_w = wellbore cross-sectional area, ft^2 [m^2], L^2
b = constant
B_o = oil FVF
B_w = water FVF
C = constant
d_{ci} = casing ID, ft [m], L
d_T = tool OD, ft [m], L
f = flowmeter response, rev/sec, rev/t
f_{100} = flowmeter response above all production zones, rev/sec, rev/t
F = correction factor
g = acceleration of gravity, ft/sec^2 [m/s^2], L/t^2
m = flowmeter-response slope, ft/rev [m/rev], L/rev
N_{GR} = gamma ray density tool count rate, cycles/sec [Hz], cycles/t
Δp = pressure differential, psi/ft [kPa·m], m/L^{2}t^2
Δp_{E_K} = pressure differential resulting from kinetic energy change, psi/ft [kPa·m], m/L^{2}t^2
Δp_f = frictional pressure differential, psi/ft [kPa·m], m/L^{2}t^2
Δp_{HH} = pressure differential resulting from hydrostatic head, psi/ft [kPa·m], m/L^{2}t^2
q = volumetric flow rate, ft^3/sec [m^3/s], L^3/t
q_g = gas volumetric flow rate, ft^3/sec [m^3/s], L^3/t
q_o = oil volumetric flow rate, ft^3/sec [m^3/s], L^3/t
q_{os} = oil volumetric flow rate at surface, ft^3/sec [m^3/s], L^3/t
q_{o100} = oil volumetric flow rate above all production zones, ft^3/sec [m^3/s], L^3/t
q_t = total volumetric flow rate, ft^3/sec [m^3/s], L^3/t
q_{t100} = total volumetric flow rate above all production zones, ft^3/sec [m^3/s], L^3/t
q_w = water volumetric flow rate, ft^3/sec [m^3/s], L^3/t
q_{ws} = water volumetric flow rate at surface, ft^3/sec [m^3/s], L^3/t
q_{w100} = water volumetric flow rate above all production zones, ft^3/sec [m^3/s], L^3/t
$\bar{v}$ = average in-situ velocity, ft/sec [m/s], L/t
v_{bs} = rise velocity of gas bubble in stagnant liquid, ft/sec [m/s], L/t
$\bar{v}_g$ = average in-situ gas velocity, ft/sec [m/s], L/t
v_{gs} = superficial gas velocity, ft/sec [m/s], L/t
$v_{\lim}$ = bubble-rise velocity in stagnant liquid, ft/sec [m/s], L/t
$\bar{v}_L$ = average in-situ liquid velocity, ft/sec [m/s], L/t
v_M = mixture velocity, ft/sec [m/s], L/t
$\bar{v}_o$ = average in-situ oil velocity, ft/sec [m/s], L/t
v_s = slip velocity, ft/sec [m/s], L/t
v_{t100} = total average velocity above all production zones, ft/sec [m/s], L/t
$\bar{v}_w$ = average in-situ water velocity, ft/sec [m/s], L/t
y_g = gas holdup
y_L = liquid holdup
y_o = oil holdup
y_w = water holdup
θ = deviation from vertical, degrees [rad]
$\bar{\rho}$ = average in-situ density, g/cm^3, m/L^3
ρ_f = fluid density, g/cm^3, m/L^3
ρ_g = gas density, g/cm^3, m/L^3
ρ_{Gr} = Gradiomanometer density reading, g/cm^3, m/L^3

ρ_L = density of liquid phase, g/cm^3, m/L^3
ρ_o = oil density, g/cm^3, m/L^3
ρ_w = water density, g/cm^3, m/L^3
$\Delta\rho$ = density difference between phases, g/cm^3, m/L^3
σ = IFT, mN/m^2, m/t^2

References

1. Carlson, N.R. and Barnette, J.C.: "The Significance of the Temperature Log in Multiphase Flows," *Proc.*, SPWLA Annual Logging Symposium, Houston (June 9-13, 1986) paper R.
2. Smolen, J.J.: "Cased-Hole Logging: A Perspective," *The Log Analyst* (March-June 1987) **28,** No. 2, 165-74.
3. Hammack, G.W., Myers, B.D., and Barcenas, G.H.: "Production Logging through the Annulus of Rod-Pumped Wells to Obtain Flow Profiles," paper SPE 6042 presented at the 1976 SPE Annual Technical Conference and Exhibition, New Orleans, Oct. 3-6.
4. Hill, A.D. and Oolman, T.: "Production Logging Tool Behavior in Two-Phase Inclined Flow," *JPT* (Oct. 1982) 2432-40.
5. Anderson, R.A. *et al.*: "A Production Logging Tool With Simultaneous Measurements," *JPT* (Feb. 1980) 191-98.
6. Taitel, Y., Barnea, D., and Dukler, A.E.: "Modelling Flow Pattern Transitions for Steady Upward Gas-Liquid Flow in Vertical Tubes," *AIChE J.* (1980) **26,** No. 6, 345-54.
7. Weiss, E.N., Taylor, J.G., and Toronyi, R.M.: "Productivity Testing Using Production Logging Techniques," paper SPE 9610 presented at the 1981 SPE Middle East Oil Conference, Bahrain, March 9-12.
8. Connolly, E.T.: "Resume and Current Status of the Use of Logs in Production," *Proc.*, SPWLA Annual Logging Symposium, Dallas (May 4-7, 1965) paper L.
9. *Production Log Interpretation,* Schlumberger Ltd. (1970).
10. *Production Log Interpretation,* Schlumberger Ltd. (1973).
11. Carlson, N.R. and Roesner, R.E.: "Water-Oil Flow Surveys with Basket Fluid Capacitance Tool," *Proc.*, SPWLA Annual Logging Symposium, Corpus Christi (July 6-9, 1982) paper F.
12. Piers, G.E., Perkins, J., and Escott, D.: "A New Flowmeter for Production Logging and Well Testing," paper SPE 16819 presented at the 1987 SPE Annual Technical Conference and Exhibition, Dallas, Sept. 27-30.
13. *Interpretive Methods for Production Well Logs,* second edition, Atlas Wireline Services, Western Atlas Intl. Inc., Houston (1982).
14. Carlson, N.R., Barnette, J.C., and Davarzani, M.J.: "Applications of the Fluid Capacitance Log in Multiphase Flows," *Proc.*, Aberdeen Well Log Analysts Society Symposium, Aberdeen (1986) paper B.
15. Curtis, M.R.: "Flow Analysis With the Gradiomanometer and the Flowmeter," Schlumberger, Houston (1966).
16. Curtis, M.R.: "Flow Analysis in Producing Wells," paper SPE 1908 presented at the 1967 SPE Annual Meeting, Houston, Oct. 1-4.
17. Nicolas, Y. and Witterholt, E.J.: "Measurements of Multiphase Fluid Flow," paper SPE 4023 presented at the 1972 SPE Annual Meeting, San Antonio, Oct. 8-11.
18. Orkiszewski, J.: "Predicting Two-Phase Pressure Drops in Vertical Pipes," *JPT* (June 1967) 829-38.
19. Carlson, N.R., Barnette, J.C., and Davarzani, M.J.: "Utilization of Surface Flow Loop for Analysis of Field Logs in Production Wells," *Proc.*, SPWLA Annual Logging Symposium, London (June 29-July 2, 1987) paper X.

SI Metric Conversion Factors

°API	141.5/(131.5+°API)		=	g/cm^3
bbl	× 1.589 873	E−01	=	m^3
cp	× 1.0*	E−03	=	Pa·s
cycles/sec	× 1.0*	E+00	=	Hz
ft	× 3.048*	E−01	=	m
ft^3	× 2.831 685	E−02	=	m^3
°F	(°F−32)/1.8		=	°C
gal	× 3.785 412	E−03	=	m^3
in.	× 2.54*	E+00	=	cm
lbm	× 4.535 924	E−01	=	kg
psi	× 6.894 757	E+00	=	kPa

*Conversion factor is exact.

Chapter 10
Noise Logging

10.1 Introduction

The noise log is a relative newcomer to production logging services. Though described by Enright[1] in 1955, noise logging was not commercially available until after McKinley *et al.*[2] demonstrated its utility in 1973. A noise log is simply a record of a passive measure of the audible sound detected by a sensitive hydrophone at a number of locations in the wellbore. Because sound is generated by fluid turbulence, high noise amplitudes indicate locations where the flow path causes additional turbulence to develop. Fluid moving through restricted channels, leaks, flow from perforations, and flow past the logging sonde are among the phenomena that can produce characteristic sounds in the wellbore and that may be detected with a noise log. Analysis of the frequency characteristics of the measured noise can distinguish between the various possible sources of high sound amplitudes, making the noise log a powerful tool for well diagnosis.

The noise log has been used primarily as a qualitative indicator of channeling behind pipe, often in conjunction with temperature logs. It can now be used quantitatively to estimate flow rates in some instances, and its application has been extended beyond channel detection.

This chapter first describes noise logging tools and their operation and then discusses use of noise logs for channel detection. The following section describes other applications of the noise log. The chapter concludes with guidelines for the running and interpretation of noise logs.

10.2 Tools and Operations

A noise logging tool is simply a sensitive microphone with a downhole amplifier. A piezoelectric crystal sound detector typically is used; the downhole amplifier arrangement depends on the particular tool manufacturer. The AC voltage signal generated by the detector is transmitted up a single-conductor wireline to a surface panel where the signal is again amplified. The noise amplitude spectrum is then typically conditioned by four high-pass filters that transmit the noise amplitudes above 200, 600, 1,000, and 2,000 cycles/sec [200, 600, 1000, and 2000 Hz]. While these four filter cuts are the standard output from a noise logging sonde, the surface equipment should also allow the full noise spectrum to be recorded because it can be helpful in the positive identification of noise sources. The surface panel is usually equipped with a speaker or headphones so that the logging operator can listen to the noises from the well to identify regions of interest more quickly. Fig. 10.1 illustrates a noise logging tool and a surface panel. Refs. 2 through 5 give further details on noise logging tool specifications.

For quantitative analysis, the noise logging signal must be corrected for cable attenuation. The raw noise amplitude signal must be multiplied by a line factor that depends on the type and length of conductor cable used to yield the amplitude that would be measured without the cable effect. Fig. 10.2 shows a line-factor chart for an Atlas Wireline Services noise logging tool. Of course, the length of cable to use in this correction is the total length spooled on the truck, not only the amount used in the well. Line factors such as those in Fig. 10.2 are specific to the particular type of cable used and thus must be supplied by the logging operator.

A noise logging sonde is run without centralizers so that the tool will rest against the casing wall, allowing better acoustic coupling to sounds generated outside the casing. Measurements must be made with the tool stationary to avoid the noise made by motions of the tool and wireline. After a desired depth for a measurement is reached, the operator should wait a short time (40 to 60 seconds) before recording the noise log to prevent the measurement of tool-motion noise. By listening to the tool output on the speaker or headphones, the operator can determine when extraneous noise has dissipated.

A noise log consists of a series of discrete noise measurements made throughout the zones of interest in the well. Standard practice is to make initial measurements at 50-ft [15-m] intervals. From this coarse log, the high-noise regions can be located; measurements are then taken every 2 to 3 ft [0.6 to 0.9 m] in these regions to locate the noise sources more precisely.

10.3 Noise Logging Theory

When a fluid expands through a constriction, turbulence that dissipates energy in the form of noise is generated. The amplitude of the noise depends on the flow rate and the pressure drop across the throttling location—e.g., 100 Mcf/D [2.83×10^3 m^3/d] of gas leaking from a high-pressure line is much noisier than 10 Mcf/D [283 m^3/d] leaking from a low-pressure separator. Furthermore, the frequency characteristics of noise generated by a throttling process depend on specifics of the flow—the flow rate, whether the flow is single- or two-phase, etc. These qualitative characteristics of noise produced by flowing fluids form the basis for noise logging and were investigated theoretically and experimentally by McKinley *et al.*[2] The following discussion is drawn primarily from their work.

10.3.1 Noise Frequency Characteristics. Flow through a constriction produces noise over a wide frequency range, depending on the particular type of flow.

Single-phase liquid flow through a throttle was found to display a noise amplitude peak at about 1,000 cycles/sec [1000 Hz] (Fig. 10.3). The high-amplitude noise below 100 cycles/sec [100 Hz] is generally not considered in noise logging because noise in this frequency range can result from pumps, motors, and other surface sources of background noise (recall that the lowest filter used on a noise log records noise above 200 cycles/sec [200 Hz]). An increase in the pressure drop across the restriction increases the frequency of the noise peak in single-phase liquid flow (Fig. 10.4). Throttling of single-phase gas flow produces a frequency spectrum similar to that of single-phase liquid flow, perhaps shifted to slightly higher frequencies (Fig. 10.5).

Two-phase gas/liquid flow generates noise at lower frequencies; therefore, the noise spreads over a frequency band wider than that observed with single-phase flow. In laboratory experiments, McKinley and his coworkers observed three distinct types of noise

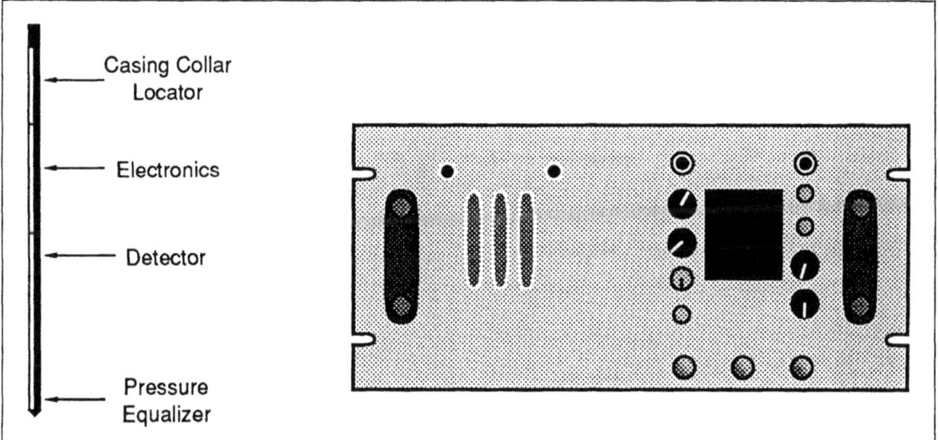

Fig. 10.1—Noise logging tool and surface panel (from Ref. 3, courtesy Society of Professional Well Log Analysts).

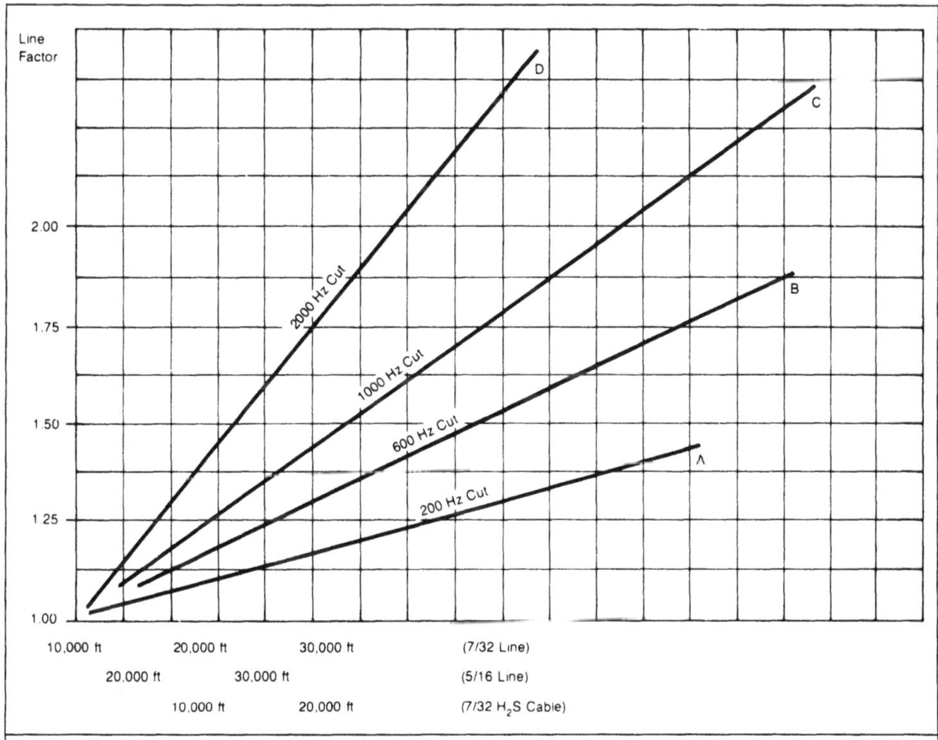

Fig. 10.2—Line factors for Atlas Wireline Services wirelines (from Ref. 6, courtesy Atlas Wireline Services, Western Atlas Intl.).

spectra in two-phase flow: discrete bubbling, mild slugging, and severe slugging. Discrete bubbling exhibits an amplitude peak in the 300- to 600-cycle/sec [300- to 600-Hz] range. With mild slugging, the "bubble" peak at 300 to 600 cycles/sec [300 to 600 Hz] is diminished, but still distinguishable. Severe slugging concentrates more sound energy at ~200 cycles/sec [~200 Hz], with noise spread across the entire spectrum to >2,000 cycles/sec [>2000 Hz]. In all cases, a small "single-phase" peak occurs at about 1,000 cycles/sec [1000 Hz]. Fig. 10.6 illustrates these spectra.

The differences in the spectral behavior of single- or two-phase flow provide a qualitative means for interpretation of flow in a channel with a noise log. An amplitude peak on the log indicates a throttling location—i.e., an entrance or exit of a channel, a constriction in a channel, a leak, or some other source of pressure drop. Whether the flow is single- or two-phase can be determined by the spectral characteristics reflected in the frequency cuts recorded. If the flow is single-phase, noise is concentrated at frequencies above 1,000 cycles/sec [1000 Hz]; thus, the 200-, 600-, and 1,000-cycle/sec [200-, 600-, and 1000-Hz] frequency cuts will have almost the same amplitude at the depth of the noise source (Fig. 10.7). Two-phase flow results in noise spread over a broad frequency band above 200 cycles/sec [200 Hz] so that the various frequency cuts have widely different amplitudes (Fig. 10.8).

10.3.2. Calculation of Flow Rate in a Channel. Because noise results from energy dissipated by fluid turbulence, it follows that the noise amplitude generated in a throttling process is some function of the energy dissipation rate, or

$$a=f(\mathrm{d}E/\mathrm{d}t), \quad \text{(10.1)}$$

where a=noise amplitude and $\mathrm{d}E/\mathrm{d}t$=rate of mechanical energy dissipation. From a mechanical energy balance, we can determine that

$$\mathrm{d}E/\mathrm{d}t \sim \Delta p_t q, \quad \text{(10.2)}$$

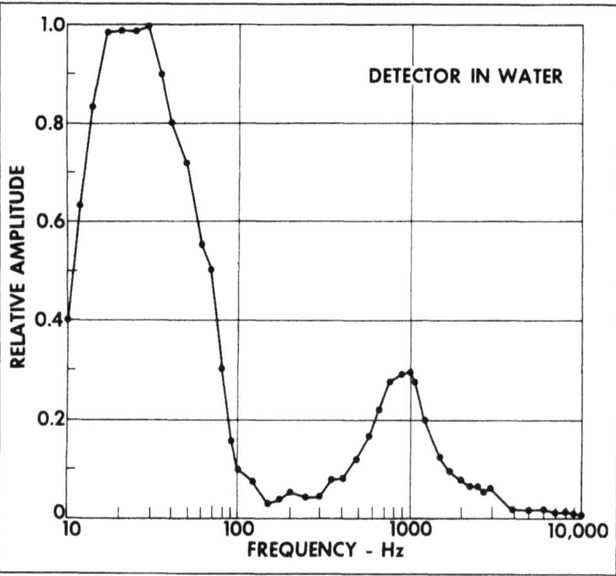

Fig. 10.3—Noise spectrum for water throttling across 10-psi/in. pressure drop (from Ref. 2).

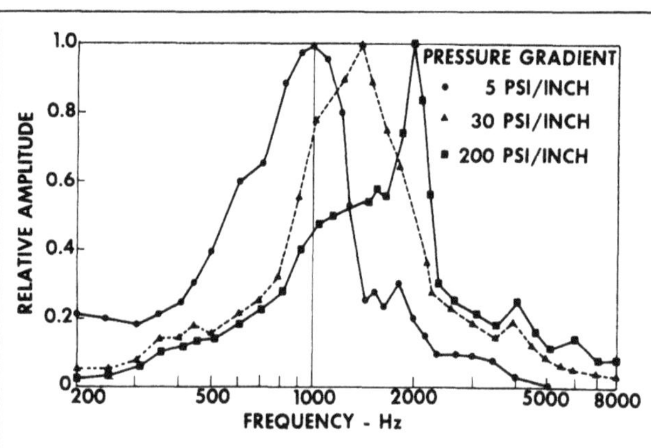

Fig. 10.4—Noise spectra for water throttling at various pressure gradients (from Ref. 2).

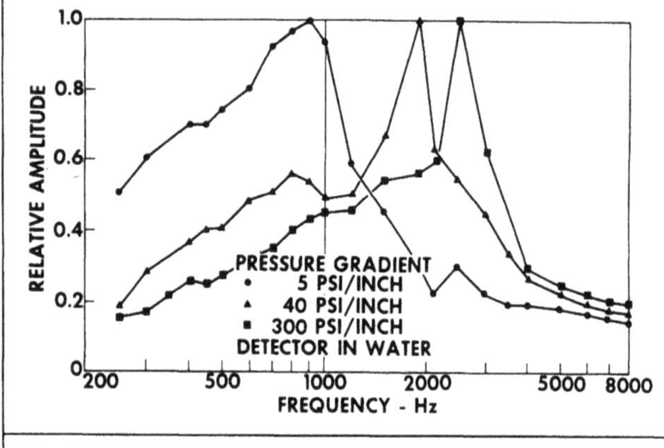

Fig. 10.5—Noise spectra for air throttling at various pressure gradients (from Ref. 2).

where Δp_t=pressure drop across the throttle and q=volumetric flow rate. Combining Eqs. 10.1 and 10.2,

$$a=f(\Delta p_t q). \quad (10.3)$$

McKinley *et al.*[2] used $\Delta p_t q$ as a correlating parameter for data taken in a leak simulator. The results (Fig. 10.9) indicate a linear

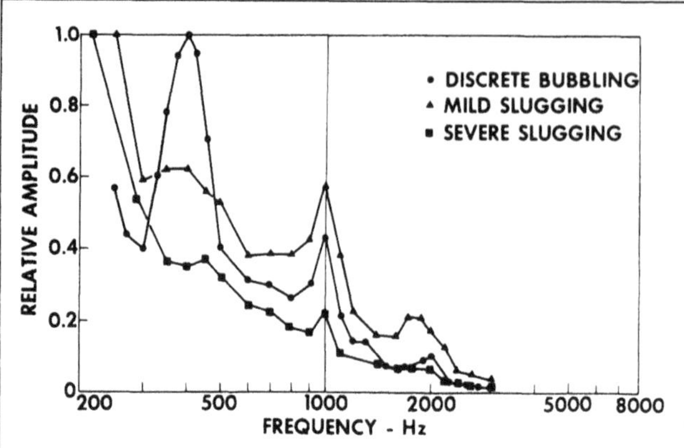

Fig. 10.6—Noise spectra for air throttling into water-filled channels (from Ref. 2).

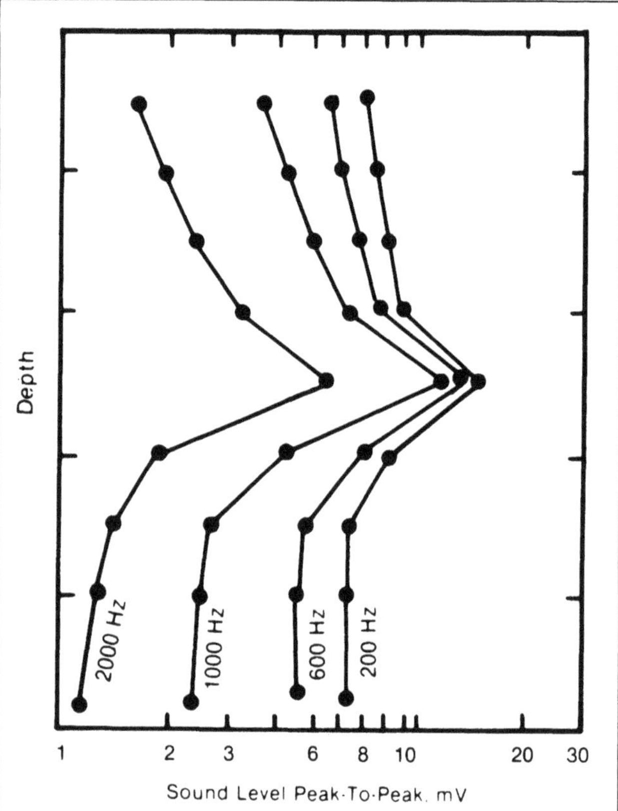

Fig. 10.7—Noise log characteristics of single-phase flow in a channel (from Ref. 6, courtesy Atlas Wireline Services, Western Atlas Intl.).

relationship between noise amplitude and the product of pressure drop and flow rate over a wide range of data. Despite the considerable scatter in the data, this correlation provides a means of estimating the flow rate in a channel from a noise log. From the amplitude of noise from the log, $\Delta p_t q$ can be obtained from the correlation line in Fig. 10.9. Then, by estimation of the pressure drop across the throttling location from independent means, the flow rate through the channel can be calculated.

Fig. 10.9 applies to single-phase flow in a channel behind one string of water-filled casing. If the channel is behind more than one string of pipe or if the casing or annulus between tubing and casing is gas-filled, the noise is more attenuated and a lower noise amplitude is measured for a given flow rate. Table 10.1 presents well-geometry factors used to correct the correlation of Fig. 10.9 for different wellbore geometries. McKinley *et al.* multiplied the amplitudes from the noise log by the well-geometry factors to obtain

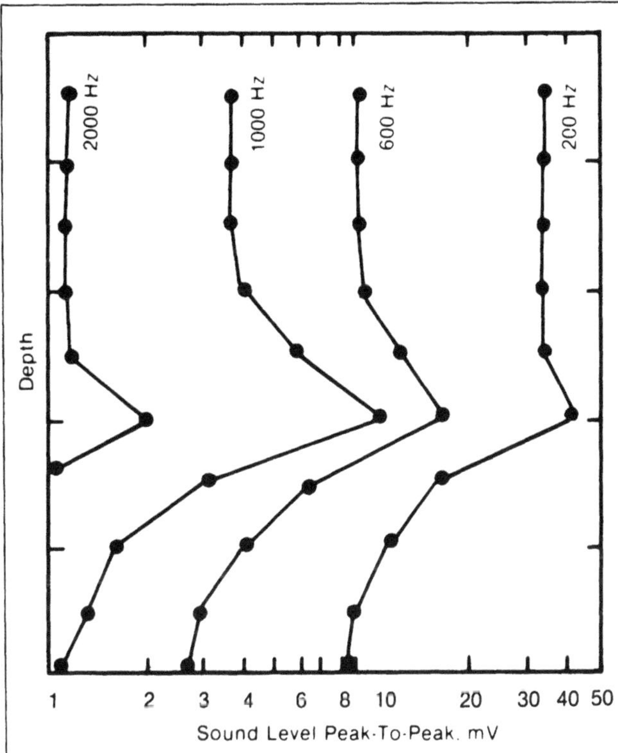

Fig. 10.8—Noise log characteristics of gas/liquid flow in a channel (from Ref. 6, courtesy Atlas Wireline Services, Western Atlas Intl.).

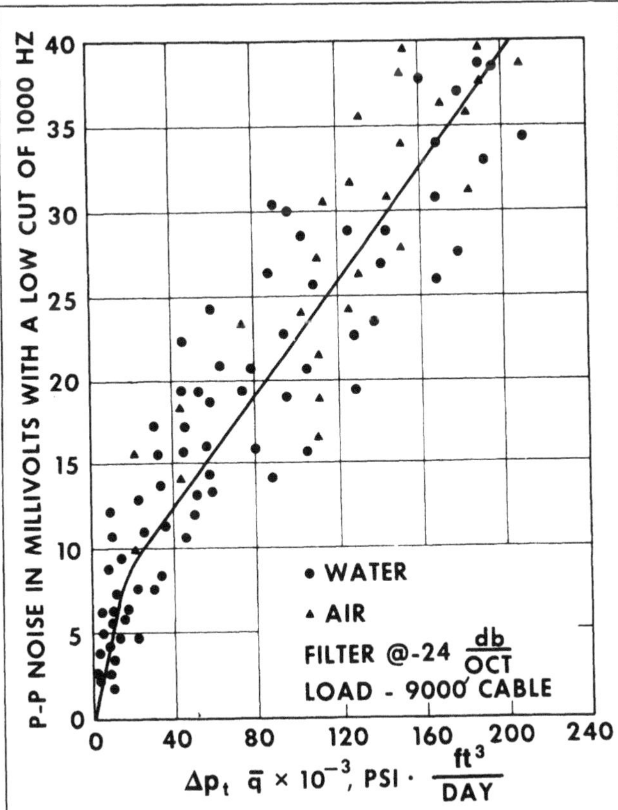

Fig. 10.9—Correlation of noise amplitude with energy dissipation rate (from Ref. 2).

TABLE 10.1—WELL GEOMETRY FACTORS

Well Type	Fluid Content	Noise Amplitude Multiplication Factor
Tubingless completion	Liquid in string	1
	Gas in string	3.5
Tubing string in casing	Liquid in tubing, liquid in annulus	4
	Gas in tubing, liquid in annulus or vice versa	10
	Gas in both tubing and annulus	30
Leak into same string as detector	Liquid in string	0.06
	Gas in string	0.20

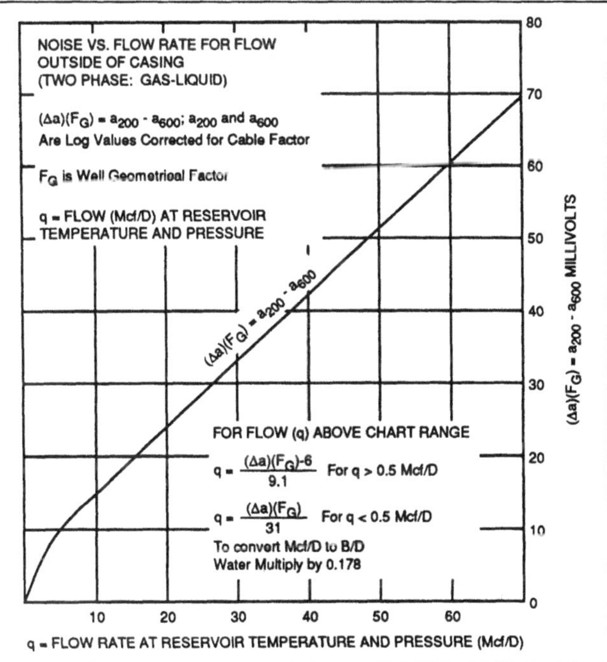

Fig. 10.10—Noise log correlation for two-phase flow in a channel (after Fig. 3 in Ref. 5, courtesy Society of Professional Well Log Analysts).

the appropriate noise amplitudes. Britt[5] presented geometry factors that differ slightly from those of McKinley *et al.* because they are specific for the Gearhart Industries Borehole Audio Tracer Survey noise logging tool. Also, McKinley *et al.*'s geometry factors are based on laboratory experiments in which the pipes were acoustically separate, whereas Britt's are derived from field results.

Britt also presented a correlation for flow rate of gas/liquid flow in a channel for a Gearhart Industries noise logging tool by fitting a curve to the two-phase data of McKinley *et al.* The correlation (Fig. 10.10) is approximated by the equations

$$q=[(a_{200}-a_{600})F_G-6]/9.1 \quad (10.4)$$

for $q>0.5$ Mcf/D [>14 m^3/d], and

$$q=[(a_{200}-a_{600})F_G]/31 \quad (10.5)$$

for $q<0.5$ Mcf/D [<14 m^3/d].

In these equations, a_{200} and a_{600} = amplitudes in peak-to-peak millivolts of the noise above 200 and 600 cycles/sec [200 and 600 Hz], respectively, F_G = well geometry factor, and q = flow rate in Mcf/D at downhole conditions. These equations may be specific to this

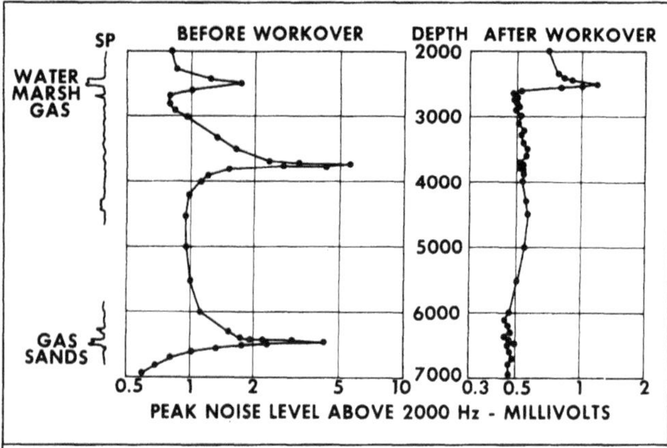

Fig. 10.11—Noise logs from tubingless completion leaking to surface (from Ref. 2).

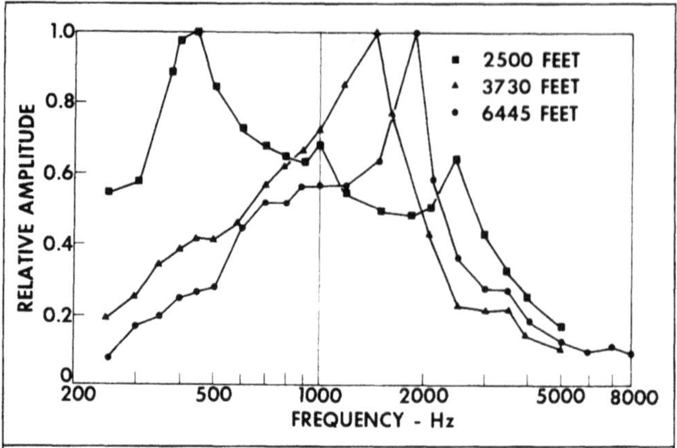

Fig. 10.12—Noise spectra at various depths, example well (from Ref. 2).

tool; however, the form of the correlation should be general to other instruments.

The correlations presented should not be expected to yield highly accurate values for flow rates behind casing, considering the scatter in the laboratory data and the uncertain geometry for sound transmission in a well. They do, however, provide a means of obtaining an estimate of the flow rate in a channel from a noise log.

Example Log—Flow in a Channel. McKinley *et al.*[2] illustrate the ability of a noise log to locate a channel and the procedure for calculating flow rate in a channel. The noise log in Fig. 10.11 was run on a well that had a surface gas flow in the annulus between the surface casing and the three inner casing strings; the surface pressure on the annulus was 1,350 psi [9.3 MPa]. A temperature log showed a slight cooling anomaly at 2,730 ft [832 m]. The temperature log, coupled with the surface shut-in pressure, suggested a shallow gas source. The noise log, however, showed three distinct peaks at 2,500, 3,730, and 6,445 ft [762, 1137, and 1964 m], with no indication of a flow source at 2,730 ft [832 m]. The noise amplitude peaks are also definitely skewed upward, indicating upward flow. The noise source at 6,445 ft [1964 m] is opposite a 2,900-psi [20-MPa] gas sand, the source of the channeling gas. At 3,730 ft [1137 m], openhole logs indicate a shale formation, so the noise generated at this depth apparently results from a restriction in the channel through the cement. The noise at 2,500 ft [762 m] is caused by some of the gas charging a water sand at this depth.

Noise spectra of the three noise sources confirm the interpretation of flow in a channel (Fig. 10.12). The spectra at 6,445 and 3,730 ft [1964 and 1137 m] exhibit single-phase gas-throttling characteristics, with a single noise peak above 1,000 cycles/sec [1000 Hz]. The higher frequency of the amplitude peak at 6,445 ft [1964 m] indicates a pressure gradient higher than that at 3,730 ft [1137 m]. The noise spectrum at 2,500 ft [762 m] looks similar to the spectrum characteristic of discrete bubbling of gas through liquid (Fig. 10.6), confirming that gas is charging the water sand at this depth.

Rough estimates of the flow rates at each throttling location can be obtained with McKinley *et al.*'s correlations. First, the noise amplitude above 1,000 cycles/sec [1000 Hz] is read from the log for each noise source, then the value of $\Delta p_t q$ (see Table 10.2) is read from the correlation line in Fig. 10.9. To calculate flow rates, we must now estimate the pressure drop at each location. With the gas source at 2,900 psi [20 MPa] and the assumption that the pressure at 2,500 ft [762 m] is close to the surface pressure of 1,350 psi [9.3 MPa], the gas experiences a total pressure drop of ~1,500 psi [10.3 MPa] as it flows from the source at 6,445 ft [1964 m] to the water sand at 2,500 ft [762 m]. Allocating this pressure drop equally among the two throttling locations and the rest of the channel yields a 500-psi [3.4-MPa] pressure differential at the gas source (6,445 ft [1964 m]) and at the constriction (3,730 ft [1137 m]). Thus, at 6,445 ft [1964 m],

$$q=\Delta p_t q/\Delta p=85{,}000/500$$

$$=170 \text{ ft}^3/\text{D at } 2{,}700 \text{ psi } [4.8 \text{ m}^3/\text{d at } 18.6 \text{ MPa}]$$

TABLE 10.2—NOISE AMPLITUDES ABOVE 1,000 cycles/sec [1000 Hz], EXAMPLE WELL

Depth (ft)	Peak-to-Peak Noise Amplitude (mV)	$\Delta p_t q$* (psi-ft^3/D)
6,445	20	85,000
3,730	30	145,000
2,500	11	30,000

*Values from Fig. 10.9

or

$$q=25{,}000 \text{ scf/D } [716 \text{ std m}^3/\text{d}].$$

Similarly, at 3,730 ft [1137 m],

$$q=\Delta p_t q/\Delta p=145{,}000/500$$

$$=300 \text{ ft}^3/\text{D at } 1{,}900 \text{ psi } [8.5 \text{ m}^3/\text{d at } 13.1 \text{ MPa}]$$

or

$$q=35{,}000 \text{ scf/D } [1002 \text{ std m}^3/\text{d}].$$

At 2,500 ft [762 m], the pressure in the annulus is assumed to be about the same as the surface pressure, 1,350 psi [9.3 MPa]. Assuming hydrostatic pressure in the water sand yields a pressure of 1,100 psi [7.6 MPa] in the water sand so that the pressure differential at 2,500 ft [762 m] is about 250 psi [1.7 MPa],

$$q=\Delta p_t q/\Delta p=30{,}000/250$$

$$=120 \text{ ft}^3/\text{D at } 1{,}200 \text{ psi } [3.4 \text{ m}^3/\text{d at } 8.3 \text{ MPa}]$$

or

$$q=10{,}000 \text{ scf/D } [286 \text{ std m}^3/\text{d}].$$

If the noise generated at 2,500 ft [762 m] is from the gas charging the water sand, then presumably the flow rate calculated is an estimate of the charging rate into this zone. In summary, the analysis shows that 20 to 40 Mscf/D [57×10^3 to 114.5×10^3 std m^3/d] of gas is produced at the gas source, with about 10 Mscf/D [286 std m^3/d] of this gas entering the water sand at 2,500 ft [762 m] and the rest being produced to the surface. An orifice meter at the surface indicated a flow rate of 10 Mscf/D [286 std m^3/d], making the estimates of downhole rates from the noise log seem reasonable. Though the interpretation of flow rates in a channel from a noise log is obviously quite approximate, it does lend more confidence to the qualitative interpretation.

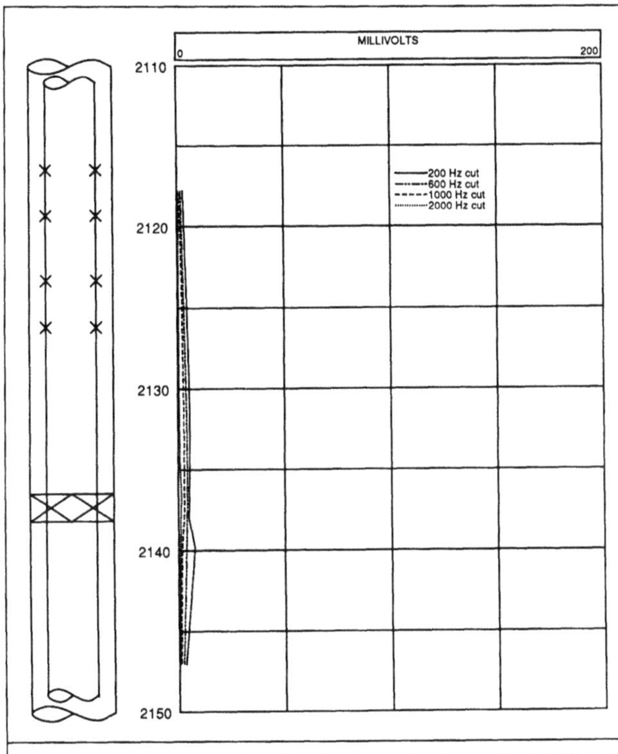

Fig. 10.13—Suspected packer leak—noise log with 1,200 psi annulus pressure.

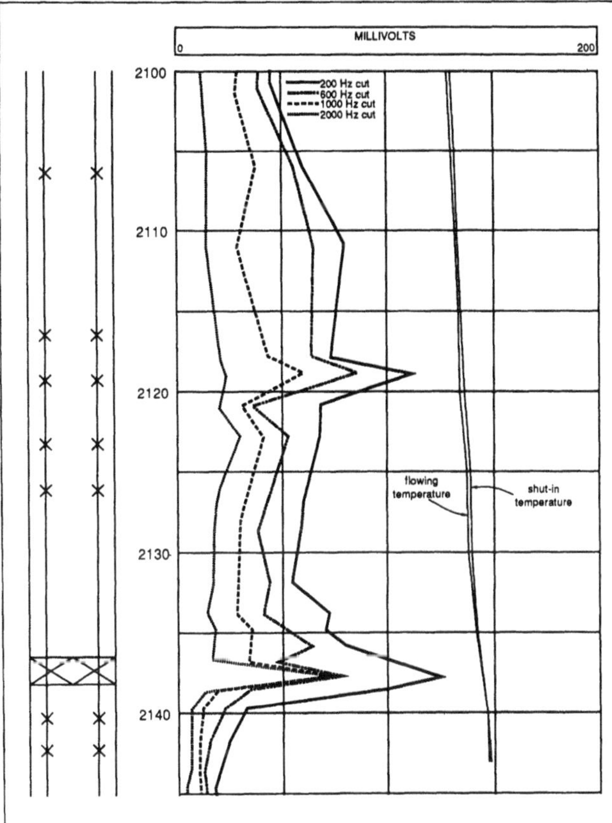

Fig. 10.14—Suspected packer leak—noise log with pressure bled off annulus.

In this example well, the remedial action was to cement squeeze above and below the gas source at 6,445 ft [1964 m], which resulted in an immediate drop in the annulus pressure to 350 psi [2.4 MPa]. The noise log run after the workover (Fig. 10.11) shows that the gas source was eliminated, with a noise peak remaining at 2,500 ft [762 m] because of the discharging of gas from the water sand at this depth. Without the noise log, a workover would have been attempted at 2,730 ft [832 m], where the temperature log showed a slight cooling anomaly. A workover at this depth would have had a slim chance of success. The cooling at 2,730 ft [832 m] was attributed to evaporative cooling of the gas as it passed a water zone.

Example Log—Packer Leak. Noise logs are commonly applied to locate leaks in or around well hardware because the flow constriction at the leak will generate noise. Fig. 10.13 shows a noise log run in a well with a suspected packer leak that allowed flow into the annulus. The log was run with 1,200-psi [8.3-MPa] surface pressure on the annulus; with this backpressure, no leak is evident because no noise anomalies occur at the packer location. Pressure was then bled off the annulus, and the noise log was repeated (Fig. 10.14). With low pressure in the annulus, a leak at the packer was clearly indicated by the noise amplitude peaks at the packer location.

Notice that the noise generated by the packer leak displays characteristics of both single- and two-phase leaks. The high amplitude at frequencies above 2,000 cycles/sec [2000 Hz] corresponds to gas flow through a constriction and is probably caused by gas leaking through the packer. There is also considerable noise at lower frequencies, as indicated by the difference in the amplitude of the 200- and 600-cycle/sec [200- and 600-Hz] cuts. This lower-frequency noise probably results from gas bubbling into the liquid column in the annulus. The temperature log reflects the presence of liquid in the annulus by the steeper slope above the packer than below.

10.4 Other Applications of Noise Logging

Noise logging has been extended to applications other than the detection of channels in completed wells. Among these are the measurement of flow rate in the wellbore, the measurement of flow rates from individual perforations, detection of liquid production from gas perforations, detection of sand production, and the use of noise logging during drilling. McKinley and Bower[7] first pointed out the utility of noise logging for most of these applications.

10.4.1. Axial Flow Rate in the Wellbore. If the flow rate is sufficiently high, flow past the noise logging sonde in the wellbore generates sufficient turbulence to make detectable sounds. Following McKinley and Bower, we can again relate the noise produced to the flow rate and pressure drop in the restriction, as was done in the analysis of flow in a channel:

$$a=f(\Delta pq), \quad (10.6)$$

where a=noise amplitude, Δp=pressure drop in annulus between the tool and the casing, and q=volumetric flow rate. For turbulent flow in an annulus, the pressure drop can be written as

$$\Delta p=K_D\rho_f(q/A_{an})^2, \quad (10.7)$$

where K_D=drag coefficient, ρ_f=fluid density, and A_{an}=annular cross-sectional area. Combining Eqs. 10.6 and 10.7 and using the amplitude of noise above 600 cycles/sec [600 Hz] yields

$$a_{600}=K_D\rho_f q^3/A_{an}^2, \quad (10.8)$$

where K_D=drag coefficient that depends on the Reynolds number. For N_{Re} above 6,000,

$$K_D=(4\pm1.5)\times10^{-6}.$$

Inverting Eq. 10.8,

$$q=63(A_{an}{}^2 a_{600}/\rho_f)^{1/3}, \quad (10.9)$$

where q is in actual ft^3/D, ρ_f is in lbm/ft^3, A_{an} is in ft^2, and a_{600} is in mV (peak-to-peak). From these equations, it is apparent that

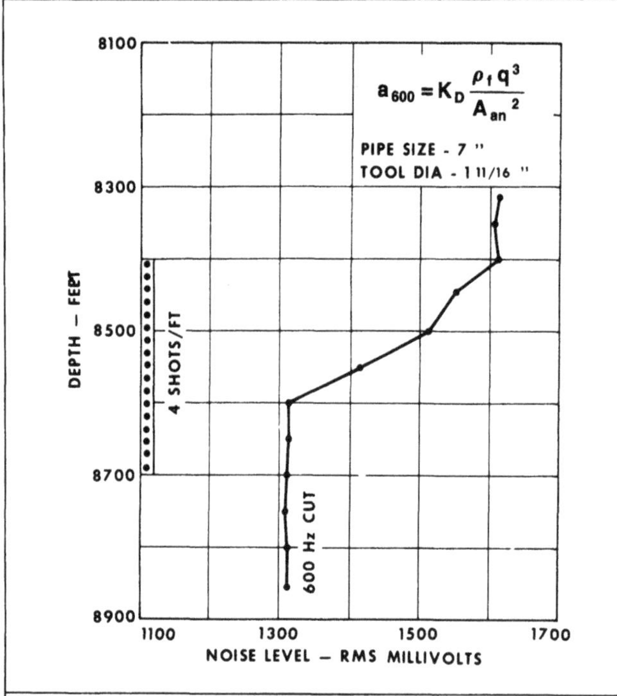

Fig. 10.15—Noise profile across a perforated interval taking 4% of flow (from Ref. 7).

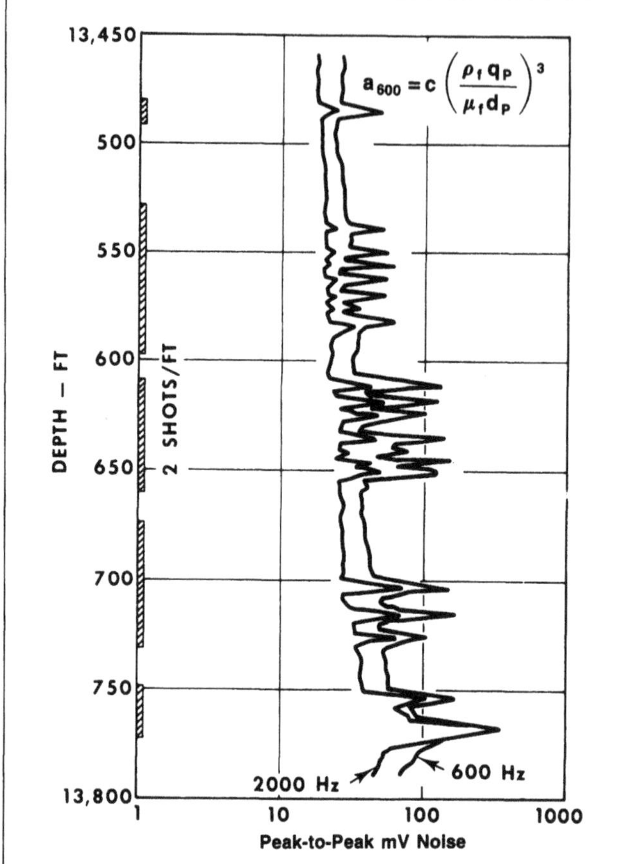

Fig. 10.16—Noise log showing flow from individual perforations (from Ref. 7).

the noise level depends on the cube of the flow rate and thus should be a sensitive measure of flow rate in the wellbore. This is indeed the case if the flow rate is high enough to generate significant noise above background level. The noise log is best applied as a flowmeter to locate small changes in flow rate in high-rate wells.

Fig. 10.15 shows a noise log across the upper perforations in a water injection well with a total injection rate of 25,000 B/D [3975 m^3/d]. This zone is taking only 4% of the total injection, which is near the limit of detection for a spinner-flowmeter log. The high flow rate would also prevent a radioactive-tracer log from accurately measuring the fluid loss to this zone. The noise log, however, shows a 19% decrease in noise amplitude across the interval resulting from the cubic dependence of noise amplitude on flow rate. Applying Eq. 10.9 to the noise log above and below the perforated interval yields a flow rate of 1,600 B/D [254 m^3/d] into the zone. The noise log also shows that all injection in this zone is into the upper two-thirds of the perforations. Note that the noise level on the log is in root-mean-square (RMS) millivolts, while the flow-rate correlation (Eq. 10.9) is in peak-to-peak millivolts, requiring that the RMS millivolt value be multiplied by 2.83 to convert to the peak-to-peak millivolt value. In this example, the noise level also was multiplied by 3.2 to account for cable attenuation and specific tool gain so that the correlation could be applied. The tool manufacturer should provide this information so that quantitative interpretation is possible.

10.4.2 Flow From Perforations. The flow from a perforation is similar to flow through an orifice—i.e., the fluid is accelerated and a substantial pressure drop occurs because the cross-sectional area of the perforation hole is smaller than the surface area in the perforation tunnel. For noise generated by flow through an orifice, McKinley and Bower derived that the amplitude would be expressed as

$$a_{600}=c[\rho_f^4 q^3/(\mu_f^3 d_P^4)], \quad \text{(10.10)}$$

where μ_f=fluid viscosity and d_P=perforation diameter. Because of the uncertainties in estimating the perforation diameter and the downhole fluid density in gas wells, McKinley and Bower simplified this equation to

$$a_{600}=cX_P^3, \quad \text{(10.11)}$$

where

$$X_P=\rho_f q/\mu_f d_P. \quad \text{(10.12)}$$

An advantage to this form is that, for gas wells, $\rho_f q$ at downhole conditions is equal to $\rho_f q$ at standard conditions, eliminating the need to estimate downhole gas density. Experiments performed with a variety of perforation sizes and formation permeabilities yielded the empirical correlation

$$a_{600}=3.84\times10^{-5}X_P^{2.86}, \quad \text{(10.13)}$$

which is close to the theoretical form derived for flow from a perforation. To interpret a noise log, Eq. 10.13 can be inverted to

$$X_P=35(a_{600})^{0.35}=\rho_f q/\mu_f d_P, \quad \text{(10.14)}$$

where q=volumetric flow rate from the perforation (Mcf/D), ρ_f= fluid density (lbm/ft^3), μ_f=fluid viscosity (cp), and d_P=perforation diameter (in.).

Fig. 10.16 is a noise log in which the flows from individual perforations show up distinctly. In this oil well, 19 distinct peaks in noise amplitude are observed. With Eq. 10.14, the rates from these perforations were calculated to total 2,530 B/D [402 m^3/d], as compared with the actual total downhole rate of 1,750 B/D [278 m^3/d]. It is interesting that in a well with about 400 perforations, 19 account for virtually all the production.

10.4.3. Liquid Entrained With Gas Flowing From Perforations. McKinley and Bower investigated the noise characteristics of gas/liquid streams produced from perforations into a gas- or foam-filled wellbore. In this situation, the liquid impacting the logging

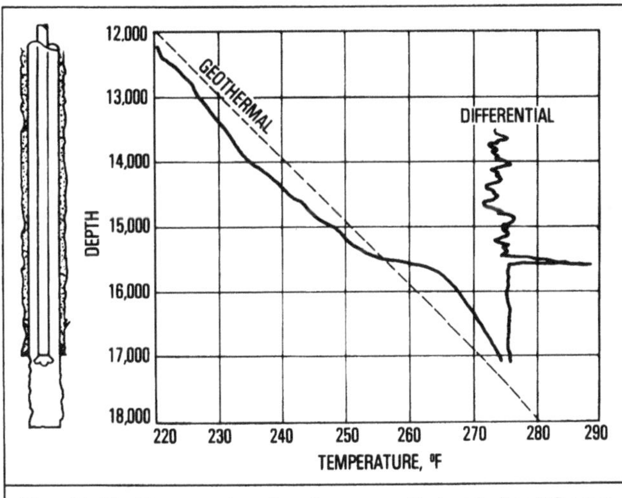

Fig. 10.17—Temperature log from a well shut in for 6 hours after a surface kick (from Ref. 9).

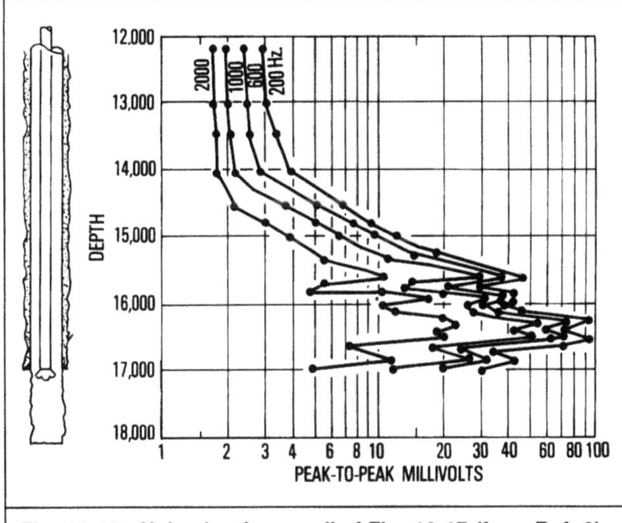

Fig. 10.18—Noise log from well of Fig. 10.17 (from Ref. 9).

tool as droplets adds a high-frequency component to the noise of the gas jetting from the perforation. They suggested the criterion

$$a_{2,000}/X_P^3 \leq 3\times10^{-5} \quad \text{.......................... (10.15)}$$

to indicate dry gas production. In this equation, $a_{2,000}$=amplitude of noise above 2,000 cycles/sec [2000 Hz] (peak-to-peak millivolts). If this condition is not satisfied, the liquid rate can be approximated by

$$q_L = 1.5 d_P a_{2,000}/\sqrt{X_P}, \quad \text{........................ (10.16)}$$

where q_L=liquid flow rate in barrels per day and d_P=perforation diameter in inches.

Example Calculation. A perforation producing 20 Mscf/D [57.3 std m³/d] of gas generates a noise amplitude above 2,000 cycles/sec [2000 Hz] of 300 RMS mV. The cable attenuation factor is 1.5. The gas has a density at standard conditions of 0.06 lbm/ft³ [961 g/m³] and a viscosity at downhole conditions of 0.02 cp [0.02 mPa·s]. The perforation is 0.35 in. [8.9 mm] in diameter. Is the perforation producing any liquid, and, if so, at what rate?

Solution. Apply Eq. 10.15 to solve this calculation.

$$X_P = (\rho_f q)_{sc}/(\mu_f d_P)$$

$$= (0.06 \text{ lbm/ft}^3)(20 \text{ Mscf/D})/(0.02 \text{ cp})(0.35 \text{ in.}) = 171.$$

The noise amplitude must be corrected for cable attenuation and converted to peak-to-peak millivolts:

$$a_{2,000} = (300 \text{ mV})(2.83)(1.5) = 1{,}274 \text{ mV}.$$

Thus, $a_{2,000}/X_P^3 = 1{,}274/(171)^3 = 2.5\times10^{-4}$.

Because $2.5\times10^{-4} > 3\times10^{-5}$, liquid is being produced. The liquid rate is calculated from Eq. 10.16

$$q_L = 1.5 d_P a_{2,000}/\sqrt{X_P} = 1.5(0.35 \text{ in.})(1{,}274 \text{ mV})/\sqrt{171}$$

$$= 51 \text{ B/D } [8.1 \text{ m}^3/\text{d}].$$

The perforation is producing about 50 B/D [7.9 m³/d] of liquid along with the gas.

10.4.4 Sand Production. Two techniques for detecting sand production with noise logs have been advanced. The laboratory results of McKinley and Bower[7] indicated that sand production results in a broad spectrum of noise above 4,000 cycles/sec [4000 Hz], while Stein *et al.*[8] suggested using the dependence of noise amplitude on average fluid density to indicate the onset of sand production.

McKinley and Bower found that the impact of sand grains on the logging sonde produces a broad-band noise spectrum above 4,000 cycles/sec [4000 Hz], while noise in this frequency range caused by fluid movement decays rapidly with frequency. They concluded that the decay rate of noise above 4,000 cycles/sec [4000 Hz] was a good indicator of sand production and defined a decay ratio, R, as

$$R = a_{4,000}/(a_{4,000} - a_{6,000}), \quad \text{..................... (10.17)}$$

where $a_{4,000}$ and $a_{6,000}$=peak-to-peak noise amplitudes above 4,000 and 6,000 cycles/sec [4000 and 6000 Hz], respectively. Sand production is indicated by the following criteria for R.

In single-phase flow, $R > 2.5$. In gas/liquid flow, $R > 3.5$ when $X_P < 100$ or $R > 2 \log X_P - 0.5$ when $X_P \geq 100$, where $X_P = X_{P_L} + X_{P_g}$. These relationships hold for fluid jetting from a perforation at sufficient rate to produce a noise level at least four times that of background noise. The sand concentration in the fluid was correlated to the noise amplitude as

$$C_{sd} = B[(a_{4,000}/X_P^3) - 1\times10^{-5}], \quad \text{................. (10.18)}$$

where

$$C_{sd} = (\text{lbm/D})/(\rho_f q/\mu_f). \quad \text{........................ (10.19)}$$

The coefficient B varies inversely with sand size and is 3.5×10^4 for 16- to 20-mesh sand.

At present, no field results for the McKinley-Bower technique for evaluating sand production have been presented in the literature. Their technique does require two additional high-pass filters to measure the frequency cuts above 4,000 and 6,000 cycles/sec [4000 and 6000 Hz]. Many logging companies provide these additional high-pass filters so that this procedure can be easily implemented.

Stein *et al.*[8] proposed a method for determining the flow rate required for the onset of sand production by obtaining average fluid density from a noise log. By assuming that the noise generated by fluids entering the wellbore is proportional to their kinetic energy, they derived that

$$\rho_f q^3 = c_2 \Sigma a_d, \quad \text{................................ (10.20)}$$

where c_2=constant and Σa_d=summation of the noise measured at a number of d intervals. According to this equation, a plot of the log of the summed noise levels vs. the log of the flow rate at a number of different flow rates would yield a straight line with a slope of 3 if the density of the flow stream remained constant. Sand production would presumably change the average density of the fluid sufficiently to change the slope appreciably. The technique, then, for determining the rate at which sand production begins to

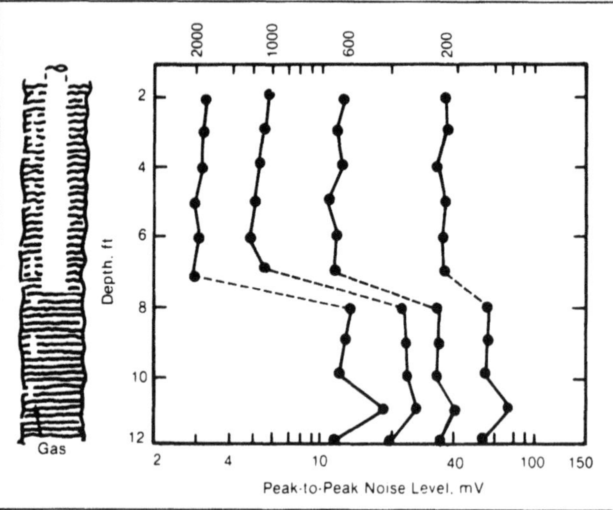

Fig. 10.19A—Location of liquid levels with a noise log, change of transmission medium (from Ref. 6, courtesy Atlas Wireline Services, Western Atlas Intl.).

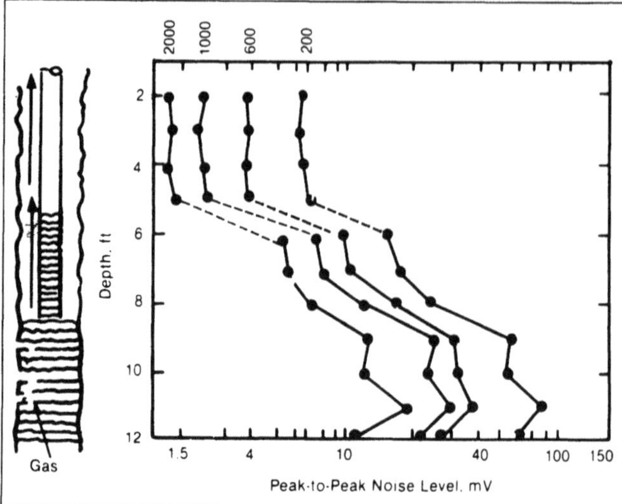

Fig. 10.19B—Location of liquid levels with a noise log, multiple changes of transmission medium (from Ref. 6, courtesy Atlas Wireline Services, Western Atlas Intl.).

occur is to increase the flow rate gradually while making noise logging measurements across the productive intervals; sand production is indicated by the rate at which the slope of the log-log plot of noise amplitude vs. flow rate increases appreciably from 3. This analysis assumes that the concentration of sand changes with flow rate as flow rate is increased after the onset of sand production; otherwise, the slope of the plot would still be predicted to be 3, though at a higher absolute noise amplitude. Of course, any other changes in average fluid density accompanying changes in total flow rate from a well (e.g., a change in gas/liquid ratio resulting from the change in pressure) would complicate this interpretation scheme. Stein *et al.* did not present any field data that indicated sand production by this method. The McKinley-Bower technique for the detection of sand production appears to be simpler and more reliable than the Stein *et al.* technique, though neither has been proven conclusively in practice.

10.4.5 Noise Logging During Drilling. Noise logs were found to be useful during well drilling to find the source and flow path of an underground blowout or crossflow.[4,9] McKinley[9] illustrated such an application (Figs. 10.17 and 10.18). In this well, a surface kick occurred during drilling below 18,000 ft [5486 m], with casing set at about 17,000 ft [5182 m]. A temperature log run after the well had been shut in for 6 hours (Fig. 10.17) indicates a crossflow from total depth up to 15,600 ft [4755 m]. Though the temperature log identifies where the flow enters the formation, it cannot distinguish between flow inside the casing exiting from a leak at 15,600 ft [4755 m] and flow outside the casing to this depth. The noise log subsequently run (Fig. 10.18) exhibits the characteristics of flow through a tortuous path because multiple peaks are observed from the casing shoe to the sink at 15,600 ft [4755 m]. Thus, the crossflow is behind the casing through channels in the cement. Because the well contains drilling mud, the noise log also shows the flow to be liquid because the crowding of the peaks of noise at different frequency cuts indicates single-phase flow.

10.4.6 Influence of Liquid Levels. The change from a gas-filled to a liquid-filled borehole is useful to recognize in noise log interpretation. Because acoustic coupling is greater in a liquid than in a gas, noise amplitudes at all frequencies are higher in a liquid-filled than in a gas-filled wellbore or tubing. Figs. 10.19A and 10.19B show two examples of liquid-level detection with a noise log. In Fig. 10.19A, the liquid level in the tubing is clearly indicated by the depth at which an abrupt change in noise amplitudes takes place. In Fig. 10.19B, two changes in the acoustic coupling occur. In the lower part of the well, the tubing and the tubing/casing annulus are liquid-filled. Above a certain depth, the annulus is gas-filled, while the tubing remains liquid-filled. This change of fluid content in the annulus results in decreased noise amplitudes, though the decrease is not as distinct as is the case for a change in the fluid occupying the pipe containing the noise logging tool (the tubing in this case). When the tubing becomes gas-filled at a higher depth, another, more abrupt, decrease in noise amplitudes occurs.

10.5 Guidelines for Running and Interpreting Noise Logs

1. A noise log is useful if the noise generated by the phenomenon of interest (flow in a channel, flow from perforations, etc.) is sufficiently greater than the background noise in the well. Thus, log quality depends on minimizing as much as possible any surface noise that could result in background noise detected by the tool.
2. The noise logging instrument should have the capability to record the full-frequency spectrum at desired depth locations to enhance interpretation of the type of flow being detected.
3. Noise logging tools should be run without centralizers to maximize acoustic coupling.
4. The operator should wait a short while (about 60 seconds) at each measurement location before recording the noise response to allow for the dissipation of noise caused by tool movement.
5. For quantitative interpretation, noise logging responses must be corrected for cable attenuation and wellbore geometry. The logging operator must supply the information required for these corrections because they are specific to the particular logging unit used.
6. A sufficient pressure drop must occur for a noise log to detect flow. Low-flow-rate channeling and channeling with no significant restrictions to flow are not detectable with a noise log.
7. Noise logs, like most production logs, are often most useful when used in conjunction with other logs; in particular, a temperature log should normally be run with a noise log.

Nomenclature

a = noise amplitude, dB, m/Lt^3
a_d = noise amplitude at location d, mV, mL^2/t^{2}q
a_{200} = noise amplitude above 200 cycles/sec [200 Hz], mV, mL^2/t^{2}q
a_{600} = noise amplitude above 600 cycles/sec [600 Hz], mV, mL^2/t^{2}q
$a_{2,000}$ = noise amplitude above 2,000 cycles/sec [2000 Hz], mV, mL^2/t^{2}q
$a_{4,000}$ = noise amplitude above 4,000 cycles/sec [4000 Hz], mV, mL^2/t^{2}q

$a_{6,000}$ = noise amplitude above 6,000 cycles/sec [6000 Hz], mV, mL^2/t^2q
Δa = $a_{200} - a_{600}$, mV, mL^2/t^2q
A_{an} = annular cross-sectional area, ft^2 [m^2], L^2
B = correlation constant
c = proportionality constant
c_2 = constant
C_{sd} = sand concentration, lbm/ft-D [kg/m·d], m/Lt
d_P = perforation diameter, in. [cm], L
E = mechanical energy, Btu [J], mL^2/t^2
F_G = well geometry factor
K_D = drag coefficient
N_{Re} = Reynolds number
Δp = pressure drop, psi [MPa], m/Lt^2
Δp_t = pressure drop across throttle, psi [MPa], m/Lt^2
q = volumetric flow rate, ft^3/sec [m^3/s], L^3/t
q_L = liquid volumetric flow rate from a perforation, B/D [m^3/d], L^3/t
R = noise decay ratio
t = time, seconds, t
X_P = $(\pi/4)N_{Re}$ for flow from perforation
X_{Pg} = $(\pi/4)N_{Re}$ for gas flow from perforation
X_{PL} = $(\pi/4)N_{Re}$ for liquid flow from perforation
μ_f = fluid viscosity, cp [Pa·s], m/Lt
ρ_f = fluid density, lbm/ft^3 [kg/m^3], m/L^3

References

1. Enright, R.J.: "Sleuth for Down-Hole Leaks," *Oil & Gas J.* (Feb. 28, 1955) 78-79.
2. McKinley, R.M., Bower, F.M., and Rumble, R.C.: "The Structure and Interpretation of Noise From Flow Behind Cemented Casing," *JPT* (March 1973) 329-38; *Trans.*, AIME, **255.**
3. Robinson, W.S.: "Recent Applications of the Noise Log," *Proc.*, SPWLA Annual Logging Symposium (June 9-12, 1976) paper Y.
4. Robinson, W.S.: "Field Results From the Noise-Logging Technique," *JPT* (Nov. 1976) 1370-76.
5. Britt, E.L.: "Theory and Applications of the Borehole Audio Tracer Survey," *Proc.*, SPWLA Annual Logging Symposium (June 9-12, 1976) paper BB.
6. *Interpretive Methods for Production Well Logs*, second edition, Atlas Wireline Services, Western Atlas Intl., Houston (1982) 101-19.
7. McKinley, R.M. and Bower, F.M.: "Specialized Applications of Noise Logging," *JPT* (Nov. 1979) 1387-95.
8. Stein, N. *et al.*: "Sand Production Determined from Noise Measurements," *JPT* (July 1972) 803-06.
9. McKinley, R.M.: "Production Logging," paper SPE 10035 presented at the 1982 SPE Intl. Petroleum Exhibition and Technical Symposium, Beijing, March 18-26.

SI Metric Conversion Factors

cycles/sec	× 1.0*	E+00	=	Hz
ft	× 3.048*	E−01	=	m
ft^3	× 2.831 685	E−02	=	m^3
°F	(°F−32)/1.8		=	°C
in.	× 2.54*	E+00	=	cm
psi	× 6.894 757	E+00	=	kPa

*Conversion factor is exact.

Chapter 11
Cement-Quality Logging

11.1 Introduction

Many production logs are aimed at evaluating the competence of a well completion. Knowledge about the completion often is gained indirectly by measuring fluid movements with such logs as radioactive-tracer or temperature logs. Acoustic logging techniques, primarily the cement-bond log, have been used for many years to try to measure directly the quality of the cement between the casing and the formation. Ultrasonic-pulse-echo techniques have recently been developed to eliminate some of the deficiencies of the cement-bond log for cement evaluation. These logs are intended to identify poorly cemented regions during the completion stage of the well so that these regions can be repaired with cement squeezes before the well is perforated for production, although their application for this purpose varies significantly from company to company.

A primary function of the cement is to prevent fluid movement between the various zones in a reservoir and between the reservoir and other zones up or down the hole. Thus, cement-quality logging is aimed at determining whether the cement is of sufficient strength and distribution to prevent fluid communication between zones. A cement-quality log ideally should indicate whether the cement is bonded to the pipe, whether the cement is bonded to the formation, and whether channels are present in the cement. A cement-quality log does not directly measure the capability of the cement to prevent fluid communication; this is inferred from the degree of acoustic coupling of the cement to the pipe and the formation as measured by the logs. For this reason, cement-quality logs are not an absolute measure of the hydraulic integrity of the cement; however, when run and interpreted properly, they are generally reliable predictors of cement placement.[1]

The two primary logs for evaluating cement quality, the cement-bond log and the ultrasonic-pulse-echo log, are both acoustic logs that differ primarily in the path taken by the sound waves between the transmitter and the detector. With the cement-bond log, sound travels axially down the casing and through the cement and formation to detectors (usually two) located below the sound source on the logging tool. An ultrasonic-pulse-echo log, on the other hand, has an array of transducers that both transmit and receive sound energy so that the sound path is radial to and from the transducers. Thus, the two types of logs make fundamentally different measurements and must be treated separately. A third logging method, the attenuation-ratio log, is essentially a borehole-compensated cement-bond log.

This chapter first discusses the cement-bond log, then the attenuation-ratio log, and finally the ultrasonic-pulse-echo log. For each log, the tools and the theory of the method are presented, and the common log presentations, interpretation procedures, and log examples are discussed. The chapter closes with guidelines for cement-quality logging.

11.2 Cement-Bond Logging

11.2.1 Tools and Operation Theory. A cement-bond log is an acoustic log with a transmitted acoustic signal detected by one or two detectors. The basic principle of cement-bond logging is that if the casing is free to move, an acoustic signal is not greatly attenuated, whereas if the pipe is held firmly in place by cement, vibrations are rapidly dampened and the received acoustic signal is low in amplitude.

Cement-bond logs were developed on the basis of the phenomenon of cycle skipping observed in sonic logs in cased wells.[2,3] In a sonic log measuring the transit time of acoustic waves in cased wells, it was observed that the transit time measured often would be longer than the time that would result from travel through the casing. The skipping of these early cycles was because of the low amplitude of the acoustic signals traveling through well-bonded casing. Thus, the sonic-log transit time gave a rough indication of the cement bond. The cement-bond log was developed to measure the amplitude of the early arriving acoustic signals to give a more complete picture of the degree of bonding. (Refs. 2 through 8 chart the early development of the cement-bond log.) It was soon recognized that more information about bonding conditions could be obtained by examining the full acoustic wave train rather than by relying on only a single amplitude measurement, so techniques were developed to display the full wave train.[9-12] Bigelow[13] presents an excellent review of current cement-bond logging practices and interpretation procedures; much of this material follows his work.

Fig. 11.1 shows a typical cement-bond log tool configuration. Acoustic signals are emitted by a crystal transmitter that is switched on and off, generating sound impulses with frequencies of about 20,000 cycles/sec [20 kHz]. The sound waves are elastic compressional waves that propagate spherically from the acoustic source. Two receivers generally are used to detect the transmitted sound waves, one spaced about 3 ft [0.9 m] below the transmitter and one spaced about 5 ft [1.5 m] below the transmitter. Some cement-bond tools have a single receiver spaced 3 to 4 ft [0.9 to 1.2 m] from the transmitter; however, experiments in a test well showed that single-receiver tools are generally inadequate.[14] The 3-ft/5-ft [0.9-m/1.5-m] dual-receiver tool is currently the standard industry configuration.

The compressional waves follow a variety of paths between the transmitter and the receivers—including down the sleeve of the instrument, axially through the borehole fluid, and radially across the borehole fluid. The waves that move radially across the borehole fluid are of most interest because some are reflected at the casing walls, while the remaining energy propagates through the casing, the cement, and the formation to the receivers by ray refraction.

Fig. 11.2 shows the signal received, which is a measure of acoustic amplitude vs. time. The closer-spaced detector is used to measure the amplitude of the first acoustic wave that arrives, because all amplitudes are higher closer to the signal source. The farther-spaced detector is used to measure the full wave train because it allows for more travel through the formation. Two important properties of the signal are the transit time (i.e., the time between signal generation and signal arrival at the detector) and the amplitude (i.e., the amplitude of the first sound wave arriving at the detector).

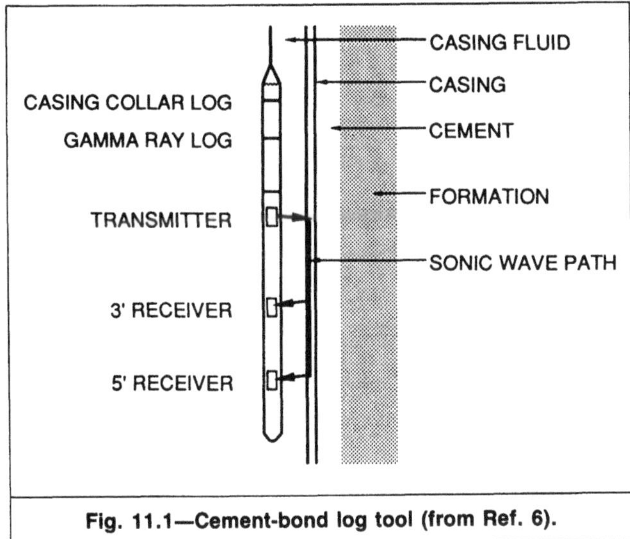

Fig. 11.1—Cement-bond log tool (from Ref. 6).

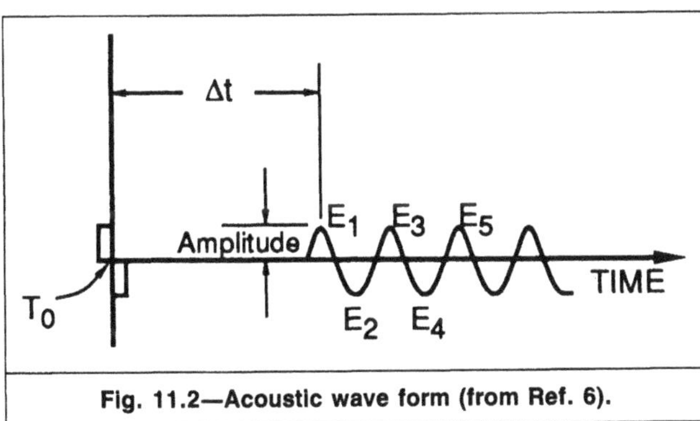

Fig. 11.2—Acoustic wave form (from Ref. 6).

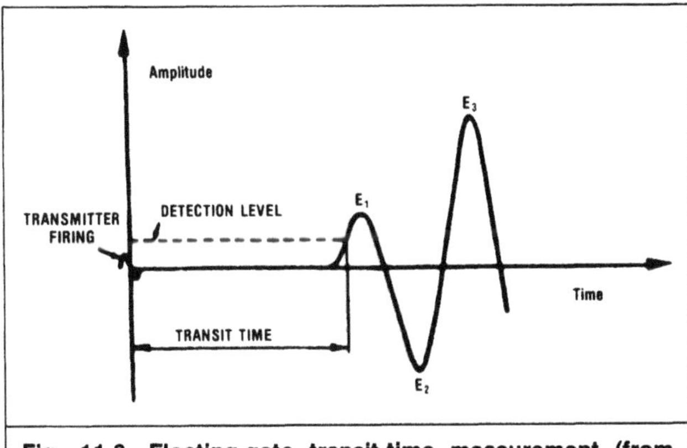

Fig. 11.3—Floating-gate transit-time measurement (from Ref. 13).

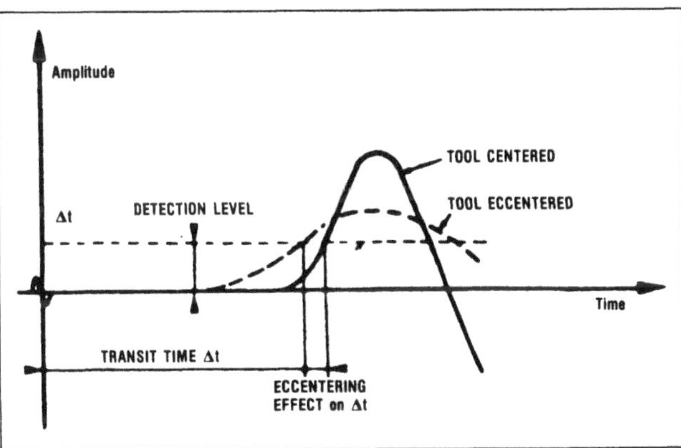

Fig. 11.4—Effect of eccentering on first amplitude-response arrival (from Ref. 13).

It is known that the sonic travel time in steel is approximately 57 μsec/ft [187 μs/m], while in mud the travel time is about 189 μsec/ft [620 μs/m]. (This value varies somewhat depending on the particular mud used.) Thus, the first signal arriving at the detector typically is from sound waves traveling through the casing. For a detector spaced 3 ft [0.9 m] from the transmitter and with the tool centered 1 in. [2.54 cm] from the casing, the pipe arrival time (i.e., the time required for the sound waves traveling through the casing to reach the first detector) is

$$\Delta t_p = (1/12 \text{ ft})(189\ \mu\text{sec/ft}) + (3 \text{ ft})(57\ \mu\text{sec/ft})$$

$$+ (1/12 \text{ ft})(189\ \mu\text{sec/ft}) = 202.5\ \mu\text{sec/ft}\ [664.4\ \mu\text{s/m}].$$

Thus, to determine the degree of bonding of the pipe to the cement, the amplitude of the acoustic waves arriving at about 200 microseconds is examined—the higher the amplitude of these pipe arrivals, the poorer the contact between pipe and cement. Even though the first pipe arrival theoretically occurs at about 200 microseconds, the first arrival measured often occurs 25 to 50 microseconds later because the amplitude of the first wave is below detection limits.

Amplitude of the pipe arrival is generally measured with a fixed electronic gate that is set to "open" shortly before the time the signal traveling through the casing is expected. The maximum amplitude measured during the gated period is recorded as the amplitude log. With some tools, a floating-gate system is also used to measure the transit time and sometimes the amplitude, though current preference is for a fixed-gate amplitude measurement. Fig. 11.3 illustrates the floating-gate transit-time measurement. The operator must set a minimum detection level, and the arrival time of the first signal exceeding this threshold amplitude is recorded as the transit time. The transit-time measurement is obviously sensitive to the detection level chosen by the operator; the detection level is usually chosen by measuring the maximum amplitude in free pipe, then setting the detection level at less than 5% of the free-pipe amplitude.

When a cement-bond log is run, proper centralization of the tool is critical. Because the sound waves are traveling through a multitude of paths along the casing, eccentricity results in signals that arrive out of phase, leading to significantly reduced amplitudes. Fig. 11.4 illustrates the effect of tool eccentering on the amplitude response.[13] This amplitude reduction is the result of the interference of waves arriving out of phase. Fig. 11.5 illustrates the theoretical change in the amplitude for a typical tool in 9⅝-in. [24.4-cm] casing as a function of the amount of eccentering.* To prevent eccentering, at least two and preferably three strong centralizers should be used on a cement-bond tool. To check for tool eccentering, it is best to run the tool in a section of uncemented casing and measure transit time. Variations in transit time of more than 4 microseconds indicate eccentering.

When some bonding exists, the sonic waves travel through the formation as well as through the casing. A measurement of these formation arrivals can tell us something about the contact between the cement and the formation. Thus, an examination of the full sonic wave train provides more information about the cement bonding than can be obtained by measuring only pipe-arrival amplitude.

Fig. 11.6 shows an idealized acoustic wave train. This wave form consists of four types of wave arrivals in the order of their arrival: the compressional wave, the shear wave, the mud wave, and the Stoneley wave.

The compressional wave is transmitted through particle motion forward and backward in the direction in which the wave travels and may be transmitted through the mud, the pipe, the cement, and the formation. Except in the case of fast formations, the first wave detected by the receiver is the compressional wave that has traveled through the casing.

*From personal communication with Gilbert L. Feather, Amoco Production Research Co., Tulsa, OK, 1987.

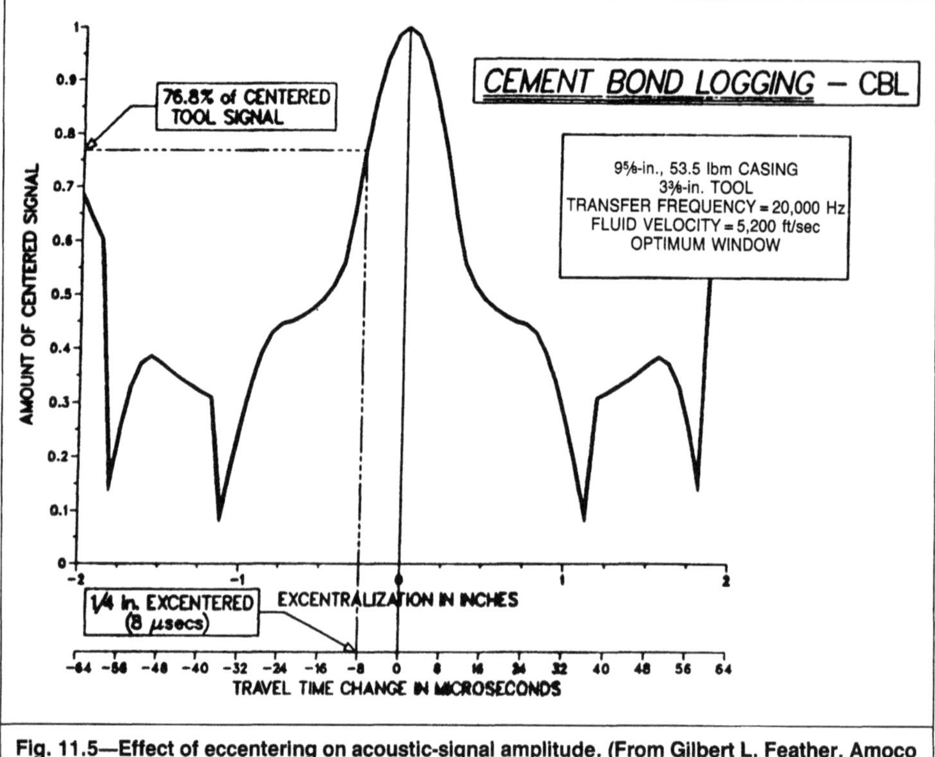

Fig. 11.5—Effect of eccentering on acoustic-signal amplitude. (From Gilbert L. Feather, Amoco Production Research Co., Tulsa, OK, 1987.)

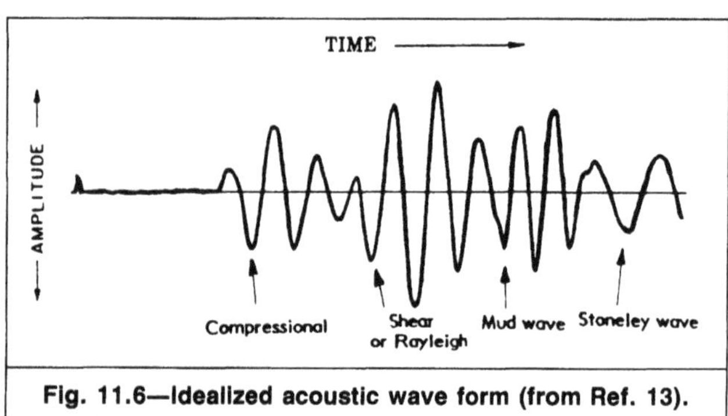

Fig. 11.6—Idealized acoustic wave form (from Ref. 13).

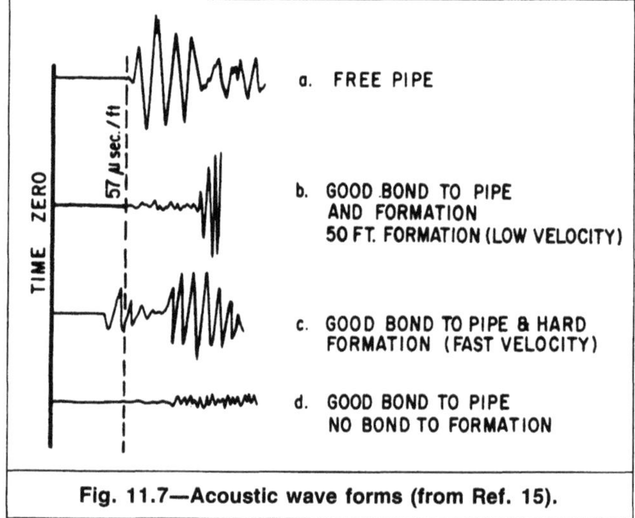

Fig. 11.7—Acoustic wave forms (from Ref. 15).

The shear wave is transmitted through particle motion perpendicular to the wave path and is supported only in solids because fluids have no shear strength. The shear waves travel 1.6 to 1.9 times slower than compressional waves in the same medium and usually have higher amplitude. Shear wave detection is generally indicative of some acoustic coupling of the pipe to the formation.

The mud wave is the compressional wave that travels through the mud from the transmitter to the receiver. Because the transit time through the mud is longer than through the pipe or formations of interest, the mud wave occurs late and usually does not interfere with subsequent log interpretation.

The Stoneley wave is a low-frequency interface wave traveling along the borehole wall and down the tool body. These waves arrive even later than the mud waves.

A careful examination of the acoustic wave train provides considerable information about the quality of the cement bonding not only to the pipe, but also to the formation. Fertl *et al.*[15] illustrated the acoustic signals characteristic of various cement-bonding and formation conditions (Fig. 11.7). In free pipe (Fig. 11.7a), a high-amplitude signal beginning at the pipe-arrival time is measured. This is caused by the pipe "ringing" with little dampening of the acoustic signal. Fig. 11.7b illustrates the sonic wave train observed with good bonding to a low-sonic-velocity formation. The amplitude of the signal is very low at pipe-arrival time, indicating the good shear contact between the pipe and the cement. The high amplitude of formation arrivals that arrived later than the pipe arrivals is indicative of the low-velocity formation and good bonding between the cement and the formation. Good bonding is also assumed in Fig. 11.7c, but the signal represents a fast formation. In this case, formation arrivals can be as early as or even earlier than pipe arrivals, yielding artificially high acoustic amplitudes at pipe-arrival time. In Fig. 11.7d, good bonding exists between the casing and the cement, but poor bonding occurs between the cement and the formation. Thus, low amplitudes are measured at pipe-arrival times because of the good bonding and also at formation-arrival times because of the limited acoustic coupling between the cement and the formation.

11.2.2 Cement-Bond-Log Presentations. The two primary measurements obtained from a cement-bond tool are the amplitude of the pipe arrivals and a display of the full acoustic wave train. In

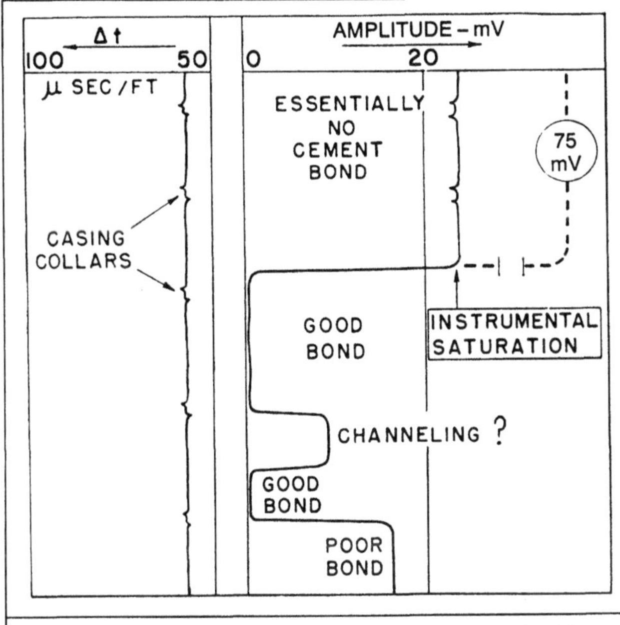

Fig. 11.8—Qualitative characteristics of amplitude log (from Ref. 2).

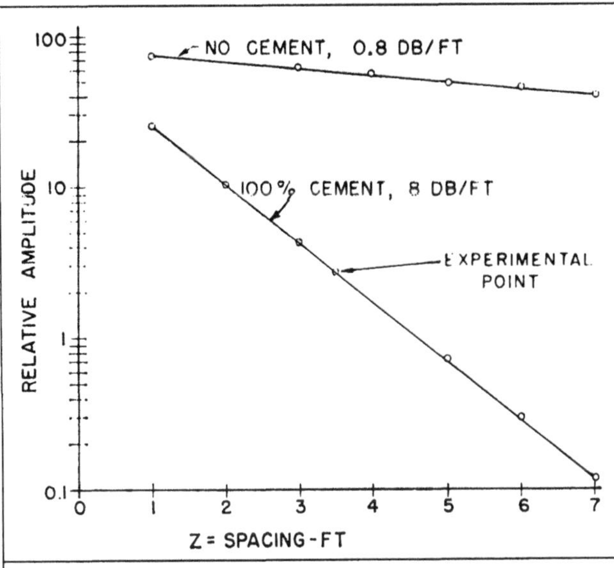

Fig. 11.9—Dependence of amplitude on transmitter-receiver spacing (from Ref. 2).

addition, the transit time of the first pipe arrival should always be displayed.

Amplitude Log. The amplitude log is a measure of the acoustic amplitude of the first pipe arrival and is usually measured by the detector closest to the transmitter. The amplitude of the pipe arrival was the first measurement from an acoustic log used as a measure of cement bonding and was referred to in the early literature as simply the cement-bond log. The amplitude of the pipe arrival is a measure of the loudness of the received acoustic signal. Unbonded pipe is free to vibrate, transmitting much of the acoustic energy of the signal from the transmitter, while in well-bonded pipe the acoustic signal is greatly attenuated. Thus, the amplitude of the sound transmitted through the casing is a measure of the bonding of the cement to the pipe. Fig. 11.8 shows the qualitative characteristics of the amplitude log. High amplitude indicates little or no bonding, low amplitude results from good bonding, and intermediate values are attributed to partial bonding around the pipe circumference. Methods have also been developed to quantify the cement bonding on the basis of the amplitude log. The amplitude log *alone*, however, should be used with caution in determining the ability of the cement to prevent fluid communication because the pipe amplitude gives no indication of the bonding between the cement and the formation. Furthermore, the quantitative analysis techniques are based on idealized conditions.

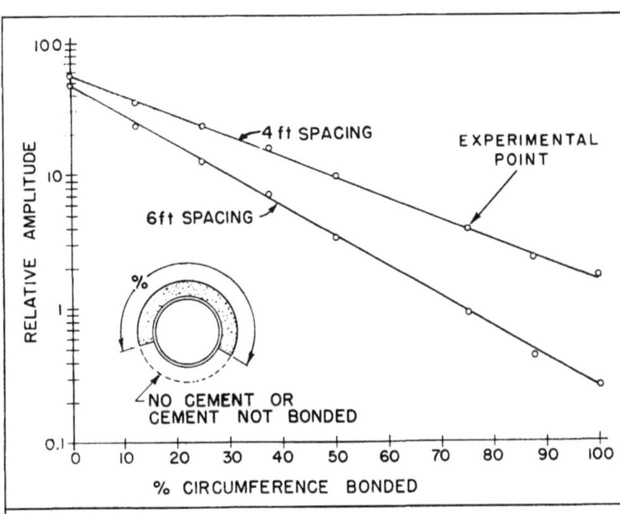

Fig. 11.10—Effect of percentage of circumference bonded on amplitude (from Ref. 2).

The amplitude is generally measured with either a fixed- or a floating-gate system. With the fixed-gate system, the method most predominantly used today, the detector is activated at a predetermined time that is shortly before the first expected pipe arrival. The tool then measures the highest amplitude detected during a brief time while the gate is open. To operate a cement-bond log in the floating-gate mode, the operator sets a trigger amplitude level. The tool then measures the amplitude of the first acoustic wave that exceeds this level. To distinguish between the measurement of pipe-arrival and formation-arrival signals, include a transit-time measurement to help distinguish a high amplitude caused by a formation arrival from high pipe amplitudes caused by poor bonding.

The amplitude of the pipe arrival decreases exponentially with distance from the transmitter,[2] or

$$a = a_0 e^{-\alpha z}, \quad \text{(11.1)}$$

where a_0 = amplitude of transmitted signal, α = composite attenuation factor, and z = distance from the transmitter. Fig. 11.9 shows that laboratory measurements confirmed this relationship. Grosmangin *et al.*[2] also demonstrated a semilogarithmic dependence of pipe amplitude on fraction of the pipe circumference that was bonded (Fig. 11.10). To compare the effect of such variables as cement compressive strength on the attenuation of the acoustic energy, the amplitudes at two locations spaced a distance z apart can be measured. From Eq. 11.1, the attenuation rate is

$$\alpha = (20/z)\log_{10}(a_1/a_2), \quad \text{(11.2)}$$

where a_1 and a_2 = amplitudes at the beginning and end of depth interval z. Pardue *et al.*[6] measured attenuation rates for several cements as they cured (Fig. 11.11) and for several casing sizes. From these experiments, a nomograph was developed (Fig. 11.12) from which the attenuation rate and the cement compressive strength could be obtained from the amplitude log. Note that this nomograph will apply strictly for a laboratory situation with a perfectly centered tool in centered casing with a continuous cement sheath of at least ¾ in. [1.9 cm]. Because these conditions are seldom all met in a well, calculation of compressive strength from the amplitude log should be considered approximate.

Another quantitative method of interpreting the amplitude log uses the bond index, defined by Brown *et al.*[12] as the attenuation in the

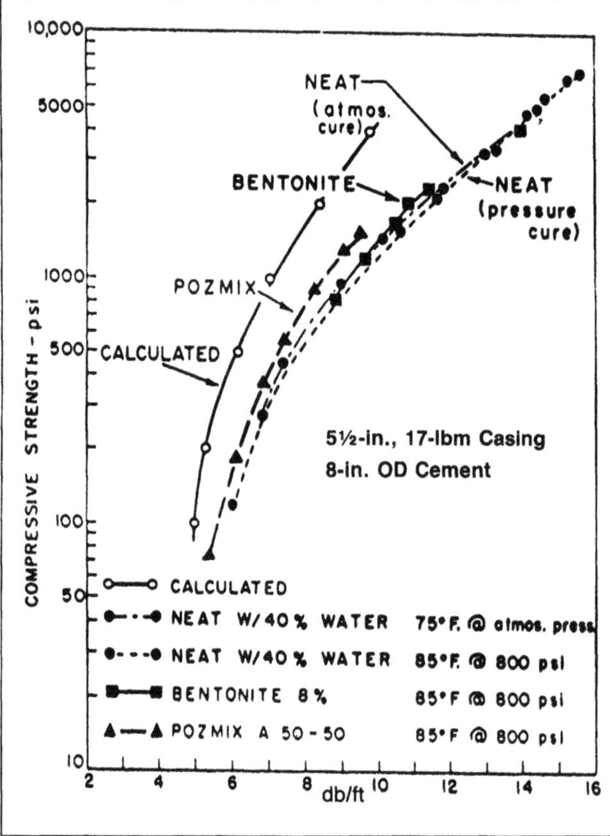

Fig. 11.11—Effect of cement compressive strength on attenuation rate (from Ref. 6).

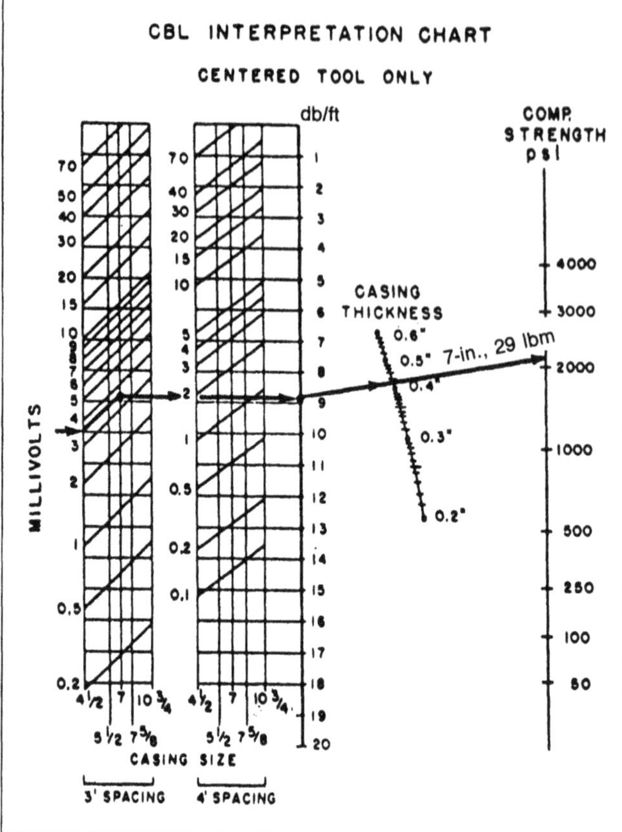

Fig. 11.12—Nomograph relating amplitude to cement compressive strength (from Ref. 6).

zone of interest minus the attenuation in free pipe, divided by the attenuation in a well-cemented section minus the attenuation in free pipe. The attenuation would be obtained from the amplitude from a correlation like the nomograph in Fig. 11.12. The bond index was assumed to represent the fraction of the pipe circumference bonded. Brown *et al.* suggested that a bond index of 0.8 was sufficient for hydraulic isolation if this degree of bonding extended a sufficient vertical distance. On the basis of field experience in South America, they presented a guideline for the interval of partial bonding needed for isolation (Fig. 11.13). Experience has shown that the 0.8 bond index cutoff is conservative and that 0.6 is generally acceptable.

The quantitative methods for interpreting amplitude logs all suffer from the fact that they are based on ideal laboratory conditions or are simply empirical in nature. The cement-bond log is better used in a qualitative way, particularly when the full wave train has been recorded. Bigelow[13] listed a number of physical conditions that can lead to erroneous amplitude interpretations, which are presented with references that discuss them in more detail.

1. *Fast formation.* When formation travel time is faster than pipe travel time, formation signals arrive earlier than or at the same time as pipe signals, leading to erroneous amplitude readings.[11,12,15-18]
2. *Floating-gate detector.* Erroneously high amplitudes can occur with floating gates because formation-arrival signals rather than pipe-arrival signals may be measured when the pipe is well bonded.[15-16] With a floating-gate detector, it is never certain what amplitude is being measured.
3. *Tool eccentering.* This condition reduces amplitude.[11,12,15-17]
4. *Insufficient cement-curing time.* This condition increases amplitude.[2,11]
5. *Less-than-¾-in. [<1.9-cm] cement sheath.* With either well-centered or poorly centered casing, this increases amplitude.[6,16,17]
6. *Microannulus.* This small gap between the casing and the cement, which is generally caused by shrinking of the casing when the pressure applied during cementing is reduced, increases amplitude.[12,16,18,19]

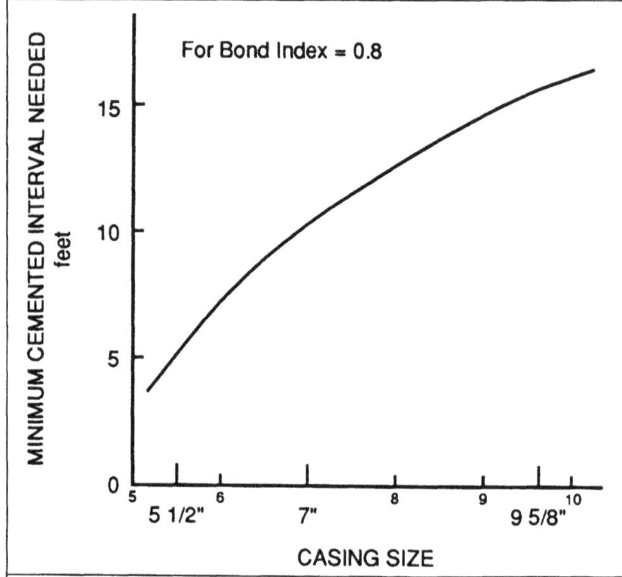

Fig. 11.13—Length of cemented interval required for zone isolation for bond index of 0.8 (from Ref. 12, courtesy Society of Professional Well Log Analysts).

7. *Gas bubbles.* The presence of gas bubbles in the borehole fluid decreases the acoustic signal.[17]
8. *Void spaces in the cement sheath.* These increase amplitude.[6,19]
9. *Pipe thickness.* Changes in pipe thickness from one joint to another cause amplitude variations.[2,6]
10. *Cement.* Cement may be bonded to the pipe, but not to the formation. This results in low pipe amplitude, even though hydraulic integrity may be poor.[19]

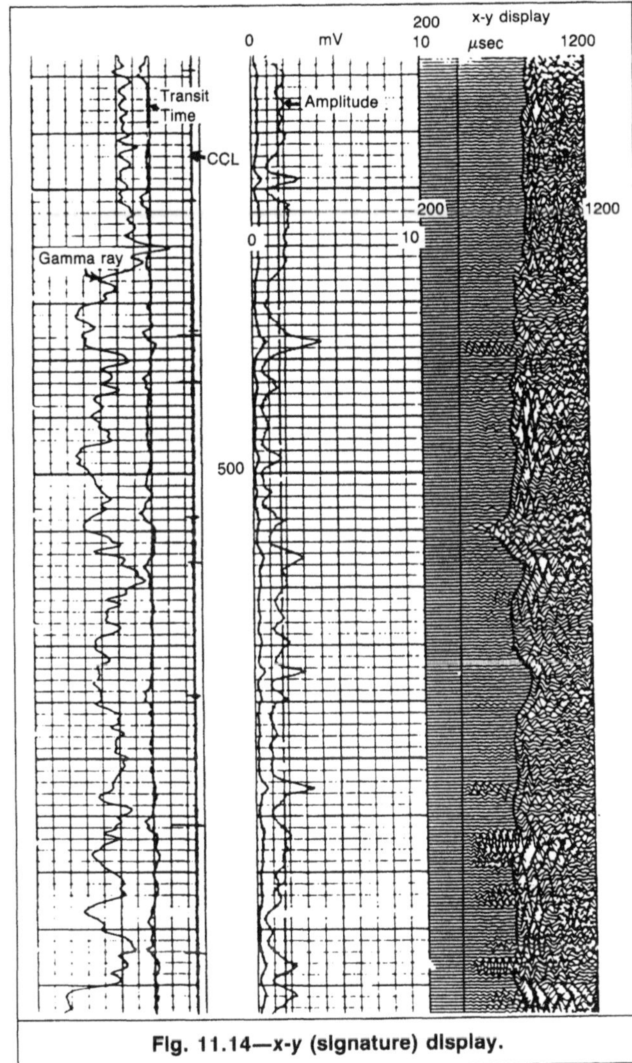

Fig. 11.14—x-y (signature) display.

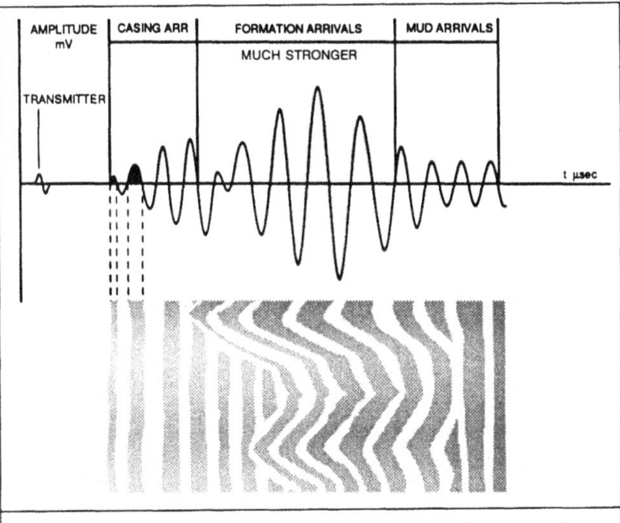

Fig. 11.15—Variable Density log construction (courtesy Schlumberger).

11. *Gas- or mud-cut cement.* The presence of gas or mud in the cement increases the amplitude, even though cement integrity may be good.

As this list shows, quantitative interpretation of an amplitude log is risky. When used qualitatively in conjunction with a full wave-train display, the amplitude log can be a good measure of bonding between the pipe and the cement.

Full Wave-Train Displays. A presentation of the full wave train of acoustic signals is the other primary display on a cement-bond log. The full wave train is usually measured with the detector spaced farthest from the transmitter, with the spacing typically set at 5 ft [1.5 m]. The two common full wave-train displays used are an *x-y* presentation (also called a signature display) and a Variable Density^{SM} log. In the *x-y* presentation,[11] the actual sonic wave train as seen on an oscilloscope is recorded on the log at selected intervals of ≥ ½ ft [0.15 m]. The full wave-train display shows the amplitude of the sonic waves arriving from the casing and from the formation as illustrated in Fig. 11.14. The main advantage of the *x-y* presentation is that it allows easy comparison of the amplitudes of the various waves received; because shear waves are typically higher in amplitude than compressional waves, the *x-y* presentation aids in identifying shear waves, which indicate acoustic coupling to the formation.

The Variable Density log (also called a microseismogram) is constructed by assigning strong positive signals a dark mark on the log, while zero-amplitude signals are bright white. All intermediate signals are shades of gray, depending on their relative amplitude. These interpreted wave signals are then recorded continuously with depth. Fig. 11.15 illustrates the construction of the Variable Density log. Like the *x-y* presentation, it displays the full wave train of acoustic signals, with the amplitude indicated by the contrast between the dark and light bands—the higher the contrast, the higher the amplitude. The Variable Density log can be displayed continuously with depth and thus highlights changes in the acoustic wave train with depth. This is particularly useful in identifying good acoustic coupling with the formation. Because the acoustic transit time in the formation varies with changes in lithology, the Variable Density log is wavy when responding to formation signals, much like an openhole sonic log. The quality of the Variable Density log is much more operator-sensitive than the full wave-train display.

Transit Time. An acoustic travel or transit time, the arrival time of the first acoustic signal, should always be displayed on a cement-bond log. The transit time is measured with a floating-gate detector as illustrated in Fig. 11.3, so the transit time is the time at which an amplitude first exceeds the bias level set by the operator. The transit-time measurement is useful for checking tool centering and for corroborating the amplitude log. A transit-time measurement is essential if the amplitude is measured with a floating-gate detector. Bigelow[13] illustrated the primary uses of the transit-time recording. For example, if the cement-bond log tool is well-centered, the transit time should normally correspond to the first expected pipe arrival. In uncemented pipe, where amplitudes are high, the transit time should be constant with depth (typically within a few microseconds), except for slight variations from one pipe joint to another caused by variations in internal diameter of individual pipe joints. Fig. 11.16 illustrates such a response. Eccentering of the logging tool results in irregular reductions in transit time of as much as 20 microseconds in unbonded pipe (approximately one-half of a wave period), as Fig. 11.17 shows. When the transit time cannot be held constant, attenuation-ratio tools should be used.

When there is bonding between the pipe and the cement, the transit time can be affected by fast formations, cycle skipping, or stretching. With good bonding to a fast formation, the measured transit time is less than the measured pipe arrival time and is helpful in interpreting the amplitude log, which is high as a result of the early formation signals (Fig. 11.18). Cycle skipping occurs when the amplitude of the first pipe arrival is less than the bias threshold set for the transit-time detector (Fig. 11.19) and results in a larger transit time than expected. Cycle skipping generally indicates good bonding between the pipe and the cement, unless the bias level was set too high. Fig. 11.20 shows a cement-bond log that illustrates cycle skipping. Transit-time stretch also indicates good bonding between the pipe and the cement because the transit time can be "stretched" by two or three microseconds when the first pipe arrival is attenuated (Fig. 11.21). The transit time and the Variable Density log or *x-y* presentation will correlate with openhole lithology logs when cement is well bonded to the formation.

11.2.3 Cement-Bond Log Interpretation. Though quantitative methods do exist for interpreting cement-bond log amplitude

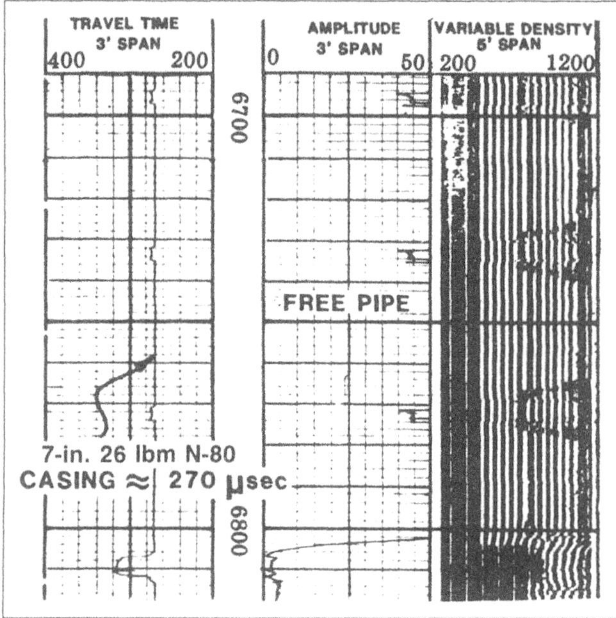

Fig. 11.16—Transit time with well-centered tool in uncemented pipe (from Ref. 13).

responses, cement-bond log interpretation is best done qualitatively, comparing the amplitude, full wave train, and, when available, transit-time displays. The following examples illustrate the cement-bond log responses expected for a variety of cement conditions.

1. *Free pipe.* In uncemented casing, the amplitude log shows high amplitude and the transit time corresponds to the casing arrival time. The Variable Density log shows strongly contrasting parallel vertical lines with no indication of formation signals. Casing collars show up distinctively on a cement-bond log in free pipe. Collar reflections, shown as chevrons on the Variable Density log, result

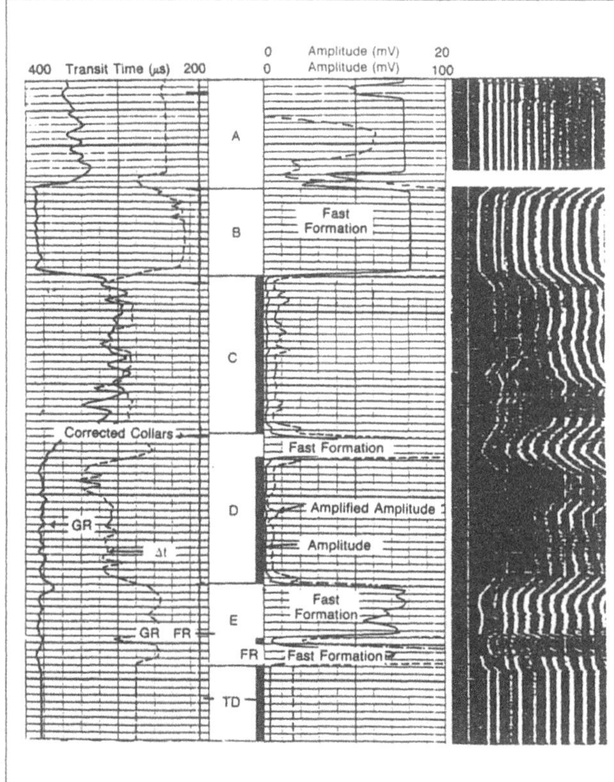

Fig. 11.18—Response to fast formations (from Ref. 20, courtesy Schlumberger).

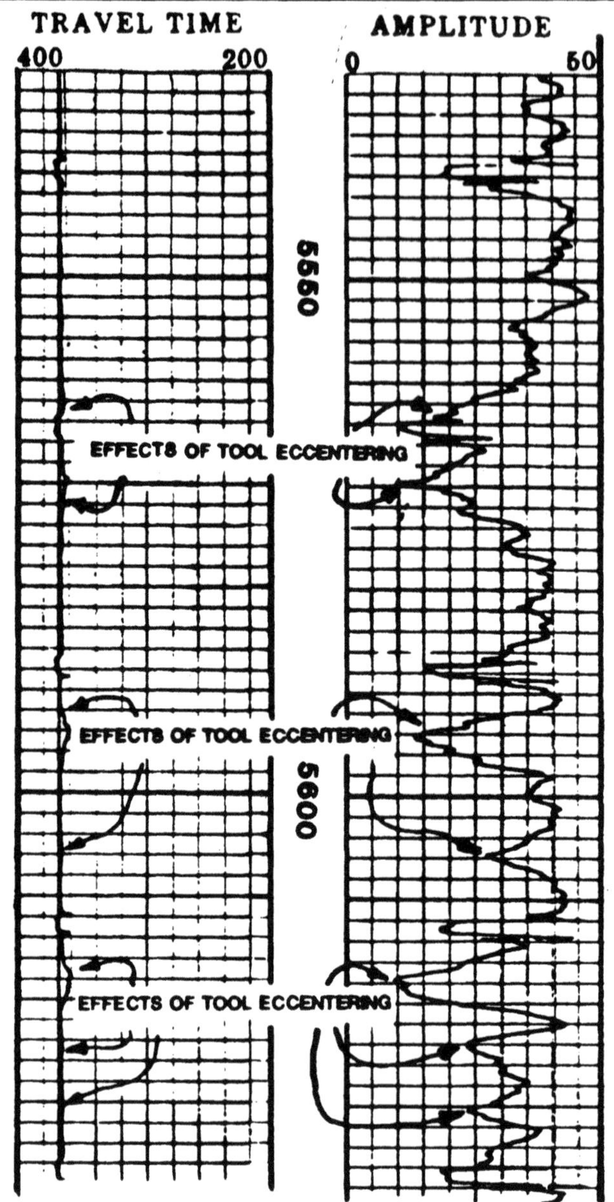

Fig. 11.17—Effect of eccentering on transit time (from Ref. 13).

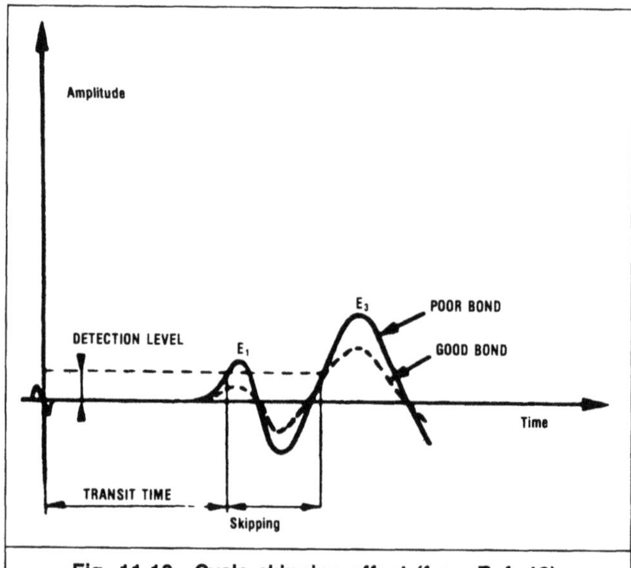

Fig. 11.19—Cycle skipping effect (from Ref. 13).

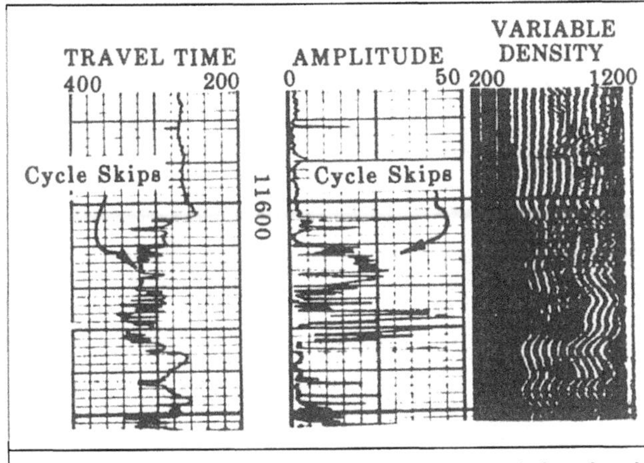

Fig. 11.20—Cycle skipping caused by attenuated pipe signal (from Ref. 13).

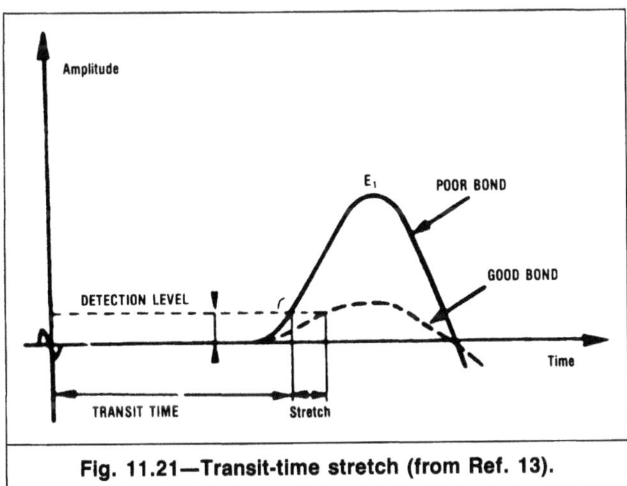

Fig. 11.21—Transit-time stretch (from Ref. 13).

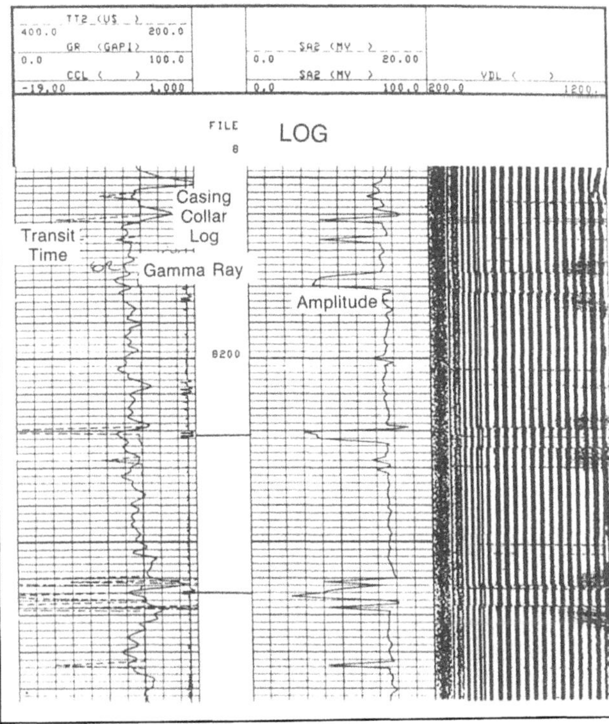

Fig. 11.22—Cement-bond log response in free pipe.

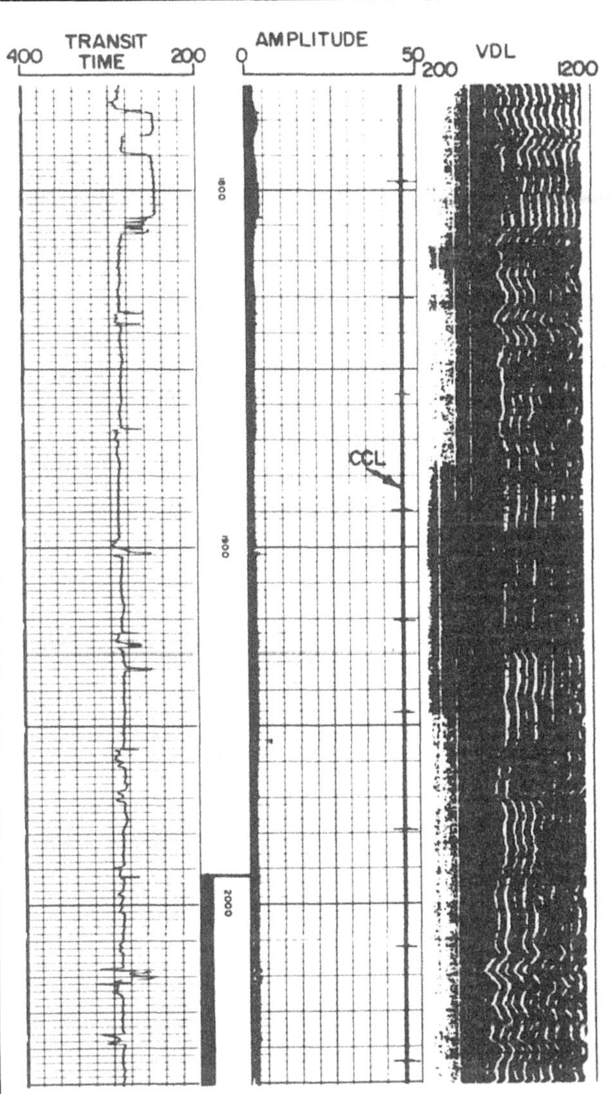

Fig. 11.23—Cement-bond log response to good bonding (from Ref. 15).

in a decrease in amplitude and an increase in transit time. Fig. 11.22 shows a cement-bond log in free pipe. It is important to log in an area of free pipe if possible when a cement-bond log is run. Any deviation from the expected response in free pipe indicates either a malfunctioning or an improperly centralized tool. Logging in free pipe calibrates the tool in a known environment under logging conditions.

2. *Good bond to formation and casing.* With good bonding, the amplitude is low. The full wave-train display shows weak or no casing signals and strong formation arrivals unless the formation attenuation is high, as would be observed for an unconsolidated gas sand, weak shales, or other low-velocity formations. Comparison of the cement-bond log with an openhole sonic log can help identify regions of high attenuation in the formation. Fig. 11.23 presents an example of good bonding to the pipe and the formation.

3. *Good casing bond, but poor formation bond.* This situation is characterized by weak casing arrivals, indicated by low amplitude and low contrast on the Variable Density log at casing arrival times, and weak formation signals on a full wave-train display. Unfortunately, these same characteristics can be caused by other factors, including high formation acoustic attenuation and tool eccentricity. Good bonding to the pipe but not to the formation can easily occur opposite permeable zones where a mud cake is built up that is not displaced by cement. Fig. 11.24 presents a case where both casing and formation amplitudes are low. Interpretation of the

bonding from the amplitude curve alone would give an erroneous picture of cement integrity—i.e., the lack of acoustic coupling to the formation indicates poor cementing, even though the pipe amplitude is quite low. This behavior alone is not sufficient to prove a lack of hydraulic seal, however, because mud occupying the space between the cement and the formation may be immobile.

4. *Microannulus.* In some instances, a small gap is present between the casing and the cement. This gap is called a microannulus if it is not large enough to allow any significant fluid migration. A microannulus can be caused by a casing pressure that is lower during production than during cementing, by heating of the casing during cement setting and subsequent cooling, or by cement dewatering. A microannulus exhibits moderate to poor bonding on a cement-bond log. The amplitude is moderate to high, and a full wave-train display exhibits weak to somewhat high casing signals and moderate formation arrivals.

The best way to distinguish a microannulus from actual poor bonding is to repeat the cement-bond log with pressure applied to the well (1,000 to 1,500 psi [6.9 to 10.3 MPa]), which reduces the size of the microannulus. If the additional pressure applied is sufficient to close the microannulus, the cement-bond log run with pressure should show lower amplitude, weaker casing signals, and stronger formation signals than the log run without surface pressure. Fig. 11.25 shows logs run with and without pressure in a microannulus region.

5. *Channeling.* Unfortunately, channeling in cement cannot be positively identified with a cement-bond log. A channel in the cement affects the cement-bond log much as a microannulus does, resulting in moderate to high amplitude, moderate to strong casing signals on a full wave-train display, and moderate formation signals. Fig. 11.26 illustrates cement-bond log responses before and after squeeze cementing in a region of channeling. Before the cement squeeze, the Variable Density log shows fairly strong pipe arrivals (note that collars can be distinguished), but strong formation signals are also evident, indicating some bonding to the formation. On this example, the "hashy" interference pattern on the presqueeze Variable Density log results from waves arriving out of phase after passing through the pipe, a gap, and the cement. The amplitude before squeezing is moderate to high. After the cement squeeze, the characteristics of good bonding are evident—low amplitude and strong formation signals on the Variable Density log. To distinguish channeling from a microannulus, the cement-bond log should be repeated with surface pressure applied. To be recognizable with a cement-bond log, a channel must occupy at least 60° of the pipe circumference.[14]

6. *Eccentric tool.* It is extremely important that a cement-bond tool be well centralized—at least three strong centralizers should be used. When the tool is eccentric, the acoustic waves travel paths of different lengths and thus arrive out of phase, resulting in reduced amplitudes. Tool eccentricity makes the Variable Density log look wavy in the pipe arrival region, yield artificially low amplitude, and exhibit a transit time less than the expected casing arrival time. Fig. 11.27 shows tool eccentering effects.

7. *Eccentered casing or thin cement sheath.* When the cement sheath is less than about ¾ in. [1.9 cm] thick, attenuation of the pipe signal diminishes greatly (Fig. 11.28), leading to a high-amplitude response. When casing is eccentered, the thin cement layer on the side of the hole where the casing is near the formation does not attenuate the pipe signal greatly. Similarly, if the pipe is well centered, but the annulus between the casing and the formation is small, pipe amplitudes are high. In such cases, the full wave-train display must be relied upon for evaluating cement integrity. If the full wave-train display shows good acoustic coupling to the formation, good bonding is possible, even though the amplitude response is high. This, however, does not rule out the presence of channels. Fig. 11.29 illustrates such a case.

As these examples show, cement-bond log interpretation is most uncertain in the cases where "intermediate" bonding is inferred. This condition, characterized by moderate to high signals from both the casing and the formation, can be caused by a number of factors, including a microannulus, channeling, poor bonding to the formation, or a thin cement sheath. To narrow the possibilities,

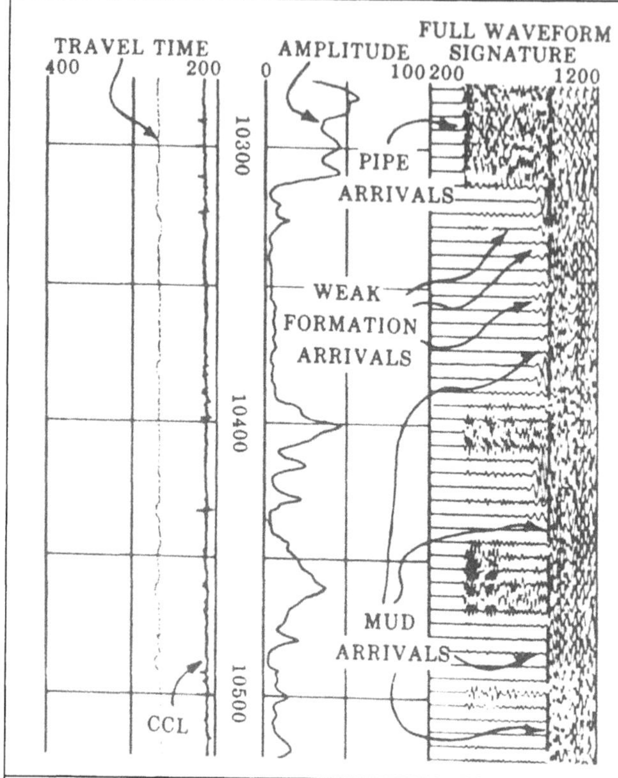

Fig. 11.24—Cement-bond log response to good pipe, poor formation bonding (from Ref. 13).

the cement-bond log should be run with additional wellbore pressure applied to eliminate the effects of microannuli and other information about the wellbore conditions—e.g., openhole sonic logs, caliper logs, and records of the casing design—should be used.

11.3 Attenuation-Ratio Logs

Borehole-compensated attenuation-ratio tools were developed[21,22] to eliminate some of the drawbacks of the cement-bond-log-amplitude measurement. These tools use three receivers spaced between two transmitters.

Gollwitzer and Masson[21] presented the principle behind measuring sound attenuation from a ratio of amplitude measurements. Consider a tool with two receivers, R_1 and R_2, spaced symmetrically between two transmitters, T_1 and T_2, as Fig. 11.30 shows. At a distance L_1 from the upper transmitter, the amplitude of the casing-borne sonic wave is attenuated and can be expressed as

$$a_{11}=p_1 S_1 \times 10^{-(\alpha L_1/20)}, \quad \text{(11.3)}$$

where a_{11}=output of Receiver R_1, p_1=pressure amplitude for $L_1=0$, S_1=receiver sensitivity, and α=attenuation rate of sonic signal. The output of Receiver R_2 can likewise be written as

$$a_{12}=p_1 S_2 \times 10^{-(\alpha L_2/20)}. \quad \text{(11.4)}$$

Similarly, when the lower transmitter is fired, the outputs of Receivers R_1 and R_2 are

$$a_{21}=p_2 S_1 \times 10^{-(\alpha L_2/20)} \quad \text{(11.5)}$$

and

$$a_{22}=p_2 S_2 \times 10^{-(\alpha L_1/20)}. \quad \text{(11.6)}$$

From Eqs. 11.3 through 11.6, the following ratio can be formed:

$$\left(\frac{a_{12}a_{21}}{a_{11}a_{22}}\right)=10^{-[2\alpha(L_2-L_1)]/20}. \quad \text{(11.7)}$$

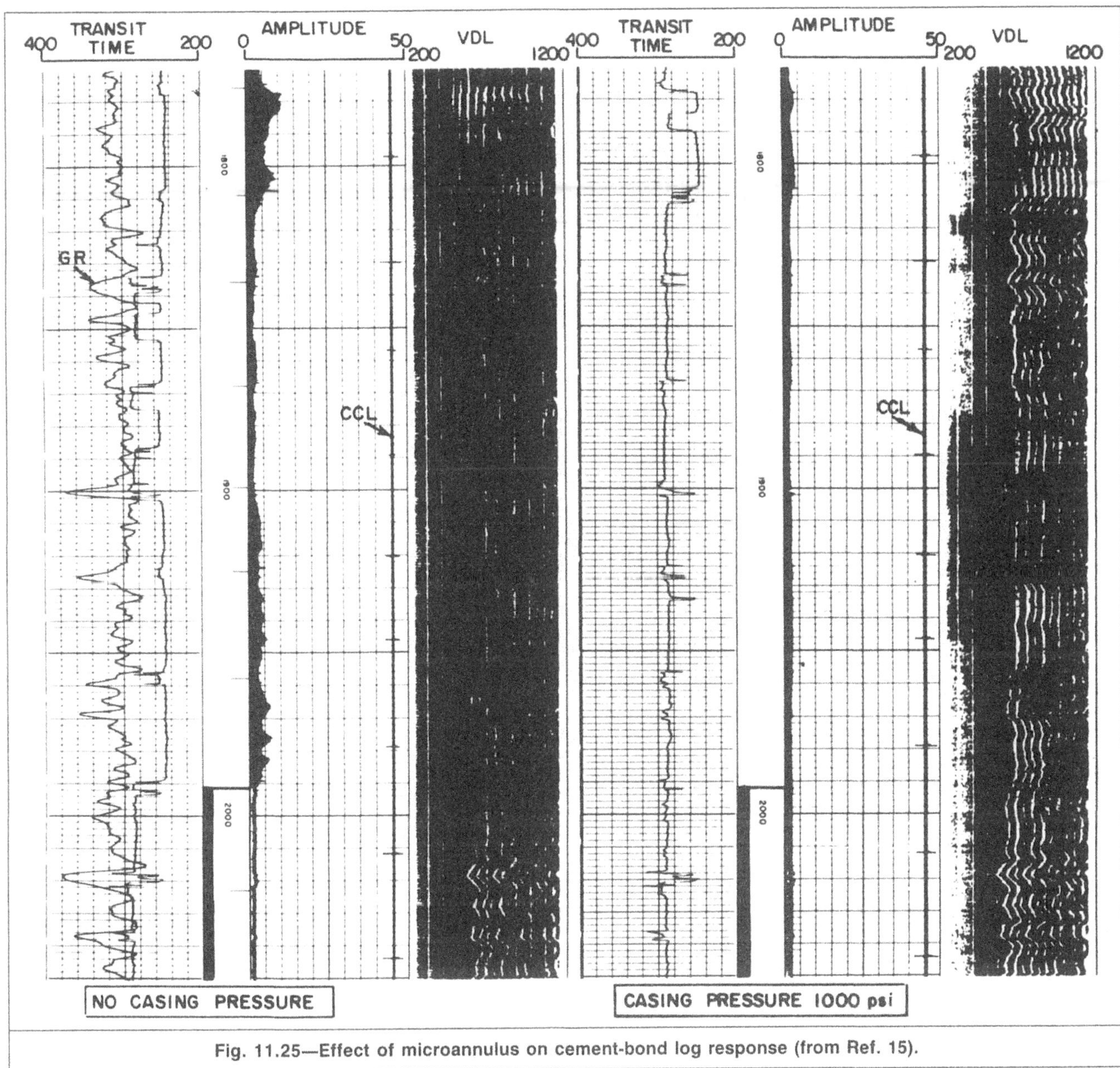

Fig. 11.25—Effect of microannulus on cement-bond log response (from Ref. 15).

Solving for the attenuation rate yields

$$\alpha = \frac{-10}{L_2 - L_1} \log_{10}\left(\frac{a_{12}a_{21}}{a_{11}a_{22}}\right), \qquad (11.8)$$

where α is expressed in decibels per foot.

As Eq. 11.8 shows, the advantage of this measurement technique is that the attenuation measurement is independent of receiver sensitivity, transmitter output power, and fluid attenuation. Thus, the attenuation-ratio tool is much less sensitive to calibration and eccentering than the cement-bond tool.

Fig. 11.31 shows a typical tool configuration for an attenuation-ratio tool. A third receiver is included so that a 5-ft [1.5-m] transmitter/detector spacing is available for a measurement of the full wave train. In addition, a 1-ft [0.3-m] amplitude can be measured by Receiver R_3 when Transmitter T_2 is fired. This 1-ft [0.3-m] amplitude measurement is useful in fast formations because the close spacing allows the pipe signal to arrive before the formation signals.

Fig. 11.32 shows a log obtained by an attenuation-ratio tool in a test well that contained built-in channels. The attenuation decreased markedly at the locations of the large channels. The interpretation of the attenuation-ratio log in this case agreed with the results of the standard cement-bond log amplitude measurement, as it should when a calibrated cement-bond tool is properly centralized. The advantage of the attenuation-ratio measurement is its reduced sensitivity to calibration and eccentering effect compared with the cement-bond log.

11.4 Ultrasonic-Pulse-Echo Logs

11.4.1 Introduction. Ultrasonic-pulse-echo techniques were developed in recent years to overcome some of the deficiencies of traditional cement-bond logs. The primary advantage of the ultrasonic devices is that they provide a circumferential picture of cement quality by use of eight transducers arrayed around the tool. In addition, they are less affected by microannuli than cement-bond tools. The ultrasonic measurements are less sensitive, however, to acoustic coupling to the formation. Pioneered by the Schlumberger Cement Evaluation Tool (CET^{SM}),[23,24] the measurements from ultrasonic-pulse-echo logs are fundamentally different from those obtained from cement-bond logs.

11.4.2 Tools and Theory of Operation. An ultrasonic-pulse-echo tool consists of an array of eight ultrasonic transducers spaced around the body of the tool, as Fig. 11.33 shows. On this tool, the transducers are arranged in a helix with a spacing of 45° between transducers, while other ultrasonic tools use a double heli-

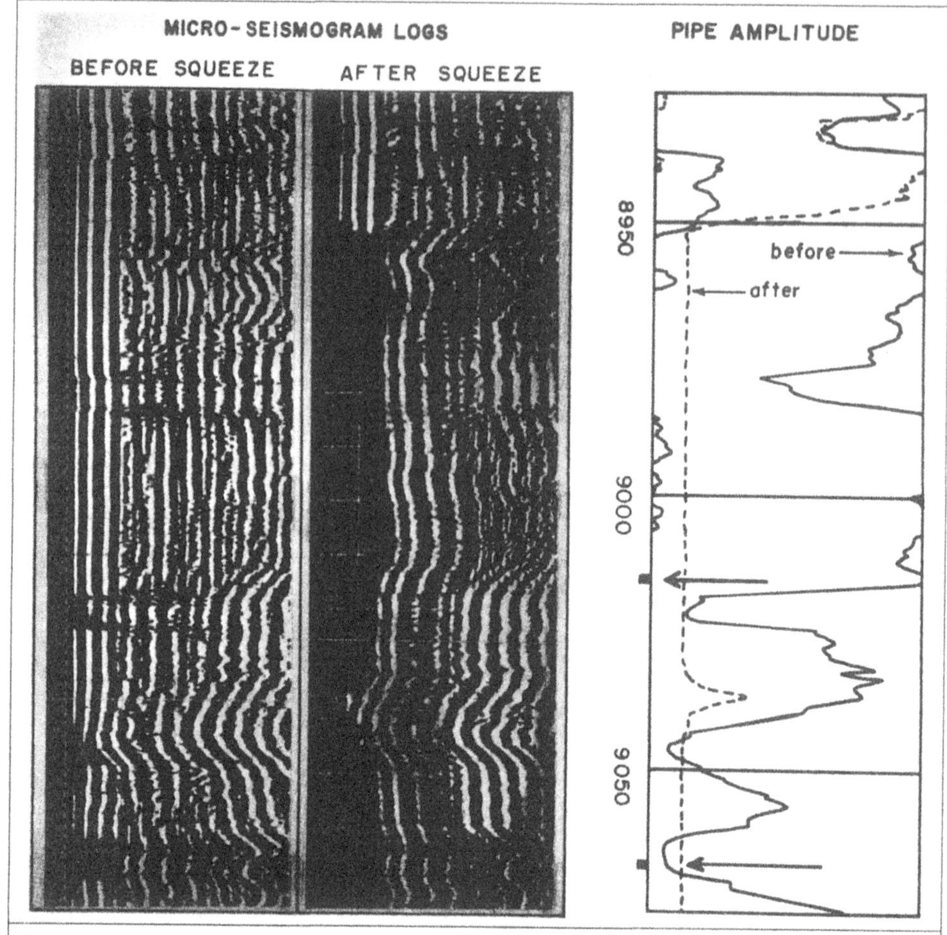

Fig. 11.26—Cement-bond logs before and after cement squeezing of a channel (from Ref. 11).

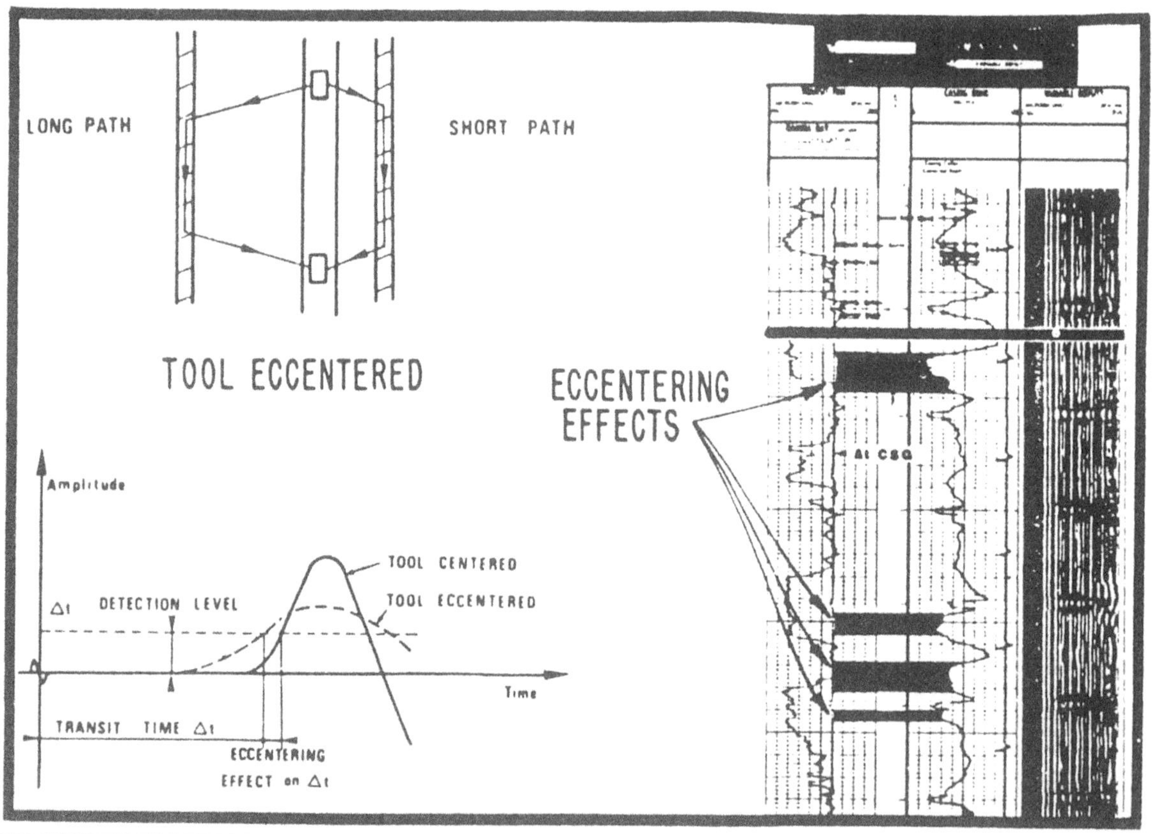

LONG PATH
SHORT PATH
TOOL ECCENTERED
ECCENTERING EFFECTS
Amplitude
Δt DETECTION LEVEL
TOOL CENTERED
TOOL ECCENTERED
Time
TRANSIT TIME Δt
ECCENTERING EFFECT on Δt
Δt CSG

Fig. 11.27—Tool eccentering effect (courtesy Schlumberger).

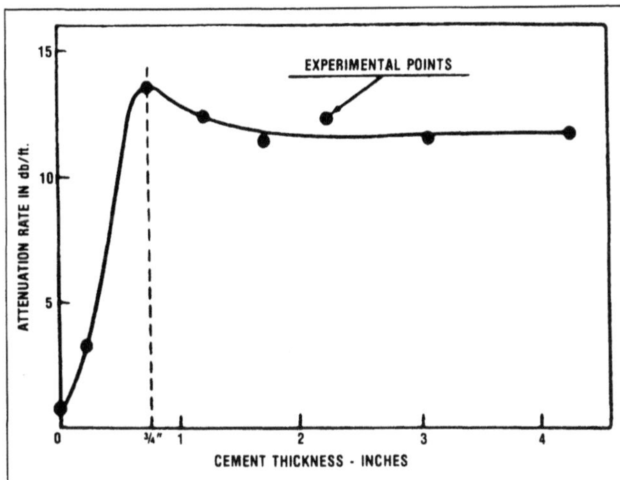

Fig. 11.28—Attenuation rate as a function of cement thickness (from Ref. 6).

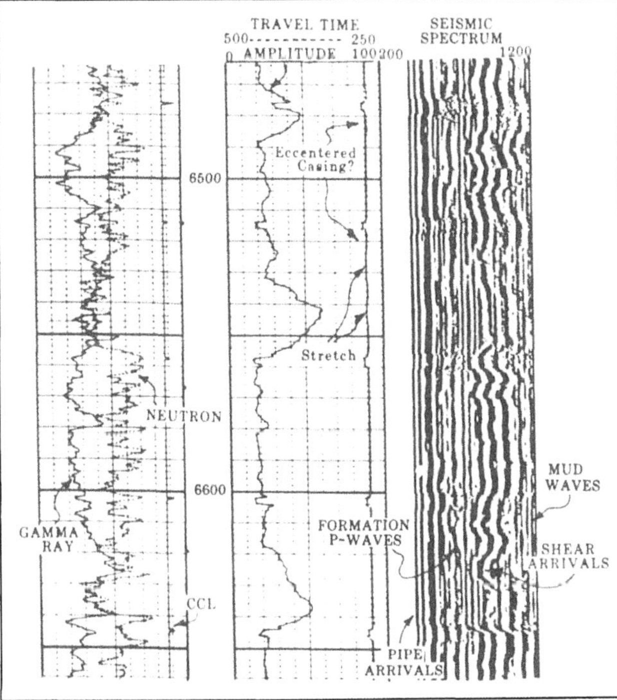

Fig. 11.29—Misleading amplitude log caused by thin cement sheath (from Ref. 13).

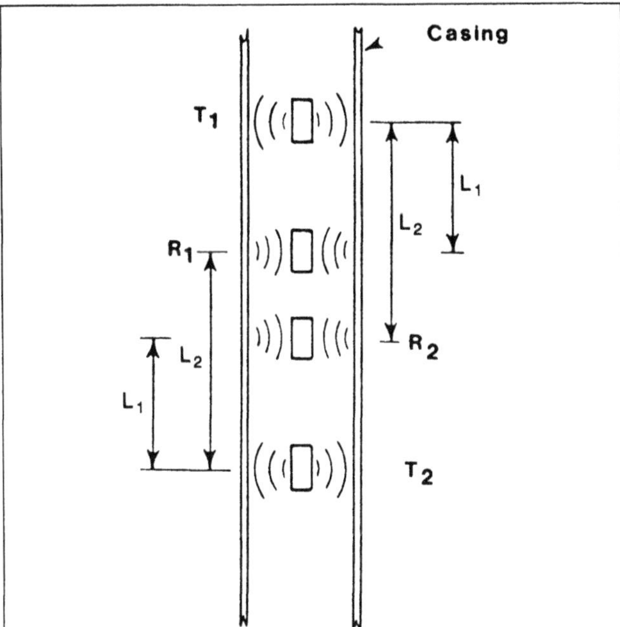

Fig. 11.30—Schematic of attenuation-ratio log transmitter and detector arrangement (from Ref. 21, courtesy Society of Professional Well Log Analysts).

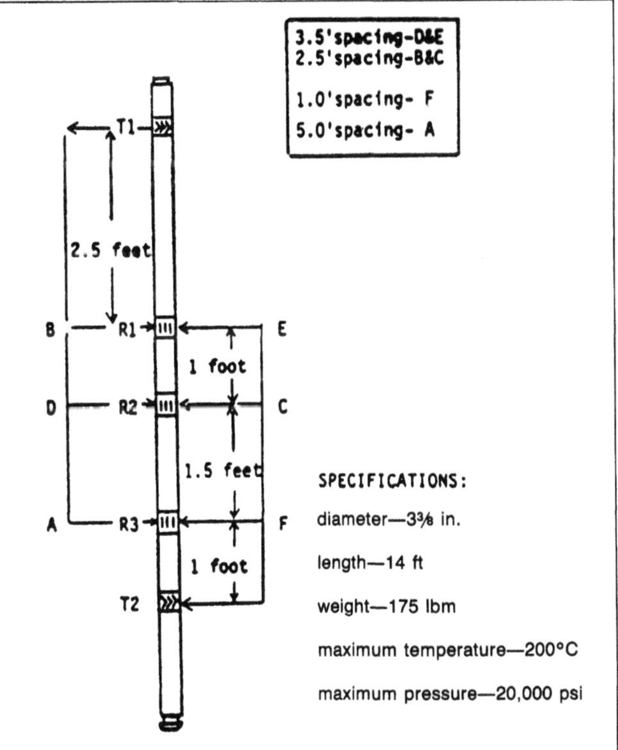

Fig. 11.31—Attenuation-ratio tool configuration (from Ref. 22).

cal array.[25] A ninth transducer is aligned axially and aimed at an acoustic mirror so that an in-situ measurement of travel time in the wellbore fluid can be made.

The ultrasonic-pulse-echo tools are designed to make the casing resonate in its thickness mode. Thus, the frequency of the acoustic signal must be in the range of the fundamental resonant frequency of the casing or one of its harmonics. This resonant frequency is[24]

$$f_0=\frac{v_c}{2h_c}, \qquad (11.9)$$

where f_0=fundamental resonant frequency, v_c=casing compressional velocity, and h_c=casing thickness. For usual casing sizes, the resonant frequency is 200×10^3 to 600×10^3 cycles/sec [200 to 600 kHz]; the ultrasonic tools thus emit a band of acoustic energy at frequencies around 500×10^3 cycles/sec [500 kHz].

Froelich *et al.*[23] illustrated the principle of the ultrasonic-pulse-echo measurement. Fig. 11.34 shows the path that would be taken by an infinitely short pulse of acoustic energy. After traveling through the fluid, the main part of the pulse is reflected back to the transducer from the inner casing wall, but a fraction of the energy enters the casing and is reflected back and forth, with part of the energy transmitted outside the casing at each reflection on the inner or outer casing wall. The impulse response measured by the transducer will thus be a succession of impulses separated by twice the travel time in the casing (Fig. 11.35); each impulse amplitude is a function of the acoustic impedance of the wellbore fluid, the casing, and the material in contact with the outside of the casing.

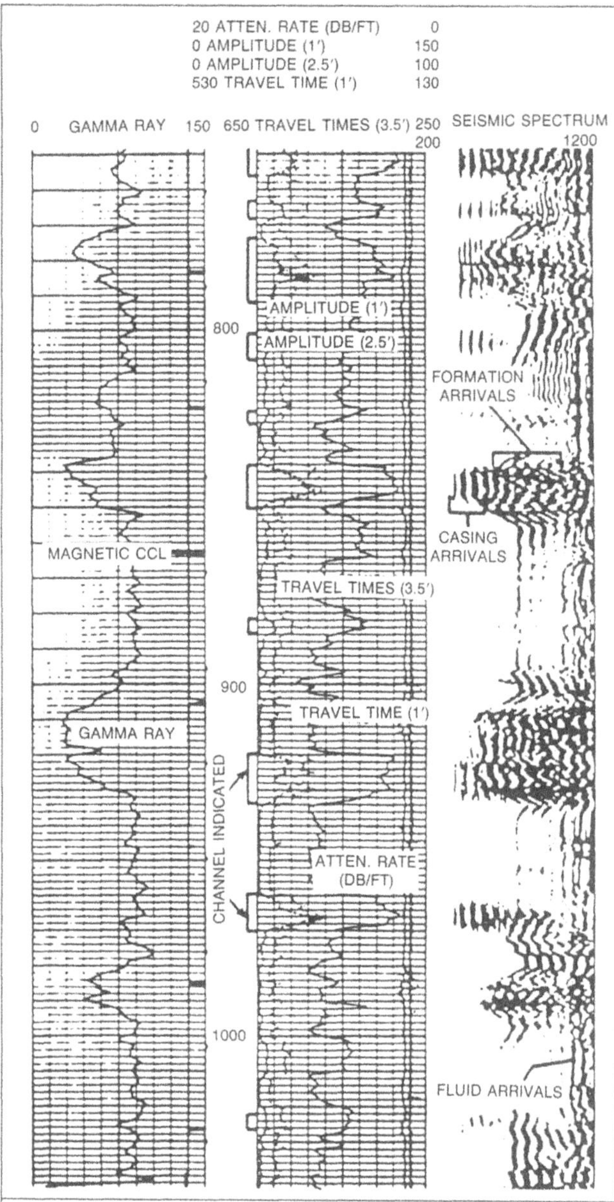

Fig. 11.32—Attenuation-ratio-tool log response in test well (from Ref. 22).

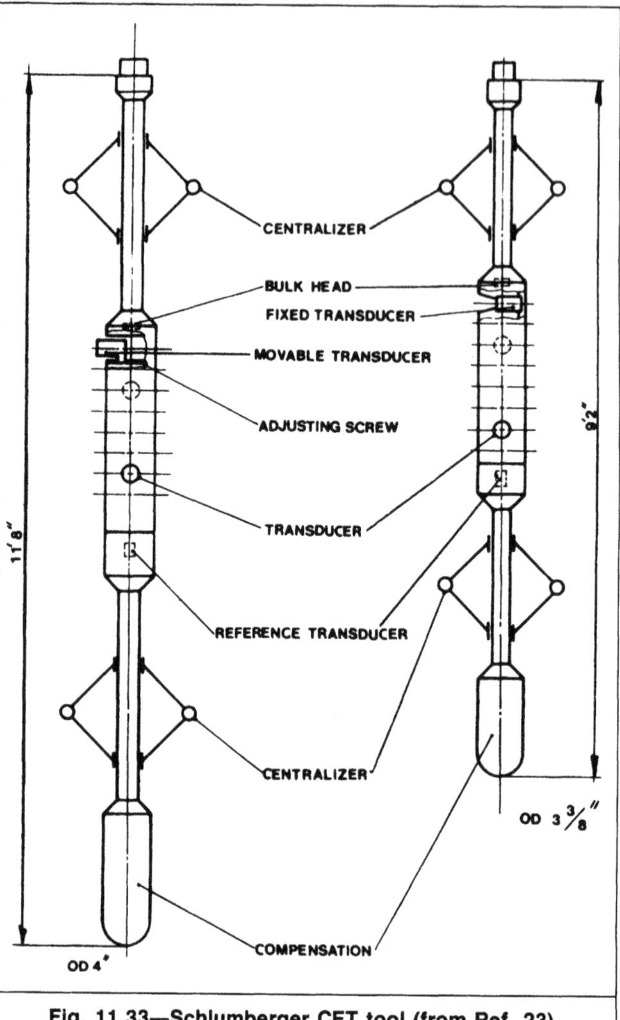

Fig. 11.33—Schlumberger CET tool (from Ref. 23).

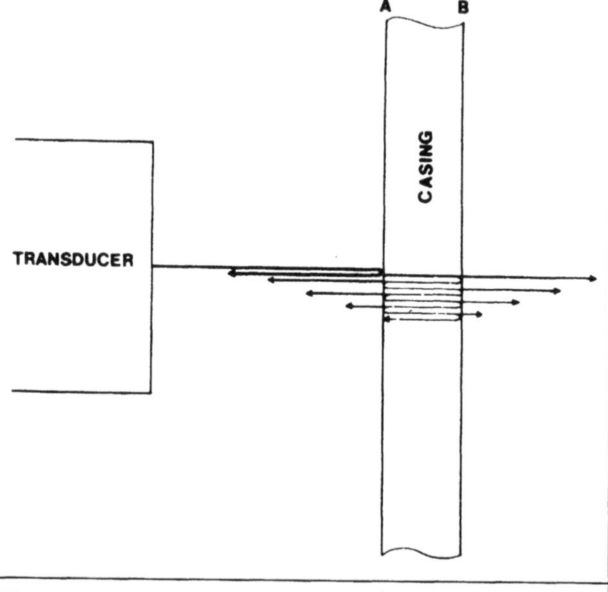

Fig. 11.34—Ultrasonic wave propagation with incident wave normal to casing wall (Ref. 23).

The first peak is approximately 10 times larger than the others. The following impulses decay exponentially, with the rate of decay depending on the material in contact with the outside of the casing. Casing surrounded by water (no cement) is relatively free to vibrate radially and the decay will be slow, while casing bonded to cement decays rapidly because there is good acoustic coupling to the surrounding medium.

The excitation of the transducer produces an acoustic pulse with a bandwidth of 300×10^3 to 600×10^3 cycles/sec [300 to 600 kHz]. The response at the transducer is like that in Fig. 11.36, except that the formation reflections are much less pronounced than those shown.

An empirical measurement of cement quality is obtained by integrating a portion of the rectified acoustic signal. Leigh *et al.*[26] illustrated the usual analysis performed on the wave form (Fig. 11.37). The two-way travel time, t_d, between the transducer and the inner casing wall is measured as the first crossing of a preset bias, and an amplitude, a_{w1}, is measured that corresponds to the inner pipe-wall reflection. All eight transducers are set in a shop calibration to give the same value of a_{w1}. The rectified wave form is then integrated over discrete time intervals (windows) to yield a_{w2} and a_{w3}, which are then normalized with respect to a_{w1} and to free pipe. These normalizations are done so that outputs from

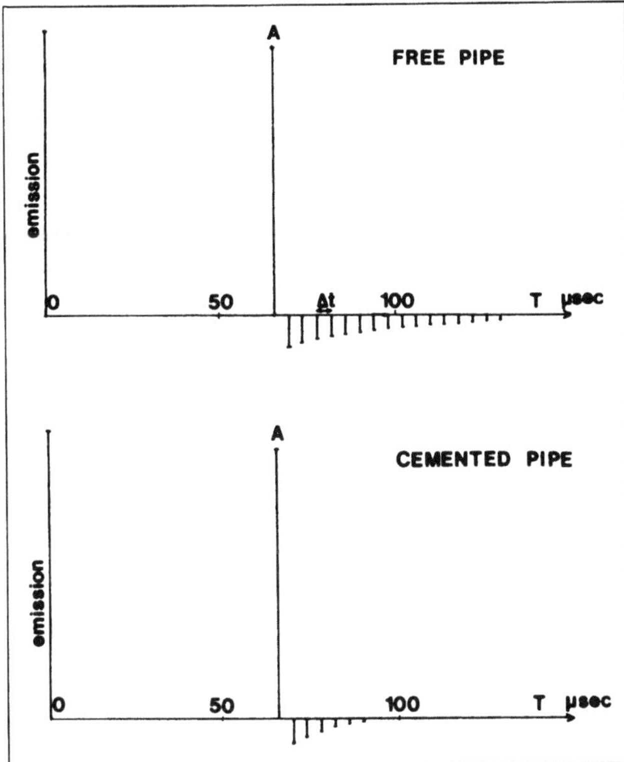

Fig. 11.35—Impulse response for free and cemented pipe (from Ref. 23).

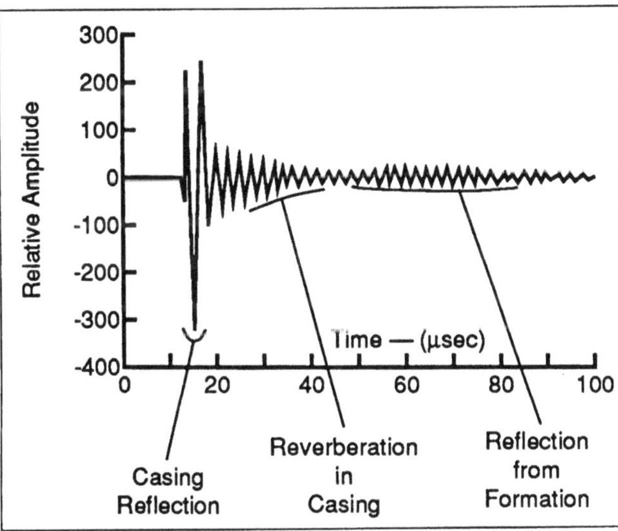

Fig. 11.36—Typical wave form (from Ref. 24, courtesy Society of Professional Well Log Analysts).

all eight transducers are comparable even if the amplitudes of the transmitted signals vary and to account for the acoustic impedance of the fluid on either side of the casing.

The normalized outputs are

$$(a_{w2})_n = \frac{(a_{w2}/a_{w1})}{(a_{w2}/a_{w1})_{fp}} \quad \text{(11.10)}$$

and

$$(a_{w3})_n = \frac{(a_{w3}/a_{w1})}{(a_{w3}/a_{w1})_{fp}}, \quad \text{(11.11)}$$

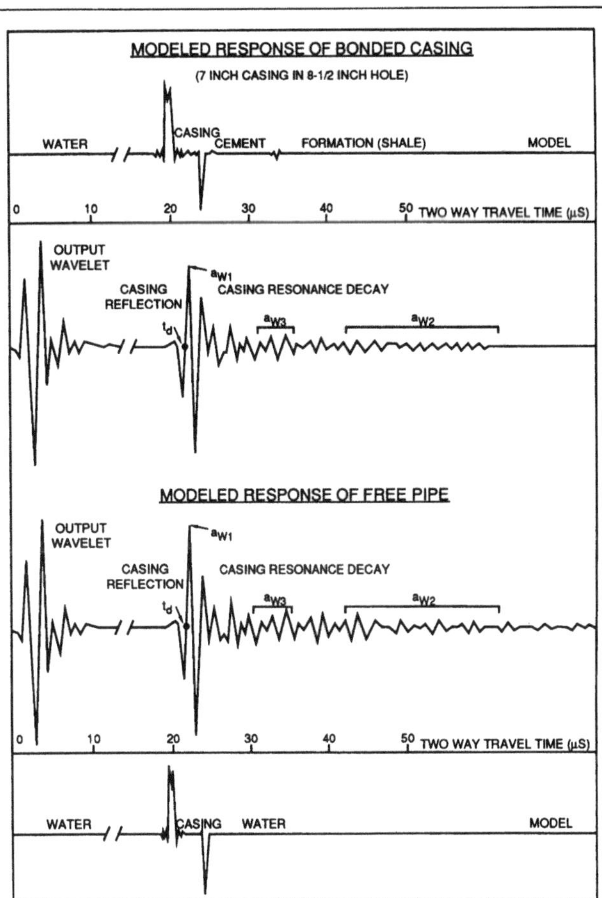

Fig. 11.37—Analysis of ultrasonic-pulse-echo log response (from Ref. 26, courtesy Society of Professional Well Log Analysts).

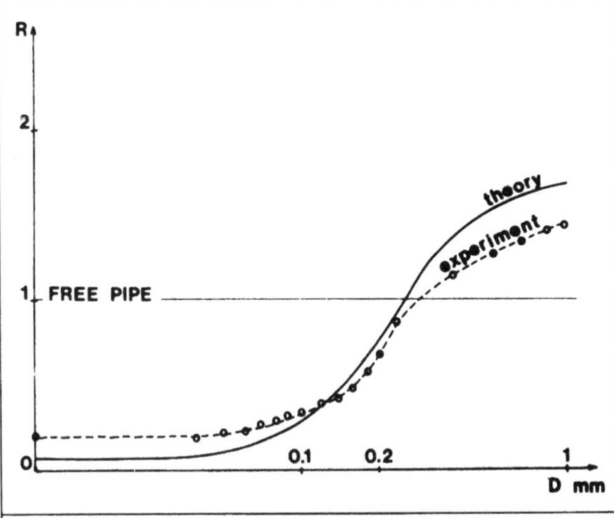

Fig. 11.38—Effect of microannulus on ultrasonic-pulse-echo log response (from Ref. 23).

where $(a_{w2})_n$ and $(a_{w3})_n$=normalized responses in Windows 2 and 3, respectively, and the subscript fp denotes values obtained in free pipe. Thus, after calibration and normalization, $(a_{w2})_n$ and $(a_{w3})_n$ for each of the eight transducers are equal to one in free pipe.

The second integral is measured primarily to check for the presence of formation signals in Window 2. If formation signals are present during the Window 2 response, the integral is high, giving

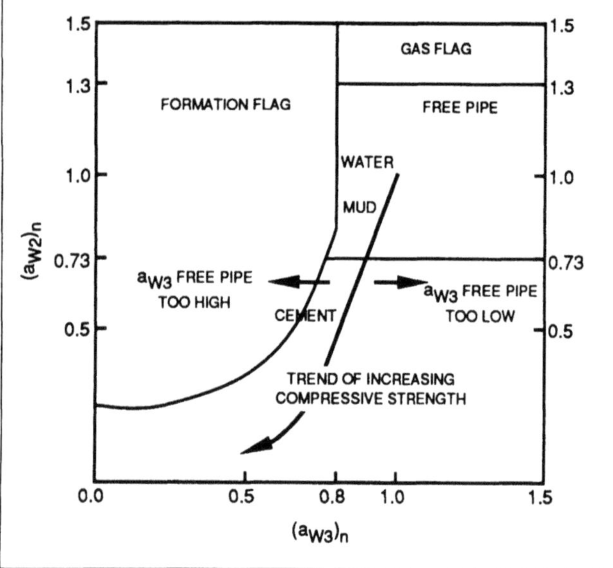

Fig. 11.39—Crossplot of $(a_{w2})_n$ and $(a_{w3})_n$ (from Ref. 26, courtesy Society of Professional Well Log Analysts).

the appearance of poor bonding conditions. With no reflections from an interface between the cement and the formation, the acoustic energy decays exponentially, while formation reflections lead to nonexponential decay. Thus, a_{w2} and a_{w3} are compared to check for nonexponential decay. When it is detected, a flag is activated to indicate on the log the presence of formation signals. Experience, however, has shown this flag to be relatively meaningless.

One purported advantage of the ultrasonic-pulse-echo technique is its insensitivity to a microannulus, which is true if the microannulus is small and liquid-filled. If the wave length of the acoustic pulse is large compared to the size of a microannulus and the contrast between the acoustic impedance of the fluid in the gap and the acoustic impedances of the surrounding solids are not too high, a microannulus does not significantly affect the tool response. Fig. 11.38 shows the effect of a water-filled microannulus on the tool response $[(a_{w2})_n]$. Sizes <0.004 in. $[<0.1$ mm$]$ do not adversely affect tool response.

When gas fills a microannulus, the steel/gas reflection coefficient is large. This results in a long, undamped resonance decay, which is similar to the response of free pipe.[26] If the responses were normalized to free pipe with liquid behind the pipe, the normalized responses with gas present behind pipe are greater than one; with some devices a gas flag is activated when $(a_{w2})_n > 1.3$.[6] This apparent gas response should be treated with caution. Fig. 11.39 summarizes the normalized response characteristics of an ultrasonic-pulse-echo device.

The transit times measured with each transducer provide eight caliper measurements of distance to the inside wall of the casing. The reference transducer is used to measure travel time in the wellbore fluid so that transit times from the eight circumferential transducers can be converted to distances. With some devices, the casing thickness is also measured by analyzing the wave form for its resonant frequency because the wave length should be twice the casing thickness. Assuming a travel time in steel of 52.55 μsec/ft [172.4 μs/m],[26]

$$h_c = 1/(2 \times 52.55)f, \quad \ldots\ldots (11.12)$$

where h_c=casing thickness (in feet) and f=response frequency.

11.4.3 Ultrasonic-Pulse-Echo Log Displays. Two fundamental measurements are made with an ultrasonic-pulse-echo tool: travel time (and hence distance from the transducers to the casing wall or walls) and a measure of the energy in a window that excludes echoes from the inside pipe wall. These basic measurements are displayed in a variety of ways.

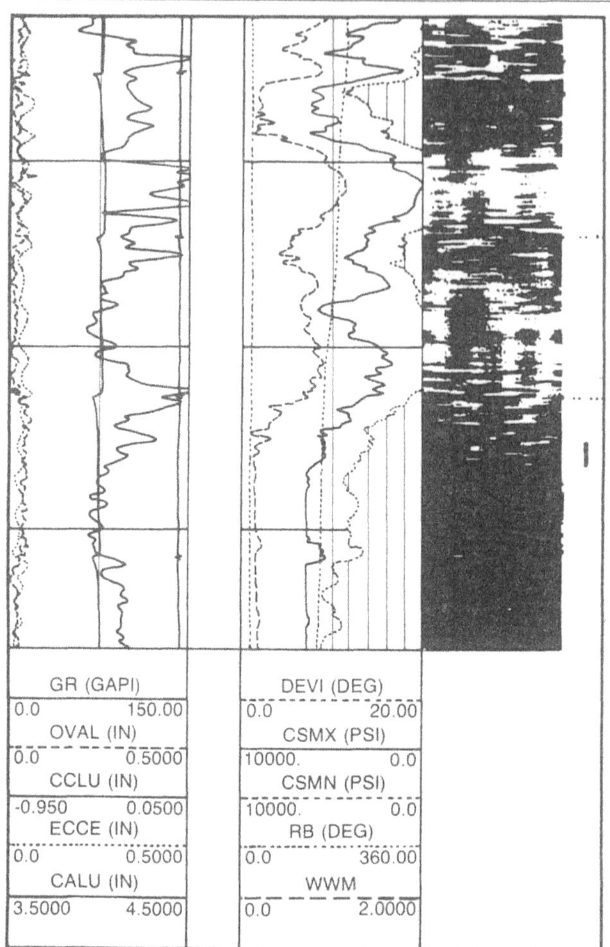

Fig. 11.40—Ultrasonic-pulse-echo log display (from Ref. 20, courtesy Schlumberger).

The measurements of distance from the transducers to the casing wall provide an accurate acoustic caliper. By addition of the responses of transducers spaced 180° apart, four measurements of casing diameter are obtained. Sometimes all four measured diameters are displayed; alternatively, a maximum and a minimum diameter or one average diameter can be presented on the log. In addition, tool eccentricity—i.e., the difference between the maximum and minimum distances from transducers to the casing wall—is often displayed.

The degree of cement bonding, as inferred from the integrated response $(a_{w2})_n$, is typically displayed in two ways. Because there are eight transducers arrayed around the logging tool, eight measures of cement integrity are obtained. A maximum and a minimum cement compressive strength are usually recorded on the basis of highest and lowest $(a_{w2})_n$ values measured by the transducers, as well as on the basis of an average $(a_{w2})_n$ for all eight transducers. The relative bonding from all eight transducers is presented by assigning a shading from white (no bonding) through shades of gray to black (excellent bonding); the relative response from each transducer is displayed on a separate track. Because the range of $(a_{w2})_n$ is small, this shading is difficult to accomplish and many logs appear to have just black and white in the display. This display gives a picture of the cement conditions circumferentially around the pipe because each band represents a 45° arc of the wellbore. This display is strongly dependent on the normalization factors and gate settings used because of the sensitivity to the offset between the tool and the pipe of this high-frequency signal.

Additional information is sometimes displayed to aid log interpretation. The relative bearing of the tool is usually recorded to assist in defining the geometry of channels in the cement. Formation and gas flags (see Sec. 11.4.2) are sometimes displayed. Fig. 11.40 shows a typical ultrasonic-pulse-echo log presentation.

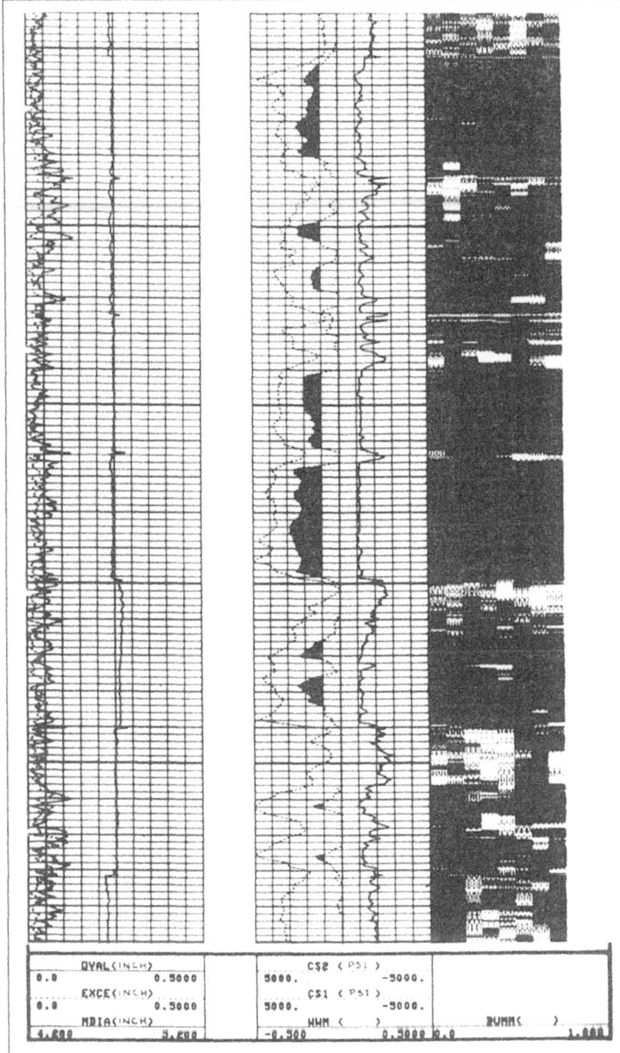

Fig. 11.41—Ultrasonic-pulse-echo log in regions of good and poor bonding (from Ref. 23).

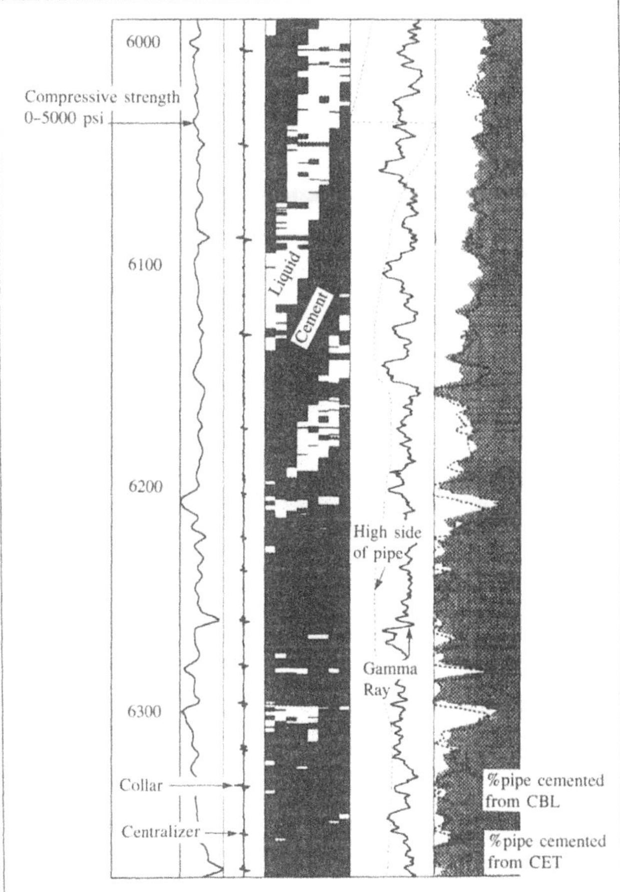

Fig. 11.42—Ultrasonic-pulse-echo response to channel (from Ref. 27).

11.4.4 Ultrasonic-Pulse-Echo Log Examples. The following examples illustrate the use of ultrasonic-pulse-echo logs.

Example 1—Alternating Good and Poor Cement Quality. Fig. 11.41 shows a typical ultrasonic-pulse-echo log in a section of well that has regions of apparently good bonding separated by regions of poor bonding. The good bonding is indicated by the solid black response from all eight transducers and by high minimum compressive strength (>1,000 psi [>6.9 MPa] where shaded). The compressive strength is obtained from an empirical laboratory relationship and should be used more in a relative sense than as an absolute measure of cement strength. In the poor bonding regions, some or all of the eight tracks exhibit light shading, which indicates poor bonding around much of the pipe circumference.

Example 2—Channeling. A primary advantage of the ultrasonic-pulse-echo log is that it can identify unsupported sections of the pipe circumference because it measures bonding conditions at eight positions circumferentially around the pipe. Fig. 11.42 shows a typical response to a channel, with a few of the tracks showing poor bonding, while good bonding is indicated around the rest of the pipe. The channel appears to be spiraling around the pipe; however, the relative bearing recording indicates that the tool was slowly rotating as the log was run, which means that the channel is consistently on one side of the pipe.

Example 3—Microannulus Effect. Figs. 11.43 and 11.44 illustrate the reduced sensitivity of the ultrasonic-pulse-echo log to a liquid-filled microannulus compared with a cement-bond log. The cement-bond log (Fig. 11.43) showed a region of intermediate bonding from 2,060 to 2,110 ft [628 to 643 m] when run with no surface pressure. When the log was repeated with a 1,500-psi [10.3-MPa] pressure, this region showed good bonding, suggesting the presence of a microannulus. The ultrasonic-pulse-echo log (Fig. 11.44) showed less change with the application of pressure, though again better bonding was indicated when surface pressure was applied. Note that the apparent channel indicated in the far track by the light shading for Transducer 8 from 2,060 to 2,110 ft [628 to 643 m] when the ultrasonic-pulse-echo log was run with no surface pressure disappeared when pressure was applied. The ultrasonic-pulse-echo log thus was noticeably affected by the microannulus.

Example 4—Presence of Gas in Cement. The presence of gas, whether in a microannulus or dispersed as bubbles in the cement, causes an ultrasonic-pulse-echo log to give an overly pessimistic measure of cement-bonding conditions. Fig. 11.45 illustrates a case where the ultrasonic-pulse-echo log shows very poor bonding, while relatively good bonding is indicated by the cement-bond log. This section of the well is opposite a gas-bearing formation, and independent tests found good zonal isolation by the cement, confirming the validity of the cement-bond log. Gas in the cement was a possible cause of the erroneous ultrasonic-pulse-echo log. Because the cement-bond log does not show weak cement, it is also possible that the gate settings were incorrect on the ultrasonic-pulse-echo log.

Example 5—Ultrasonic Caliper. The transit-time measurements from the eight transducers provide excellent caliper results. Fig. 11.46 shows four pipe diameters from an ultrasonic-pulse-echo log from a well containing 9⅝-in. [24.4-cm], 36-lbm/ft [53.6-kg/m] casing. The variation in internal diameter of the different pipe joints is clearly seen with the ultrasonic caliper measurements.

11.5 Cement-Quality Logging Guidelines

The following guidelines summarize factors that should be considered when cement-quality logs are run and interpreted.

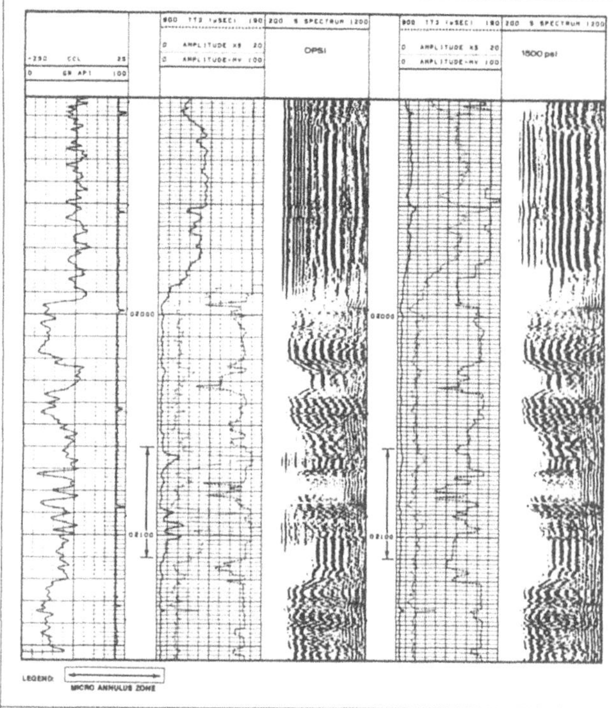

Fig. 11.43—Sensitivity of cement-bond log to a microannulus (from Ref. 25).

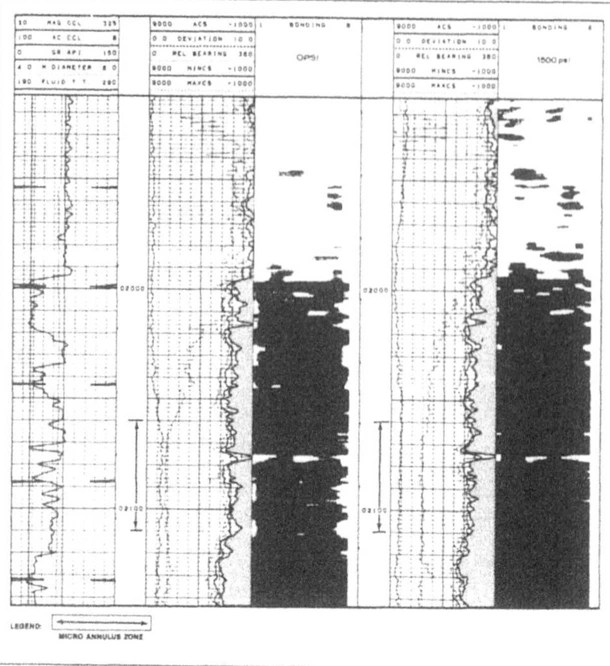

Fig. 11.44—Reduced sensitivity of ultrasonic-pulse-echo log to microannulus (from Ref. 25).

11.5.1 Cement-Bond Logs.

1. Proper centralization is critical in obtaining a useful cement-bond log. In a vertical hole, use at least three strong centralizers, with centralizers directly above and below the transmitter-receiver and one centralizer on the top of the tool string. In deviated wells, pay even more attention to centralization.
2. Make repeat runs through at least part of the logged section to check for proper tool operation and centralization.
3. Always record a full wave-train display, either a Variable Density log or an *x-y* (signature) log, because it provides the only information about bonding to the formation and aids in the interpretation of the amplitude log.
4. Consider quantitative interpretation of the amplitude log to be approximate because it is based on empirical correlations and is affected by many factors (e.g., microannulus, eccentering, and fast formations). The amplitude log is best used in conjunction with a full wave-train display.
5. Repeat a cement-bond log with pressure applied at the wellhead when intermediate bonding is indicated to distinguish between a microannulus and poor cement hydraulic integrity.

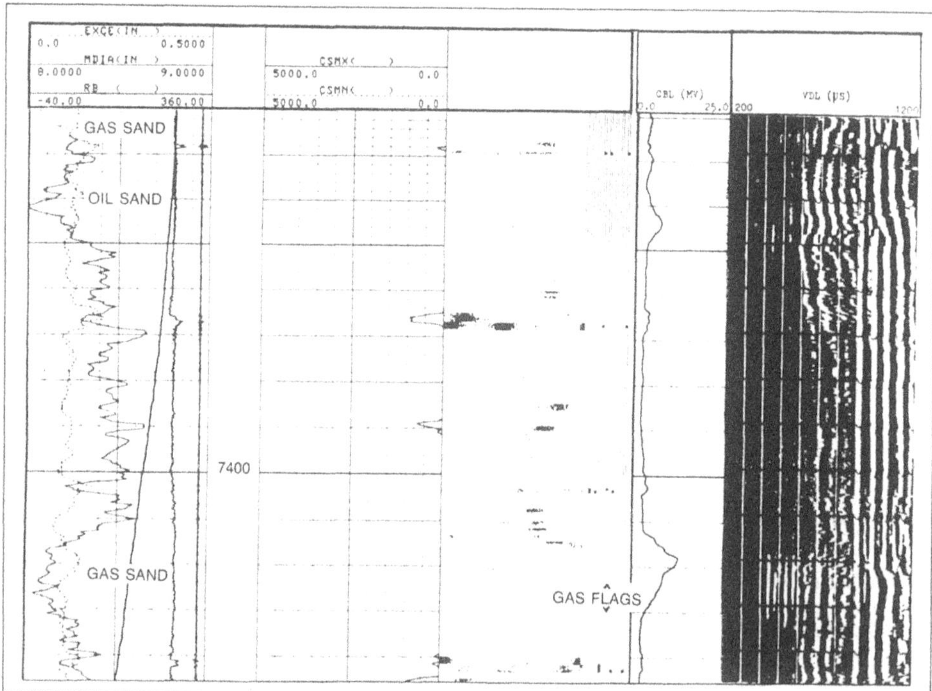

Fig. 11.45—Ultrasonic-pulse-echo log shows poor bonding because of presence of gas in cement (from Ref. 26, courtesy Society of Professional Well Log Analysts).

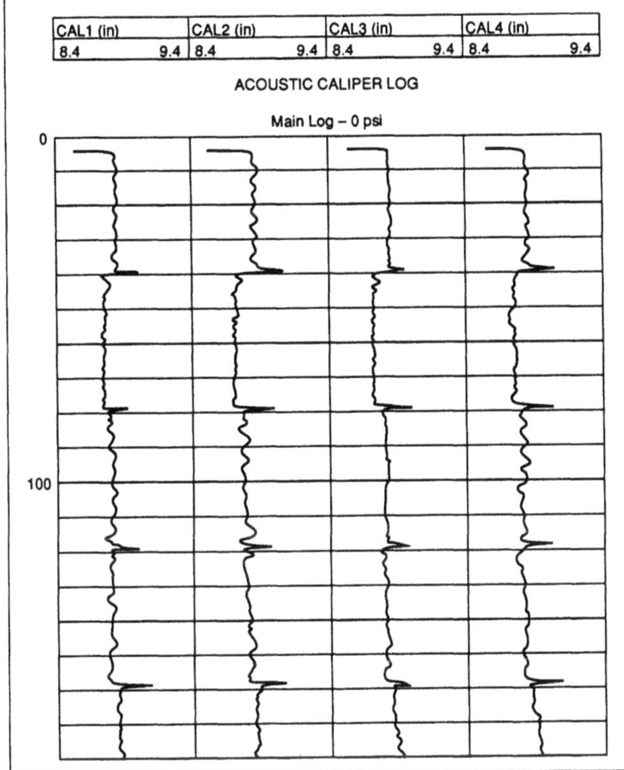

Fig. 11.46—Ultrasonic caliper showing different diameters of different joints of pipe.

6. Run the cement-bond log through the cement top whenever possible. A section of log in free pipe allows one to check tool calibration and centralization.
7. Detection of shear waves in the full wave-train display is generally a reliable indication of good bonding to the pipe and to the formation.
8. A transit-time measurement is essential in detecting eccentering and the presence of fast formations.
9. A cement-bond log can clearly indicate regions of free pipe and of very good bonding. Even in these cases, and certainly when intermediate bonding conditions are found, the cement-bond log does not provide conclusive evidence of the ability of the cement to provide hydraulic integrity.
10. An attenuation-ratio log is often preferable to a cement-bond log amplitude measurement because of its lower sensitivity to calibration errors and eccentering. Use the attenuation-ratio log in wells where centralization is difficult.

11.5.2 Ultrasonic-Pulse-Echo Logs.

1. Ultrasonic-pulse-echo logs provide an excellent acoustic-caliper log, with a measure of the inside and outside diameters of the casing in some cases.
2. The ultrasonic-pulse-echo log can clearly identify partial contact between the pipe and the cement because of the circumferential picture of cement bonding provided. A channel between the cement and the formation is not normally detectable.
3. The ultrasonic-pulse-echo log has relatively low sensitivity to bonding to the formation.
4. An ultrasonic-pulse-echo log is less sensitive to a liquid-filled microannulus than is a cement-bond log.
5. The presence of gas in a microannulus causes apparently poor bonding on an ultrasonic-pulse-echo log, even though cement integrity may be good.
6. The ultrasonic-pulse-echo log is particularly useful when intermediate bonding is indicated by a cement bond because the log can determine whether partial contact is the cause of the intermediate cement-bond log response.

Nomenclature

a_{w1} = amplitude of reflection off inside casing wall, mV, mL^2/t^2q
a_{w2} = integrated response from Window 2, mV-sec, mL^2/tq
a_{w3} = integrated response from Window 3, mV-sec, mL^2/tq
$(a_{w2})_n$ = normalized a_{w2}
$(a_{w3})_n$ = normalized a_{w3}
a_0 = amplitude of transmitted signal, mV, mL^2/t^2q
a_1 = amplitude at Location 1, mV, mL^2/t^2q
a_2 = amplitude at Location 2, mV, mL^2/t^2q
a_{11} = amplitude response of Receiver 1 to Transmitter 1, mV, mL^2/t^2q
a_{12} = amplitude response of Receiver 2 to Transmitter 1, mV, mL^2/t^2q
a_{21} = amplitude response of Receiver 1 to Transmitter 2, mV, mL^2/t^2q
a_{22} = amplitude response of Receiver 2 to Transmitter 2, mV, mL^2/t^2q
f = frequency, cycles/sec [kHz], 1/t
f_0 = fundamental resonant frequency, cycles/sec [kHz], 1/t
h_c = casing thickness, ft [m], L
L_1 = distance between transmitter and near receiver, ft [m], L
L_2 = distance between transmitter and far receiver, ft [m], L
p_1 = pressure amplitude of Transmitter 1 signal, bar [Pa], m/Lt^2
p_2 = pressure amplitude of Transmitter 2 signal, bar [Pa], m/Lt^2
S_1 = sensitivity of Receiver 1, mV/bar [mV/Pa], L^3/q
S_2 = sensitivity of Receiver 2, mV/bar [mV/Pa], L^3/q
t_d = two-way travel time between transducer and inner casing wall, seconds, t
t_p = pipe arrival time—i.e., time required for sound waves traveling through casing to reach first detector, microseconds, t
Δt = transit time—i.e., time between signal generation and signal arrival at detector, microseconds, t
z = distance from the transmitter, ft [m], L
α = sound attenuation, dB/ft [dB/m], 1/L

References

1. McNeely, W.E.: "A Statistical Analysis of the Cement Bond Log," *Proc.*, SPWLA 14th Annual Logging Symposium, Lafayette, LA (May 6-9, 1973).
2. Grosmangin, M., Kokesh, F.P., and Majani, P.: "A Sonic Method for Analyzing the Quality of Cementation of Borehole Casings," *JPT* (Feb. 1961) 165-71; *Trans.*, AIME, **222**.
3. Tixier, M.P., Alger, R.P., and Doh, C.A.: "Sonic Logging," *Trans.*, AIME (1959) **216**, 106-14.
4. Anderson, W.L. and Walker, T.: "Research Predicts Improved Cement Bond Evaluations With Acoustic Logs," *JPT* (Nov. 1961) 1093-97.
5. Riddle, G.A.: "Acoustic Wave Propagation in Bonded and Unbonded Oil Well Casing," paper SPE 454 presented at the 1962 SPE Annual Meeting, Los Angeles, Oct. 7-10.
6. Pardue, G.H. *et al.*: "Cement Bond Log—A Study of Cement and Casing Variables," *JPT* (May 1963) 545-55; *Trans.*, AIME, **228**.
7. Bade, J.F.: "Cement Bond Logging Techniques—How They Compare and Some Variables Affecting Interpretation," *JPT* (Jan. 1963) 17-22.
8. Winn, R.H., Anderson, T.O., and Carter, L.G.: "A Preliminary Study of Factors Influencing Cement Bond Logs," *JPT* (April 1962) 369-72.
9. Pickett, G.R.: "Acoustic Character Logs and Their Applications in Formation Evaluation," *JPT* (June 1963) 659-67; *Trans.*, AIME, **228**.
10. Chaney, P.E. Jr., Zimmerman, C.W., and Anderson, W.L.: "Some Effects of Frequency Upon the Character of Acoustic Logs," *JPT* (April 1966) 407-11.
11. Walker, T.: "A Full-Wave Display of Acoustic Signal in Cased Holes," *JPT* (Aug. 1968) 811-24.

12. Brown, H.D., Grijalva, V.E., and Raymer, L.L.: "New Developments in Sonic Wave Train Display and Analysis in Cased Holes," *The Log Analyst* (Jan./Feb. 1971) 27-40.
13. Bigelow, E.L.: "A Practical Approach to the Interpretation of Cement Bond Logs," *JPT* (July 1985) 1285-94.
14. Thornhill, J.T. and Benefield, B.G.: "Injection Well Mechanical Integrity," U.S. Environmental Protection Agency report EPA/625/9-87/007 (Sept. 1987).
15. Fertl, W.H., Pilkington, P.E., and Scott, J.B.: "A Look at Cement Bond Logs," *JPT* (June 1974) 607-17.
16. "The Essentials of Cement Evaluation," Schlumberger, Houston (March 1976).
17. Fitzgerald, D.D., McGhee, B.F., and McGuire, J.A.: "Guidelines for 90% Accuracy in Zone-Isolation Decisions," *JPT* (Nov. 1985) 2013-22.
18. McGhee, B.F. and Vacca, H.L.: "Guidelines for Improved Monitoring of Cement Operations," *Proc.*, SPWLA 21st Annual Logging Symposium, Lafayette, LA (July 8-11, 1980).
19. "Acoustic Cement Bond Log," Atlas Wireline Services, Western Atlas Intl., Houston (1981).
20. Morriss, S.L.: *Cement Evaluation*, Schlumberger, Houston (1989) C-20.
21. Gollwitzer, L.H. and Masson, J.P.: "The Cement Bond Tool," *Proc.*, SPWLA 23rd Annual Logging Symposium, Corpus Christi (July 6-9, 1982).
22. Albert, L.E. *et al.*: "A Comparison of CBL, RBT, and PET Logs in a Test Well With Induced Channels," *JPT* (Sept. 1988) 1211-16; *Trans.*, AIME, **285**.
23. Froelich, B., Pittman, D., and Seeman, B.: "Cement Evaluation Tool: A New Approach to Cement Evaluation," *JPT* (Aug. 1982) 1835-41.
24. Havira, R.M.: "Ultrasonic Cement Bond Evaluation," *Proc.*, SPWLA 23rd Annual Logging Symposium, Corpus Christi (July 6-9, 1982).
25. Sheives, T.C. *et al.*: "A Comparison of New Ultrasonic Cement and Casing Evaluation Logs With Standard Cement Bond Logs," paper SPE 15436 presented at the 1987 SPE Annual Technical Conference and Exhibition, New Orleans, Oct. 5-8.
26. Leigh, C.A. *et al.*: "Results of Field Testing the Cement Evaluation Tool," *Proc.*, SPWLA 25th Annual Logging Symposium, New Orleans (June 10-13, 1984).
27. Catala, G.N., Stowe, I.D., and Henry, D.J.: "A Combination of Acoustic Measurements to Evaluate Cementations," paper SPE 13139 presented at the 1984 SPE Annual Technical Conference and Exhibition, Houston, Sept. 16-19.

SI Metric Conversion Factors

cycles/sec	× 1.0*	E+00	=	Hz
ft	× 3.048*	E−01	=	m
°F	(°F−32)/1.8		=	°C
in.	× 2.54*	E+00	=	cm
lbm	× 4.535 924	E−01	=	kg
psi	× 6.894 757	E+00	=	kPa

*Conversion factor is exact.

Chapter 12
Other Logs To Measure Behind Casing

12.1 Introduction

This chapter discusses a few less-commonly-applied techniques for measuring fluid movement or the competency of the completion behind casing. These techniques include the use of pulsed-neutron logs for flow profiling, application of the radial-differential-temperature log for channel detection, and the use of the unfocused-gamma-ray-density log for gravel-pack evaluation. Though these logging methods may not be considered production logs in the classic sense, they are used to obtain the same sort of information as traditional production logs.

12.2 Application of Pulsed-Neutron Logs for Flow Profiling

The pulsed-neutron log is a commonly applied cased-hole log for formation evaluation that measures primarily the water saturation in the formations surrounding the wellbore. This application is beyond the scope of this monograph. The log is sometimes used, however, as a means of obtaining a qualitative flow profile, and in this application plays the same role as such traditional production logs as temperature, spinner, or radioactive-tracer logs. This section reviews the theory of pulsed-neutron logging and presents examples of its use for flow profiling. (See Ref. 1 for a comprehensive treatment of the theory and applications of pulsed-neutron logging and Refs. 2 and 3 for the fundamentals of pulsed-neutron logging.)

12.2.1 Theory. The basic principle of a pulsed-neutron log is that thermal neutrons traveling through a material will be captured by the atoms of the material and will emit gamma rays when captured. The capacity of elements to capture neutrons, called the capture cross section, varies, with chlorine having a high capture cross section compared with other elements commonly found in formation rocks and fluids—e.g., hydrogen, carbon, oxygen, and silicon. Thus, the pulsed-neutron log can measure the relative amount of chlorine (salt water) in the formation near the wellbore. The radiation emitted by neutron capture events is proportional to the number of events (and thus is proportional to the number of live or uncaptured neutrons) and can be measured with a gamma ray detector.

A pulsed-neutron tool emits a short burst of neutrons, then monitors their decay by measuring the gamma rays emitted. The number of thermal neutrons decreases exponentially and is given by

$$n_u = n_e e^{(-\Sigma vt)}, \qquad (12.1)$$

where n_u=number of live neutrons, v=neutron velocity (about 7,220 ft/sec [2200 m/s]), n_e=number of neutrons emitted, and Σ=macroscopic capture cross section of the medium. Because the gamma radiation emitted is proportional to the number of live neutrons, a semilog plot of gamma ray intensity vs. time should yield a straight line with a slope of Σv. Pulsed-neutron tools measure the slope of the response curve during this logarithmic decay period by measuring gamma ray intensity during multiple gating periods after the pulse of neutrons has been emitted.

The macroscopic capture cross section, Σ, is a simple average of the capture cross sections of the individual elements composing the medium. Thus,

$$\Sigma = N_1\sigma_1 + N_2\sigma_2 + \ldots + N_i\sigma_i + \ldots + N_n\sigma_n, \qquad (12.2)$$

where N_i=number of atoms of Element i per unit volume and σ_i=capture cross section of Element i. For example, for a formation saturated with oil and water,

$$\Sigma = S_w\phi\Sigma_{br} + (1-S_w)\phi\Sigma_o + (1-\phi)\Sigma_f. \qquad (12.3)$$

Thus, if the capture cross sections of the brine, oil, and rock and the porosity are known, the water saturation can be calculated from the macroscopic capture cross section. Table 1 lists the capture cross sections of some common reservoir constituents.[4]

12.2.2 Flow Profiling. The injection profile in a well can be estimated by running a baseline pulsed-neutron log in a well, injecting a fluid with a higher capture cross section than that of the normal injection fluid, then repeating the pulsed-neutron log. The portions of the formation receiving injected fluid are identified by their higher capture cross section after injection of the high-Σ fluid. The capture cross section of the fluid is typically increased by addition of boron, which has a very high Σ, or by raising the salinity.[4,5]

This method does not provide a quantitative measure of the injection profile because once the high-Σ fluid has penetrated the formation beyond the depth of investigation of the pulsed neutron tool, typically about 1 ft [0.3 m], no further change in the measured capture cross section takes place. All zones that received this amount of fluid appear similar, regardless of the rate at which fluid was injected into the different zones. The pulsed-neutron log for flow profiling is most valuable in completions where spinner or radioactive-tracer methods cannot be used with accuracy.

Tesarek and Heysse[4] presented a case where the completion prevented the use of conventional production logging tools for detailed flow profiling. In the field considered, water was being injected into three different zones, with the water injection regulated by tubing mandrels with packers separating the zones (Fig. 12.1). If a spinner-flowmeter or radioactive-tracer log was run in the tubing, the gross injection into the tubing mandrels could be measured, but no detail about the distribution of injection in a particular zone could be obtained.

Fig. 12.1 illustrates the use of the pulsed-neutron log to refine the flow profile. Normal injection into this well was 1,700 B/D [270 m^3/d] of 4,000-ppm NaCl brine. A radioactive-tracer log run in the tubing showed 80% of the flow leaving the tubing through the upper mandrel and 20% leaving through the lower mandrel. A baseline pulsed-neutron log was run, 180 bbl [29 m^3] of 132,000-ppm NaCl brine was injected, the well was relogged with the pulsed-neutron log, another 180 bbl [29 m^3] of the same brine with 1 wt% borax added was injected, and the final logging pass was run.

TABLE 1—MACROSCOPIC CAPTURE CROSS SECTION OF SELECTED MATERIALS (from Ref. 4)

Material	Σ (c.u.)
Common formations	
Sandstones, carbonates	7 to 10
Shales	20 to 40
H_2O Solutions	
0 ppm NaCl	22
1% borax	71
150,000 ppm NaCl	79
150,000 ppm NaCl + 1% borax	134
Oil	
GOR = 10 ft^3/bbl	22
GOR = 3,000 ft^3/bbl	18
Methane	
3,000 psi at 160°F	6
5,000 psi at 180°F	9

1. Σ_w of H_2O + NaCl solutions is estimated by $\Sigma_w = 22 + 0.3413\ C_w + 0.00025\ C_w{}^2$, where C_w is the salinity in 1,000 ppm NaCl.
2. Σ_w of H_2O + borax solution is estimated by $\Sigma_w = 22 + 48.8$ (% borax).

The increase of the capture cross section after injection of the higher-Σ fluid indicates injection zones. Comparison of the three pulsed-neutron logs shows injection into Intervals A through E. The amount of increase in the capture cross section in the various zones is a function primarily of their porosities, with the high-porosity zone (Zone E, where $\phi = 17.3\%$) exhibiting the largest change. The magnitude of the change in Σ is not proportional to the flow rate into the various zones. For example, the mandrel supplying Zone E receives only 20% of the total flow, yet Zone E exhibits the largest change in Σ. Another interesting feature of this log is that it shows injection into unperforated Zone A, suggesting that fluid is channeling behind the casing to this zone from lower perforations.

12.3 Radial-Differential-Temperature Logs

The radial-differential-temperature log was developed specifically for the detection and remedial treatment of channels.[6,7] The basis for this log is the fact that the fluid moving in a channel is likely to be slightly cooler or warmer than the surroundings. Thus, the temperature along the casing wall next to a channel would be slightly different from the casing-wall temperature away from the channel. The log consists of a measurement of the temperature difference on opposite sides of the casing at a given depth as a function of angular position, providing a measure of the temperature difference ($T_{w1} - T_{w2}$ in Fig. 12.2) for the entire pipe circumference.

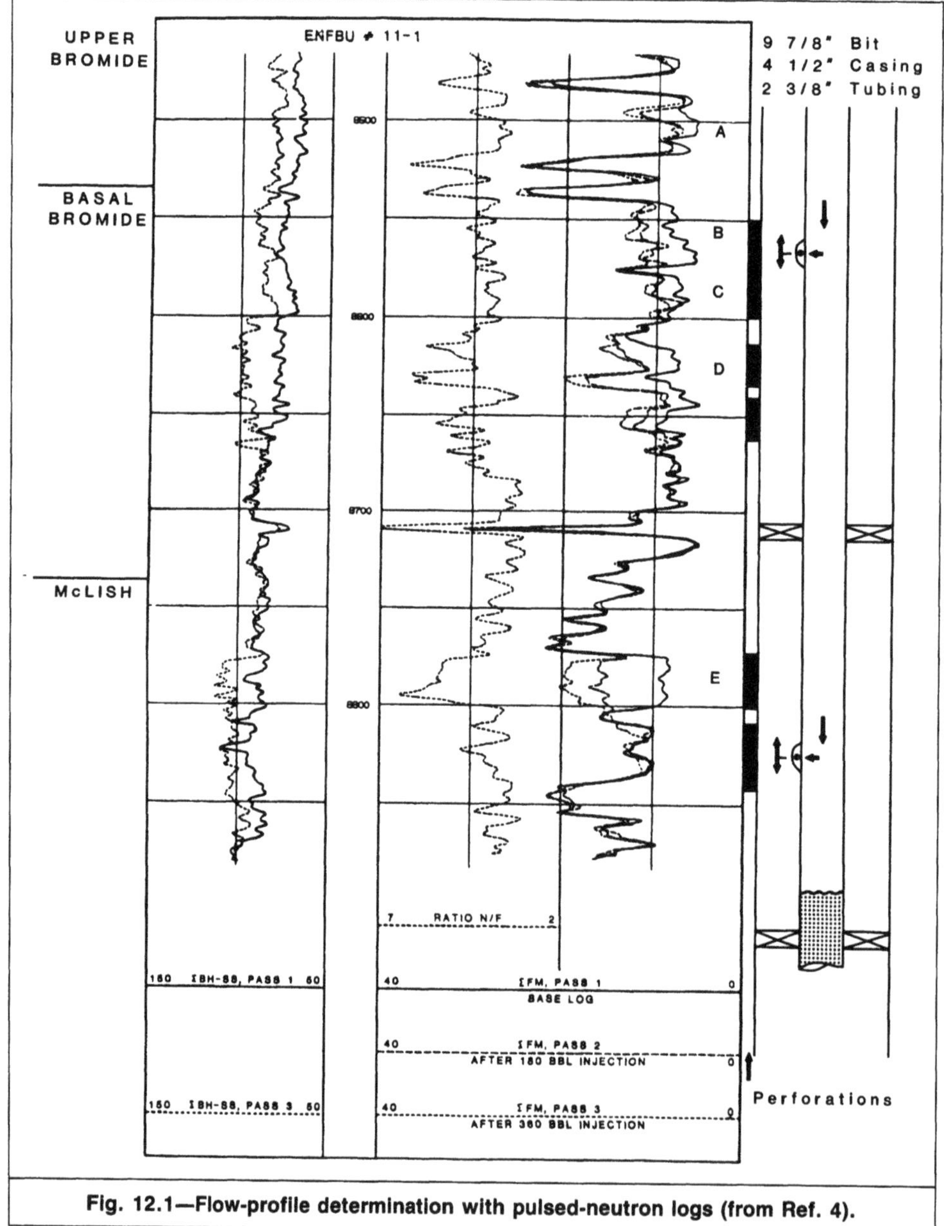

Fig. 12.1—Flow-profile determination with pulsed-neutron logs (from Ref. 4).

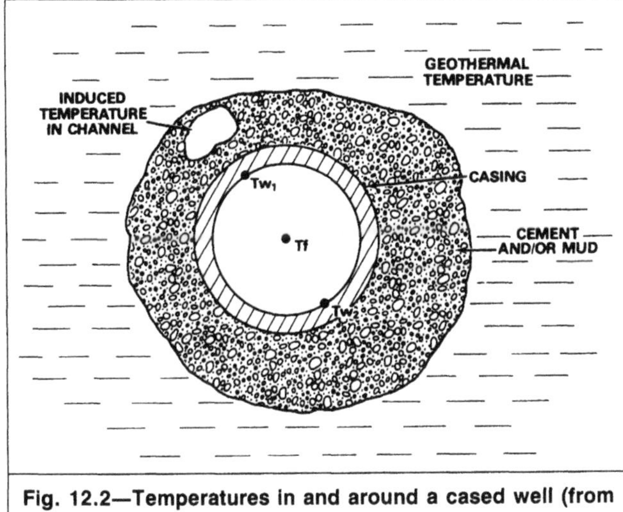

Fig. 12.2—Temperatures in and around a cased well (from Ref. 6).

The logging sonde has two temperature probes situated on arms extended against the casing wall. At the depth of a suspected channel, the tool is stopped, the arms are extended, the assembly is rotated through 360°, and the radial differential temperature is recorded. Note that the radial differential temperature is completely different from the differential temperature recorded on a conventional temperature log. The differential temperature is a measure of the vertical temperature gradient, dT/dz, while the radial differential temperature is the temperature difference on opposite sides of the casing wall at a fixed depth.

Where a channel exists, the radial-differential-temperature log should show a sinusoidal variation in differential temperature as a function of angle as the tool is rotated. Maximum and minimum temperatures are recorded when one of the probes is closest to the channel, and the differential temperature is zero when the probes are at nearly equal distances from the channel.

A sensitive measure of the temperature difference is required for this log; the highest sensitivity reported was 0.025°F/in.[0.005°C/cm] of chart deflection.[6] The radial-temperature difference caused by downhole fluids flowing between zones is typically small, usually ranging between 0.005 and 0.05°F [0.003 and 0.03°C].[6] The radial-temperature anomaly caused by a channel can often be enhanced by injecting a cooler fluid from the surface. With this technique, Cooke[6] observed radial-differential temperatures as high as 3°F [−16°C].

The radial-differential-temperature tool can be run with perforating guns so that the guns can be oriented in the direction of the detected channel. This procedure should increase the chances of perforating into a channel before a cement squeeze is attempted, particularly when the channel is fairly small. There is typically a lag in the tool response, however, and the tool cannot be placed close to the perforating gun.

The following examples illustrate the responses of the radial-differential-temperature log.

Example 1—Gas Channeling Between Zones.[6] In this well, crossflow of gas was occurring between sands in the 6,440- to 6,740-ft [1963- to 2054-m] interval, according to temperature and noise logs. The radial-differential-temperature log was run primarily to orient perforating guns to perforate into the channel for a subsequent cement-squeeze operation.

The conventional temperature log (Fig. 12.3) showed cooling beginning at about 6,450 ft [1966 m] and extending to just above 6,700 ft [2042 m], with a maximum cooling of about 20°F at 6,500 ft [11°C at 1981 m], indicating significant gas flow. To test for channeling into Sand L, the radial-differential-temperature log was first run at 6,675 ft [2035 m], just above Sand L. The radial-differential-temperature log exhibited the sinusoidal pattern typical of flow in a channel at this depth (Fig. 12.3a). A run made at 6,700 ft [2042 m] (Fig. 12.3b), just below Sand L, detected no differential temperature across the pipe. Thus, the gas flow was either originating or ending in Sand L.

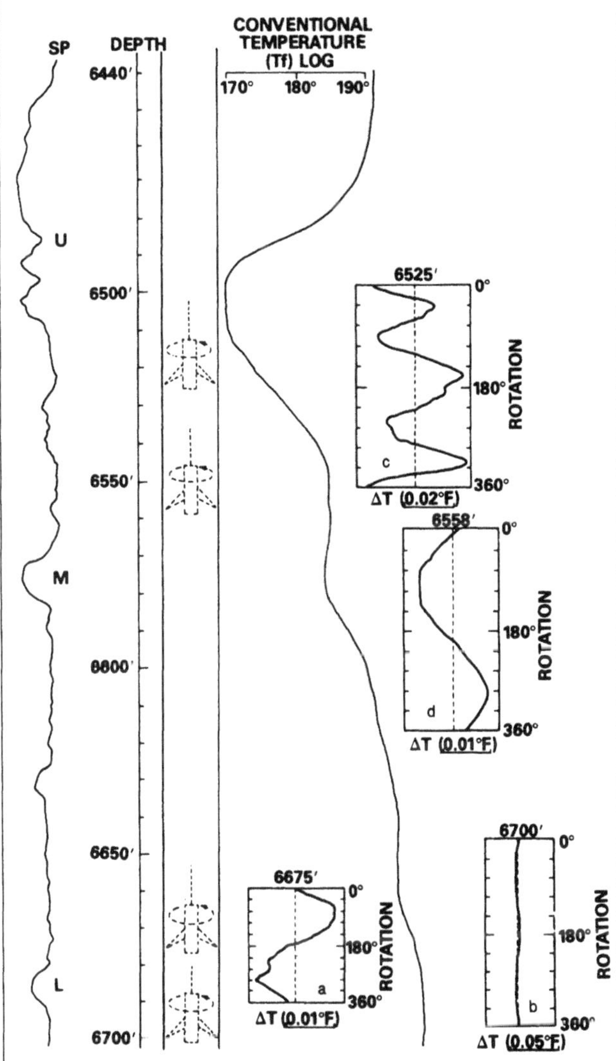

Fig. 12.3—Temperature and radial-differential-temperature logs in well with gas channel (from Ref. 6).

Radial-differential-temperature logs were then run at depths of 6,514, 6,516, and 6,525 ft [1985, 1986, and 1989 m]. Large differential temperatures were measured at these depths, but the responses did not repeat with each revolution. Fig. 12.3c shows the response obtained for one revolution at 6,525 ft [1989 m]. This behavior was assumed to result from flow through a large area outside the casing rather than from flow through a discrete channel. A measurement was then made at 6,558 ft [1999 m], where the radial-differential-temperature response again indicated flow through a channel (Fig. 12.3d). At this depth, the radial-differential-temperature log was used to orient a perforating gun. The well was also perforated at 6,525 ft [1989 m]. A cement squeeze eliminated the channel.

Example 2—Water Channeling to a Gas Zone.[7] This well (Fig. 12.4) initially produced 3 MMscf/D [90×10^3 std m^3/d] of gas from perforations at 9,093 to 9,106 ft [2772 to 2776 m], but water production increased rapidly and the well died. The well was block-squeezed above and below the productive sand and reperforated; this decreased water production and significantly decreased gas production. The well was then acidized with 15% HCl, which caused water production to increase to such an extent that the well died again.

A radial-differential-temperature log was then run, with water injection to enhance the differential temperature anomalies. An upper sand at about 8,920 ft [2719 m] was suspected as the water

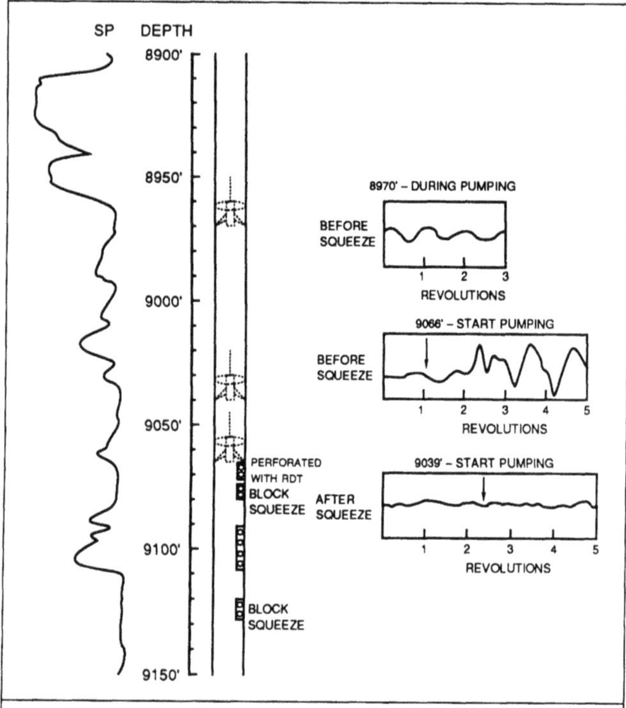

Fig. 12.4—Radial-differential-temperature logs in well with water channeling to gas zone (from Ref. 7).

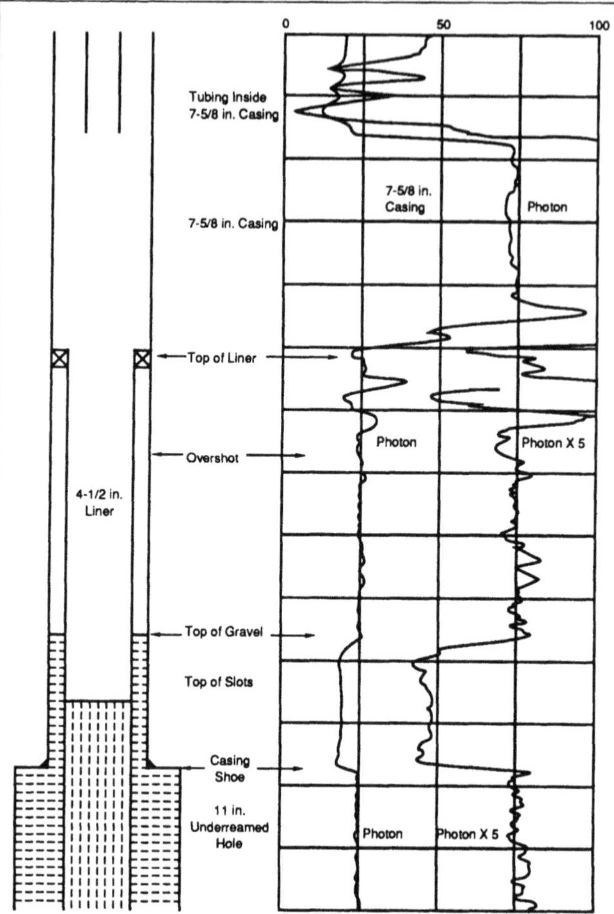

Fig. 12.5—Gravel-pack evaluation log—good gravel pack (from Ref. 8).

source, so the radial-differential-temperature measurements were made at 8,970 ft [2734 m], near the upper sand, and at 9,066 ft [2763 m], just above the producing interval. At both depths, the radial-differential-temperature log indicated flow through a channel. The well was then perforated at 9,071 to 9,074 ft [2765 to 2766 m], cement-squeezed, and reperforated at 9,100 to 9,106 ft [2774 to 2776 m]. After being swabbed for 5 days, the well produced only water. The radial-differential-temperature log was repeated to determine whether the channel had been repaired. As Fig. 12.4 shows, the radial-differential-temperature log indicated a very slight channel, if any. After another week of swabbing, the well began to produce gas, eventually reaching a rate of 1.2 MMscf/D [191×10^3 std m^3/d]. So much water channeled to the gas zone before the channel was eliminated that a large amount of water had to be removed from the gas sand to restore productivity.

12.4 Gravel-Pack Evaluation

In recent years, unfocused gamma-ray-density logs have been adapted for evaluation of gravel-pack completions.[8-11] They are used to detect voids in gravel packs by measuring the density of the materials in the region of the well completion. The tools used are gamma-gamma density devices that have gamma ray sources, typically cesium 137, and single gamma-ray detectors. The amount of radiation reaching the detector is a function of the density of the material through which it has traveled because gamma ray absorption is proportional to density. In a gravel-pack well, the material through which the gamma rays travel should be constant except for the amount of gravel present in the annular space between the liner and the casing or formation. Thus, the gamma ray intensity measured at the detector should provide at least a qualitative measure of the amount of gravel present. Where the annulus is completely filled, the detector response is a minimum, while void spaces yield higher count rates.

Fig. 12.5 illustrates the response of an unfocused gamma-ray-density log to a good gravel pack. This well has a 4½-in. [11.4-cm] slotted liner hanging in 7⅝-in. [19.4-cm] casing and an 11-in. [28-cm] -diameter openhole section. Below the casing shoe, the gamma ray response is low and relatively constant, indicating a uniform gravel pack. Farther up the well, the response decreases after the casing shoe is reached because of the absorption of gamma rays by the casing. The top of the gravel is indicated by the increase in radiation a short distance above the casing shoe; another larger increase in radiation occurs when the top of the liner is reached. Interpretation of an unfocused-gamma-ray-density log is facilitated by running a base log before gravel placement. Fig. 12.6 shows the gravel-pack evaluation log for the well of Fig. 12.5 before gravel packing. When the two logs are compared, it is clear that the gamma ray response decreased markedly in the region where gravel was placed.

An unfocused-gamma-ray-density log displays regions of high gamma ray intensity in the packed zone when the gravel pack contains voids (see Fig. 12.7). Even though a portion of the gravel-packed region produces a low count level, the gamma ray intensity is high through a large region of the completion, indicating an incomplete gravel pack.

Nomenclature

n_e = number of thermal neutrons emitted
n_u = number of uncaptured thermal neutrons
N_i = number of atoms of Element i per unit volume, 1/ft^3 [1/m^3], 1/L^3
S_w = water saturation
t = time, seconds, t
T_f = fluid temperature, °F [°C]
T_{w1} = casing wall temperature 1, °F [°C]
T_{w2} = casing wall temperature 2, °F [°C]
ΔT = radial differential temperature, °F [°C]
v = neutron velocity, ft/sec [m/s], L/t
σ_i = capture cross section of Element i, 1/ft [1/m], 1/L
Σ = macroscopic capture cross section, 1/ft [1/m], 1/L
Σ_{br} = brine capture cross section, 1/ft [1/m], 1/L
Σ_f = rock capture cross section, 1/ft [1/m], 1/L
Σ_o = oil capture cross section, 1/ft [1/m], 1/L
ϕ = porosity

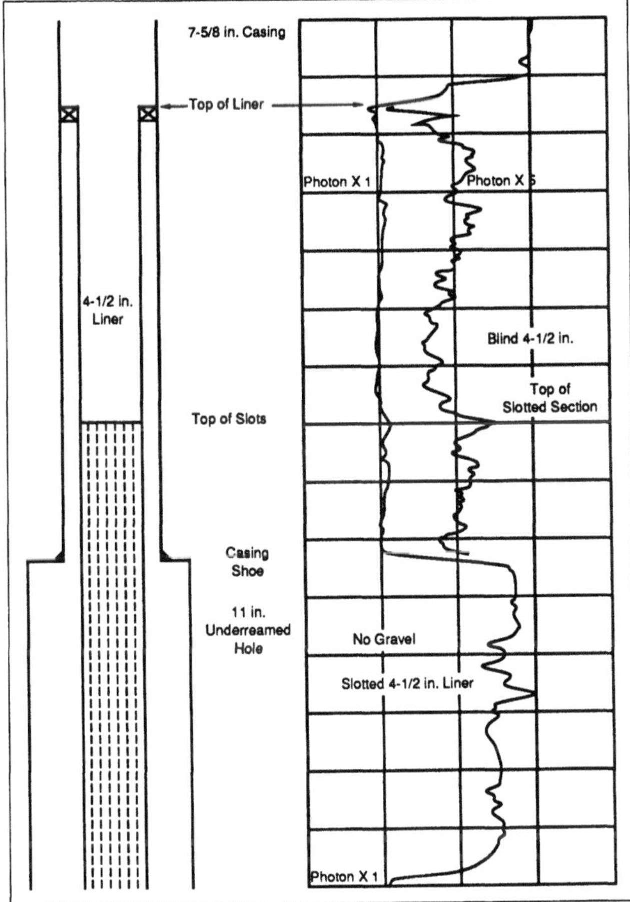

Fig. 12.6—Gravel-pack evaluation log before gravel placement (from Ref. 8).

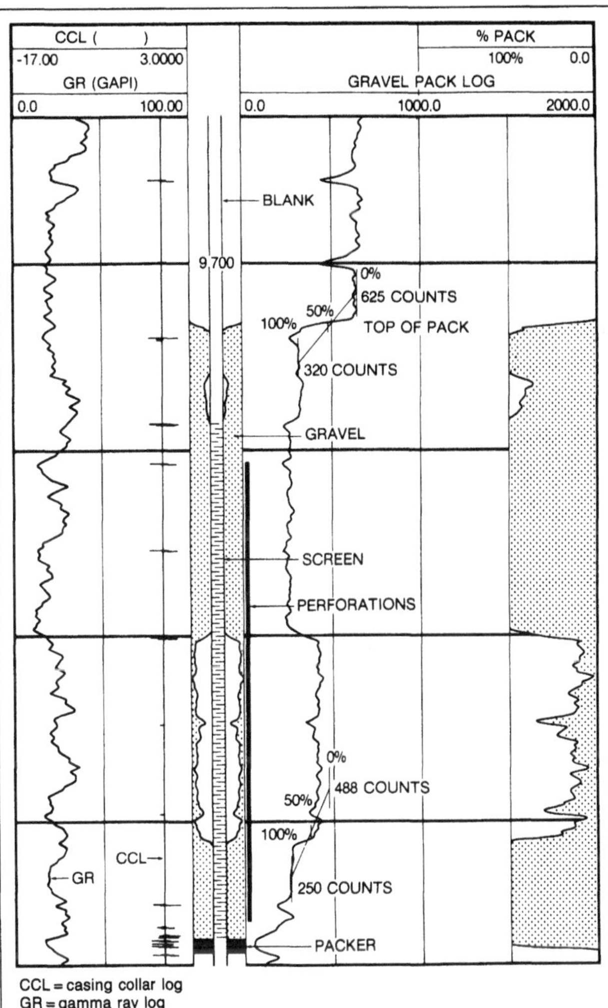

Fig. 12.7—Gravel-pack evaluation response to voids in gravel pack (from Ref. 10)

References

1. *Pulsed Neutron Logging*, Reprint Volume, SPWLA, Houston (Aug. 1979).
2. Youmans, A.H. *et al.*: "Neutron Lifetime, a New Nuclear Log," *JPT* (March 1964) 319-28; *Trans.*, AIME (1964) **231**.
3. Wahl, J.S. *et al.*: "The Thermal Neutron Decay Time Log," *SPEJ* (Dec. 1970) 365-79.
4. Tesarek, P.B. and Heysse, D.R.: "Water Injection Profiles in Eola Field Determined by Changes in Thermal Neutron Capture Cross Section," *SPEPE* (Aug. 1989) 258-64.
5. Matiisen, A.: "Defining Injection Profiles Using the Neutron Lifetime Log and High Sigma Fluids," *Trans.*, 14th SPWLA Annual Logging Symposium, Lafayette, LA (May 6-9, 1973) Paper V.
6. Cooke, C.E. Jr.: "Radial Differential Temperature (RDT) Logging—A New Tool for Detecting and Treating Flow Behind Casing," *JPT* (June 1979) 676-82.
7. Cooke, C.E. Jr. and Meyer, A.J.: "Application of Radial Differential Temperature (RDT) Logging to Detect and Treat Flow Behind Casing," *Proc.*, 20th SPWLA Annual Logging Symposium, Tulsa (June 3-6, 1979) Paper UU.
8. Chudy, S.: "Photon Log," *Proc.*, 22nd SPWLA Annual Logging Symposium, Mexico City (June 23-26, 1981) Paper GG.
9. Neal, M.R.: "Gravel Pack Evaluation," *JPT* (Sept. 1983) 1611-16.
10. Neal, M.R. and Carroll, J.F.: "A Quantitative Approach to Gravel Pack Evaluation," *JPT* (June 1985) 1035-40.
11. *Interpretive Methods for Production Well Logs*, second edition, Atlas Wireline Services, Western Atlas Intl., Houston (1982) 77-84.

SI Metric Conversion Factors

ft	× 3.048*	E−01	=	m
ft^3	× 2.831 685	E−02	=	m^3
°F	(°F−32)/1.8		=	°C
in.	× 2.54*	E+00	=	cm
psi	× 6.894 757	E+00	=	kPa

*Conversion factor is exact.

Author Index

Subject Index

www.ingramcontent.com/pod-product-compliance
Ingram Content Group UK Ltd.
Pitfield, Milton Keynes, MK11 3LW, UK
UKHW051138260726
13967UKWH00010B/3118